Urban Nexus Approach for Sustainable Cities

Wanglin Yan
Editor

Urban Nexus Approach for Sustainable Cities

Principles, Methods and Applications

Springer

Editor
Wanglin Yan
Faculty of Environment and Information Studies/Graduate School of Media and Governance
Keio University
Fujisawa, Kanagawa, Japan

ISBN 978-981-95-6155-1   ISBN 978-981-95-6156-8 (eBook)
https://doi.org/10.1007/978-981-95-6156-8

This Springer imprint is published by the registered company Springer Nature Singapore Pte Ltd.
The registered company address is: 152 Beach Road, #21-01/04 Gateway East, Singapore 189721, Singapore

Preface

Over the past decade, the academic community has made significant progress in advancing nexus thinking and its applications, deepening our understanding of the interconnections between resource provision, management, and sustainability. Among these developments, urban nexus research has sought to enhance resource-use efficiency, recognizing that cities are the primary sites of consumption and CO_2 emissions. Yet, due to the inherent complexity of modern urban systems, applying nexus thinking to promote sustainable urbanization remains a major challenge. To date, most food–energy–water (FEW) nexus studies have emphasized conceptual policy proposals or the development of analytical tools focused on the production and supply side of resource management. Far less attention has been given to the deeper nexuses that emerge from patterns of urban demand and consumption—perhaps the most defining dimensions of the urban nexus.

Building on the foundation laid by the previous volume, *Resilient and Adaptive Tokyo: Towards Sustainable Urbanization in the Perspective of the Food–Energy–Water Nexus*, this book again takes Tokyo as its focal case to showcase cutting-edge research on the urban nexus for sustainable urbanization. It integrates the fundamental principles of nexus thinking with empirical analyses grounded in quantitative and design-based methods. In contrast to most existing publications, this volume reconstructs urban FEW nexuses from the perspective of demand and consumption, offering a rigorous, data-driven structure of *policy design-led nexus approach*.

This work addresses several critical questions in contemporary nexus studies:

- How can the *urban nexus* be conceptualized within the dynamics of sustainable urbanization?
- How can *design-led nexus methods* be applied to support policy goals in complex urban nexuses?
- How can resource-use efficiency be enhanced and CO_2 emissions reduced through analysis of the food–energy–water supply and demand?

By investigating the FEW supply chain from production to consumption, this book examines nexus effects across multiple scales—from building/household and neighborhood to city and regional levels—drawing upon extensive empirical datasets. The scientific and practice-based insights presented here are intended to foster transdisciplinary dialogue and advance the operationalization of nexus thinking in real-world contexts.

This volume compiles recent research on the urban FEW nexuses and is organized into four main parts and the fifth part for conclusion:

Part I outlines the principles of nexus thinking and their relevance to urban sustainability. It includes two chapters: Chap. 1 traces the evolution, principles, and frameworks of nexus thinking; Chap. 2 introduces the nexus principle of investing in nature for the conservation of farmland ecosystems amid demographic transitions from growth to decline.

Part II presents a landscape of nexus methodologies across three chapters: Chap. 3 reviews urban nexus methods, with particular attention to the *design-led movable approach*; Chap. 4 offers a comprehensive review of system-dynamics models applied to urban FEW nexuses; Chap. 5 employs environmental input–output analysis to evaluate the impacts of disasters in Tokyo by examining shifts in energy structure and carbon emissions in 2011, 2015, and 2020.

Part III provides detailed analyses of FEW supply and demand systems across four chapters: Chap. 6 investigates city dwellers' access to ecosystem services through urban agriculture; Chap. 7 analyzes store-level logistics and structural challenges influencing emissions, proposing strategies for a more sustainable food system; Chap. 8 presents the first integrated assessment of CO_2 emissions across the entire water-supply chain—from intake to tap; Chap. 9 introduces two integrated carbon auditing methodologies—Service Point-Based Carbon Accounting (SPBCA) and the BEMS-Driven Energy Audit Method (BDEAM)—applied to central-city contexts of buildings; and Chap. 10 proposes a method for evaluating physical accessibility to grocery stores, accounting for variations in age, geography, and mobility in suburban settings.

Part IV demonstrates the broader potential of nexus thinking for urban planning and environmental action across five chapters: Chap. 11 explores the nexus among compact cities, transit-oriented development (TOD), and urban ecology; Chap. 12 challenges the foundational premises of TOD through a case study of Odawara City; Chap. 13 introduces the Suburban Industrial Community (SIC) nexus model to evaluate the role of industrial parks in sustainable urbanization; Chap. 14 reviews Tokyo's waste generation and treatment system, highlighting challenges and emerging policy responses; Chap. 15 presents an integrated approach combining a CO_2 emission database with two educational tools: food design cards, which enable calculation of both emissions and nutritional balance, and the Tangible Food Design Game, which renders invisible emissions physically perceptible.

Finally in Part V, Chap. 16 synthesizes the findings of this volume and identifies key challenges and directions for future research.

This book serves as a valuable reference for researchers in urbanization studies, nexus research, urban design, environmental science, disaster risk reduction, and climate change. It aims to help scholars grasp current trends and advanced methodologies in the field, while offering practitioners and policymakers actionable insights for collaborative research, evidence-based design, and informed decision-making.

Fujisawa, Kanagawa, Japan Wanglin Yan

Acknowledgments

This volume presents a collection of advanced research conducted at Shonan Fujisawa Campus (SFC), Keio University, in recent years. The contributors are primarily doctoral students together with several master's students, from the Graduate School of Media and Governance, Keio University. The chapters included here are either concise versions of published journal articles or selected excerpts from dissertations that have not yet been published. The editor extends heartfelt gratitude for their dedication, creativity, and perseverance—especially during the challenging period of their degree research.

The origins of nexus research at SFC can be traced back to the *Sustainable Urbanisation Global Initiative on Food–Water–Energy*. With the implementation of the funded project *Moveable Nexus for Urban–Energy–Water Management Innovation in New Boundary Conditions (M-NEX)*, our journey toward sustainable urbanization through the lens of nexus thinking began. While the Japan team focused on the Tokyo Metropolitan Area, international partners contributed fresh perspectives from Amsterdam, Belfast, Detroit, Doha, and Sydney. Through a series of global workshops and shared research practices, professors and students were inspired from the outset, and this collaborative process continuously nurtured the growth of the Japan team.

Although the project formally concluded in March 2022—one year later than planned due to the COVID-19 pandemic—our intellectual journey has never ceased. Since then, the principles of the nexus have been applied across multiple spatial scales within the metropolitan region. Nexus methods have informed regional and community-level investigations, while the *policy design-led nexus approach*—developed within our laboratory—has been applied at the scales of households, buildings, and neighborhoods to respond to the spatial, social, and political dynamics of sustainable urbanization.

These achievements would not have been possible without the generous support and collaboration of our research partners, including *Ryoko Okumura*, Senior Researcher at Tokyu Research Institute; *Yoshifumi Murakami* and *Akinobu Yasuda* at Ad-Sol Nissin Ltd.; and *Kentaro Kitanami* at Tokyo Tatemono Ltd., among others. The editor also expresses sincere appreciation to the officials of *Yokohama City*, *Odawara City*, and numerous municipal officers across the Tokyo metropolitan region, who kindly shared their time and insights through interviews and discussions.

Their contributions—together with the creativity and persistence of the researchers and collaborators involved—have made this collective work possible and have advanced the evolution of the *policy design-led nexus approach* as a practical framework for achieving sustainable and resilient urban futures.

Wanglin Yan
30 September 2025, Tokyo

Contents

Editor and Contributors

About the Editor

Wanglin Yan is a Professor in the Faculty of Environment and Information Studies and Head of the Program for Environmental Design and Governance at the Graduate School of Media and Governance, Keio University, Shonan Fujisawa Campus, Japan. He received his Bachelor of Engineering from the Wuhan Technical University of Surveying and Mapping in 1982, followed by Master's and Doctoral degrees in Civil Engineering from the University of Tokyo. Over the past three decades, Dr. Yan has become an influential figure in the fields of environmental science, geoinformation technology, and sustainable urban and regional planning. Dr. Yan was formerly President of the Geographic Information Systems Association (Japan) and received the *Gijuku Award* of Keio University in 2024.

At Keio University, Dr. Yan has pioneered the development of education and research that integrate environmental design, spatial information science, and governance innovation. His academic leadership has shaped a new interdisciplinary paradigm bridging geography, engineering, urban planning, and social policy. As Head of the Program for Environmental Design and Governance, he has trained numerous doctoral and master's students, encouraging them to combine scientific analysis with creative design and evidence-based policy formulation to address complex environmental challenges.

Dr. Yan's research and practice have increasingly centered on what he defines as the *Policy Design-Led Nexus Approach*—a framework that integrates nexus thinking with design methodologies and policy innovation. This approach views cities as dynamic socio-ecological systems in which the interdependencies of food, energy, and water can be understood, visualized, and governed through co-creation among scientists, designers, citizens, and policymakers. By linking conceptual frameworks with tangible spatial and policy design outcomes, the Policy Design-Led Nexus Approach bridges the long-standing gap between scientific knowledge and urban practice.

Internationally, Dr. Yan leads the Sub-Regional Node for the Asia-Pacific Adaptation Network (APAN) under the United Nations Environment Programme (UNEP), where he contributes to advancing climate change adaptation and disaster risk reduction across Japan and Northeast Asia. His expertise encompasses spatial analysis, resilience planning, and post-disaster reconstruction design, with a focus on climate-responsive urban transformation and regional sustainability.

From 2018 to 2022, Dr. Yan directed the *Moveable Nexus Project* (M-NEX) under the *Sustainable Urbanisation Global Initiative (SUGI)/Food–Water–Energy Nexus*, a joint program of the Belmont Forum and the Joint Programming Initiative Urban Europe. This multinational collaboration—with partners in Amsterdam, Belfast, Detroit, Doha, Sydney, and Tokyo—demonstrated how design-led nexus thinking can inform policy innovation at multiple spatial scales, from households and neighborhoods to metropolitan regions.

Dr. Yan has authored and edited numerous academic books and journal papers addressing resilience, sustainability, and the transformation of cities under climatic and social change. His recent book, *Resilient and Adaptive Tokyo: Towards Sustainable Urbanisation in the Perspective of the Food–Energy–Water Nexus* (Springer Singapore, 2024), synthesizes the results of the M-NEX project and establishes a foundation for the ongoing development of the Policy Design-Led Nexus Approach.

Through his interdisciplinary vision, international collaboration, and commitment to bridging science, design, and governance, Dr. Yan continues to advance the frontiers of sustainable urbanisation. His work exemplifies how nexus thinking can evolve into a practical framework for co-creating resilient, low-carbon, and inclusive cities.

Contributors

Amane Fujita SFC Research Institute, Keio University, Fujisawa, Kanagawa, Japan

Shuang Gan Graduate School of Media and Governance, Keio University, Fujisawa, Kanagawa, Japan

Xujie Hu Graduate School of Media and Governance, Keio University, Fujisawa, Tokyo, Japan

Takahiro Kanamori SFC Research Institute, Keio University/PASCO Corporation, Fujisawa, Kanagawa, Japan

Keidai Kishimoto School of Human Science and Environment, University of Hyogo, Himeji City, Hyogo, Japan

Hanyu Liu Graduate School of Media and Governance, Keio University, Fujisawa, Kanagawa, Japan

Shun Nakayama Graduate School of Media and Governance, Keio University, Fujisawa, Kanagawa, Japan

School of Network and Information, Senshu University, Chiyoda, Tokyo, Japan

Graduate School of Media and Governance, Keio University/School of Network and Information, Senshu University, Tokyo, Japan

Ryoko Okumura Tokyu Research Institute, Shibuya, Tokyo, Japan

Haichen Wei Data Intelligence Team, JTB Corporation, Tokyo, Japan

Weiyao Yang School of Urban Planning and Design, Peking University, Shenzhen, China

Wanglin Yan Faculty of Environment and Information Studies, Graduate School of Media and Governance, Keio University, Fujisawa, Kanagawa, Japan

Ayase Yonamine Nikken Sekkei Research Institute, Fujisawa, Kanagawa, Japan

Ruiyi Zhang School of Media and Governance, Keio University, Fujisawa, Kanagawa, Japan

Qiannan Zhuo SFC Research Institute, Keio University, Fujisawa, Kanagawa, Japan

Part I
Understanding Urban Nexus

Chapter 1
Nexus Thinking and Its Principles for Sustainable Cities

Wanglin Yan

Abstract This chapter explores the evolution, frameworks, and practical implications of the food–energy–water (FEW) nexus in the context of sustainable urban development. It synthesizes recent advances in nexus assessment, modeling, governance, design, and policy integration. By examining multiple analytical approaches—including input–output analysis, life cycle assessment, and participatory modeling—the author highlights the strengths and limitations of current methods, particularly their often-fragmented treatment of cross-sectoral dynamics and lack of spatial resolution. The role of governance is emphasized through stakeholder engagement and codesign processes that support system understanding and decision-making. Conceptually, the FEW nexus is advanced through boundary objects such as urban metabolism and traits frameworks, which help translate complex interdependencies across disciplines and scales. Entry points—such as food systems, water management, or urbanization—are critical for operationalizing the nexus in place-based contexts. Key themes discussed include the need for transdisciplinary integration, improved data infrastructure, and adaptive governance structures. The chapter argues that embedding the nexus in urban planning and policy can unlock synergistic solutions that enhance resource efficiency, resilience, and equity. The findings offer a roadmap for moving beyond sectoral silos toward a more integrated, systemic approach to sustainability in the Anthropocene.

Keywords Urban Nexus · Nexus principles · Nexus interest · Nexus approach · Nexus effects

---

W. Yan (✉)
Faculty of Environment and Information Studies/Graduate School of Media and Governance, Keio University, Fujisawa, Kanagawa, Japan
e-mail: w_yan@keio.jp

W. Yan (ed.), *Urban Nexus Approach for Sustainable Cities*,
https://doi.org/10.1007/978-981-95-6156-8_1

1 Introduction

1.1 Nexus Thinking

The concept of "nexus" has emerged as a valuable framework for interdisciplinary academic research, enabling scholars to explore the interconnectedness of various systems. The term of nexus itself dates back approximately 140 years, first appearing in the context of an "organic nexus between the motor and tactile centers" in an 1876 issue of *Nature* (Nature 2016). Its contemporary relevance resurfaced in the late 1980s when researchers began to recognize the close relationship between water and food security. Despite this recognition, little theoretical progress was made over the subsequent three decades in understanding or formalizing these interdependencies (Allan et al. 2015).

The modern momentum of nexus gained significant traction following the 2011 Bonn Conference (Hoff 2011), which catalyzed global interest in the water–energy–food (WEF) nexus. It soon emerged as a potential solution to the interconnected challenges in complex society (Cairns and Krzywoszynska 2016). Only 5 years from 2011 and 2015, nearly 300 organizations worldwide initiated projects under the food–energy–water (FEW) nexus framework (Endo et al. 2017). This surge in attention is driven by three main factors: (1) increasing interdependencies among resources due to mounting scarcities, (2) recent disruptions in resource supply, and (3) the limitations of siloed, sector-specific management approaches (Al-Saidi and Elagib 2017).

Nexus thinking advocates for the integrated management of WEF systems to minimize trade-offs and enhance synergies across sectors, contrasting sharply with conventional sectoral governance models. At its core is the interconnectedness of natural resource flows and integrated strategies for resource management (Al-Saidi and Elagib 2017; Hoff et al. 2019). It shares significant conceptual ground with resilience thinking. Both of them is rooted in systems thinking approaches that aim to understand socio-ecological interlinkages (Glibo and Koenigstorfer 2023; Alcamo 2017; Laspidou et al. 2020). Nonetheless, they are frequently applied independently. Scholars have cautioned that systems approach often suffer from a lack of theoretical rigor or sufficient engagement with political, historical, and relational dimensions (Leck et al. 2015). The final report from the Bonn Conference underscored the urgency of this agenda, warning of the substantial challenges that humanity would face if resource management systems were not improved in the coming decades (Lazaro et al. 2022).

1.2 Urban Nexus

Given the deeply embedded relationships between cities and their surrounding environments, urban governance must account for both internal dynamics and external interdependencies (Zhang et al. 2019).

Nexus thinking aims to dismantle sectoral silos by promoting a systemic understanding of how multiple resource flows—particularly water, energy, and food—interact within specific spatial and temporal contexts, such as cities (Covarrubias et al. 2019). This cross-sectoral governance and planning methodology applied within urban settings is often referred to as the "Urban Nexus." Despite the growing body of FEW nexus literature, operational definitions that effectively bridge research and practice—particularly in urban systems—are still lacking (Romero-Lankao et al. 2017). Most studies remain focused on national and regional scales (Covarrubias 2019). Research at the urban scale—where FEW systems interact directly with population dynamics, social structures, and economic activities—remains relatively scarce (Tye et al. 2022). The application of the food–water–energy (FEW) nexus specifically within urban contexts remains relatively underexplored (Tye et al. 2022; Francisco et al. 2023).

The Urban Nexus represents a forward-looking approach to designing sustainable urban development strategies. It responds to the urgent need for governance mechanisms and implementation tools that can overcome silo-based thinking in key urban sectors (Vogt et al. 2014).

A key innovation within the Urban Nexus approach is the emphasis on "designing for circular abundance," a concept proposed by McDonough and Braungart, which integrates water and energy-efficient technologies and practices in synergistic ways (Vogt et al. 2014). The model promotes decentralized, closed-loop systems of urban planning and service provision that align environmental sustainability with economic vitality and social equity. It envisions transitions from fossil fuels to renewables, the reuse of wastewater for water and fertilizer recovery, and the harvesting of rainwater and stormwater runoff. In this vision, the Urban Nexus aims to produce food using recycled water and energy, pursue zero-waste objectives, and ensure the affordability and accessibility of essential services—particularly for disadvantaged populations. These goals align with broader sustainability paradigms such as the "circular economy" and "circular abundance" (Vogt et al. 2014).

Another critical dimension of the Urban Nexus research involves modeling and forecasting FEW resource demands as a pathway toward resilient urban systems. This aspect supports the development of tailored governance models for rural–urban linkages and resource interdependencies, informed by comprehensive spatial and systems-based perspectives. The Urban Nexus thus contributes significantly to global urban sustainability agendas, including the Habitat III framework and the New Urban Agenda. Ultimately, the evolution of nexus thinking into effective resource governance is critical for achieving the United Nations Sustainable Development Goals (SDGs) (Lazaro et al. 2022).

1.3 What Matters in Nexus

The nexus concept has generated substantial interest across academic, policy, and planning communities, yet confusion persists regarding its precise meaning and application. Despite big number of publications, distinctions between "nexus thinking," "nexus approach," and the operationalization of nexus through tools and methodologies remain blurry. Nexus is often treated as the analytical endpoint itself—abstract and detached from practical outcomes. Several critiques point to the lack of conceptual novelty, insufficient clarity, and limited actionable utility of the nexus frameworks, which is often perceived as facilitating discussion without driving implementation (McGrane et al. 2018). This ambiguity raises a key epistemological question: What is the essence of nexus thinking? Does the nexus thinking primarily represent a philosophical perspective or a practical set of tools? (Micheli et al. 2019).

As global challenges intensify, there is an urgent call to move beyond conceptualization toward actionable outcomes—to truly "walk the nexus walk" (McGrane et al. 2018). The imperative to transform nexus thinking into practical, inclusive, and evidence-based action becomes ever more pressing. That is the transition from "nexus thinking" to "nexus doing" (Newell et al. 2019). This transition necessitates the dismantling of institutional silos and the cultivation of collaborative environments where data, knowledge, and resources are shared openly among stakeholders (Purwanto et al. 2021). It also requires a shift in disciplinary focus: while "nexus thinking" has largely been situated within discourses on resource security and policy analysis, "nexus doing" emphasizes practical domains such as planning, design, development, and implementation.

A key challenge in operationalizing nexus lies in the application of disciplinary tools not originally designed to capture cross-sectoral dynamics and feedback loops (Albrecht et al. 2018). The absence of integrated and coherent datasets further hampers efforts by researchers and policymakers to trace interdependencies and evaluate trade-offs or synergies. Research groups are encouraged to assess their methodological needs before constructing bespoke models, with emphasis placed on integration and interoperability across systems. Rather than continuously developing new context-specific tools, there is a growing call for critical reflection on existing frameworks.

Additional methodological challenges persist in defining system boundaries, capturing internal mechanisms within nexus systems, and assessing overall system performance. As Zhang et al. (2018) argue, existing models often lack comprehensive descriptions of internal feedback and suffer from inadequate performance metrics tailored to nexus evaluation. Another major gap in the current nexus landscape is the underrepresentation of nonacademic stakeholders. Private sector actors—including supermarkets, utilities, and farms—are not only significantly affected by nexus dynamics but also possess valuable data and implementation capacity. Building stronger transdisciplinary partnerships could enhance the relevance and applicability of nexus-based research.

Cairns and Krzywoszynska (2016) warn against treating the nexus as a mere "matter of fact"—a problem to be solved solely through technocratic or science-driven solutions. Instead, they argue that the nexus should be viewed as a "matter of concern"—a dynamic and socially contested issue that demands critical inquiry.

This perspective calls for deeper engagement from the social sciences to challenge the political, institutional, and epistemological assumptions embedded in nexus frameworks. That is, a holistic nexus perspective is required to account for both ecosystem services (like aquifer recharge or carbon sequestration) and public services (such as irrigation, wastewater treatment, or potable water delivery).

To bridge the gaps between theory and practice, four strategic imperatives have emerged from recent research (Lazaro et al. 2022): (1) enhancing relevance—making nexus frameworks meaningful for stakeholders and aligned with policy needs; (2) ensuring data reliability—addressing gaps in data quality, availability, and interoperability; (3) creating adaptable frameworks—designing flexible models that can be tailored to different spatial, temporal, and governance contexts; (4) focusing on applicability—prioritizing tools and approaches that are accessible and usable by practitioners. Meanwhile, quantitative indicators and standardized assessment criteria specific to nexus analysis must be developed and widely adopted. This challenge has been partly addressed through initiatives in the world such as Urban Nexus (Urban_Nexus 2013a, b, c) and SUGI Nexus, which aim to develop a nexus-based horizon for resource management and sustainable development.

2 Nexus Interests

Nexus thinking is mostly advocated as tools for promoting integrated governance and cross-sectoral coordination across systems and scales (Allouche et al. 2014). Or it is often treated as nexus approach for quantifying the linkages and trade-offs between system components.

Despite growing academic and policy interest of nexus, there remains no universally accepted definition of the term. Interpretations vary widely across sectors, geographical contexts, and disciplinary perspectives. Etymologically, the term of nexus derives from the Latin term *nectere*, meaning "to bind," which implies a relationship or connection between two or more entities (Rezaei Kalvani et al. 2024). The Oxford English Dictionary defines "nexus" as "a connection or series of connections linking two or more things" (Fig. 1.1). The dictionary definition evokes the term of nexus can be a configuration of nexus such as nodes, links, and networks. Scholar interpretations of nexus have predominantly emphasized integration and interlinkage. It can be also operationalized as a set of interactions among subsystems or sectors (e.g., water, energy, food), emphasizing dynamic interdependencies (Zhang et al. 2018). In principle, any study investigating interlinkages among multiple components could be categorized as a nexus study (Sušnik and Staddon 2022).

This section overviews the understandings and operations of nexus in concept and practice.

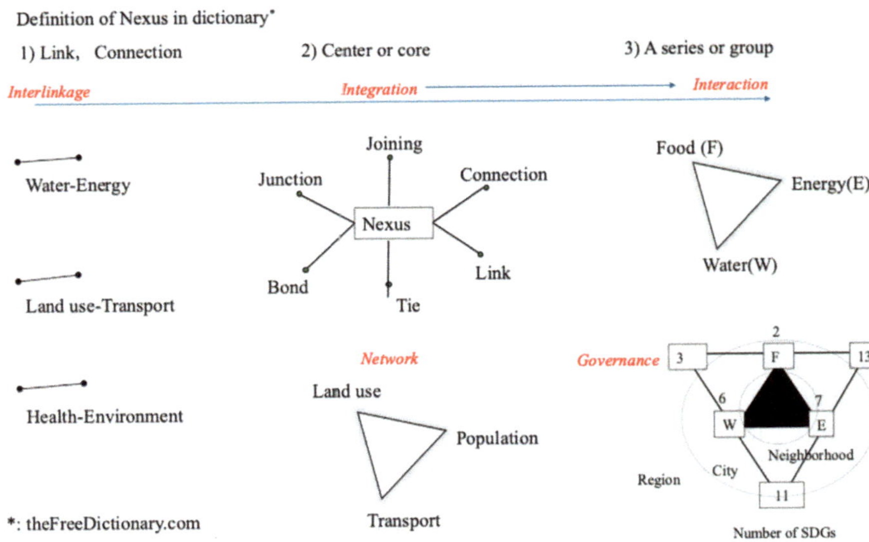

Fig. 1.1 Terminology of nexus

2.1 *Nexus as Integration*

Integration lies at the heart of nexus thinking and has been reinforced through key international policy frameworks. Within the academic literature, integration is widely acknowledged as a key tenet of the water–energy–food (WEF) nexus. Notably, the outcomes of early nexus discussions contributed to the United Nations Conference on Sustainable Development, "Rio+20," held in 2012. Its final document, *The Future We Want*, explicitly called for the adoption of "holistic and integrative approaches to sustainable development" (paragraph 40) (Lazaro et al. 2022). The same document further advocated for integrated planning approaches in the context of sustainable urban development, emphasizing the role of local authorities, public awareness, and participatory decision-making (paragraph 135) (United Nations 2012). Expanding on this perspective, Sperling and Berke (2017) proposed a practical framework for "urban nexus science" to identify integrated pathways for achieving urban sustainability.

The interconnections among food, energy, and water systems become particularly salient in the face of growing economic demand, climate variability, and rapid urbanization (Hoff 2011). These challenges have intensified the need for integrated approaches that consider the synergistic and competitive relationships among resource flows. Building upon this foundation, the United Nations' *Agenda 2030*, adopted in 2015, formalized the "integrative approach" as a guiding principle. This approach underscores the interconnectedness of social progress, economic development, and environmental sustainability, urging the design of policies and interventions that acknowledge and strengthen these linkages (United Nations 2015).

Al-Saidi and Elagib (2017) identify three primary drivers necessitating integration: (1) increasingly interlinked resource systems due to growing scarcity, (2) recent crises in resource supply chains, and (3) the limitations of sector-specific management paradigms. These conditions collectively point to the urgent need for systemic approaches that transcend sectoral silos.

Therefore, integration as a core of nexus thinking implies nexus as a pool to collect resources and services through the enhancement of connections. On the other hand, there remains little consensus on what integration within the nexus actually entails. There remains a lack of integration of critical sectors, particularly ecosystems (Sušnik and Staddon 2022). Definitions and operationalizations vary significantly depending on context, scale, and sectoral focus. This lack of clarity has implications for both research coherence and policy implementation.

2.2 *Nexus as Interlinkage*

Achieving long-term sustainable development requires a nuanced understanding of the interlinkages among critical sectors and resources, particularly within the constraints of availability, time, and spatial distribution (Weitz et al. 2014). The water–energy–food (WEF) nexus framework explicitly foregrounds these interlinkages to both conceptual understanding and practical governance (Hoff 2011; Zhang et al. 2019; Tan et al. 2025).

Lazaro et al. (2022) provide a longitudinal analysis of how the WEF nexus has evolved conceptually and in practice. Their study traces the emergence of key themes, maps the evolving interlinkages between WEF topics, and offers insight into the shifting paradigms that have shaped nexus research over time.

For cities, the interlinkages of FEW present a strategic opportunity to enhance service provision and optimize metropolitan systems in support of sustainable development at local, national, and global scales (Vogt and Brekke 2014; Avellán et al. 2017; Bellezoni et al. 2021). As urban areas grow, ensuring the sustainable governance of essential resources—energy, food, and water—becomes increasingly vital. These three resources are not only essential for human activity but are intrinsically interdependent (Zhang et al. 2019). A practical example of such interdependence is seen in wastewater treatment plants: the volume of incoming wastewater directly affects the amount of recoverable waste heat, illustrating a clear linkage between water and energy systems (Lazaro et al. 2022).

Beyond technical systems, the nexus has also been conceptualized as a dual coupling between human society and the natural environment. The Food and Agriculture Organization (FAO 2014) interprets the FEW nexus as comprising external resource components bound together by internal interlinkages that are essential for driving sustainable economic development. This perspective underscores the socio-ecological dimension of the nexus, highlighting the interplay between resource flows and human well-being.

Therefore, nexus as interlinkages focuses on the connections among sectors, spatial scales, or institutional frameworks. The interconnectedness is not only a defining feature of the nexus but also a powerful lens through which sustainability transitions can be understood and pursued.

2.3 *Nexus as Interactions*

Nexus is also increasingly understood as a framework that captures the interactions among resource systems, supply chains, and socioeconomic processes. The energy–food–water (EFW) nexus is typically a valuable conceptual model that captures the dynamic interactions among system components—whether within a single system or across multiple, interconnected systems (Paavola 2020). White et al. (2018), for instance, interpret the nexus as the interaction between FEW resources and economic activities, examining how regional, national, and supranational demands are met through global trade mechanisms. Owen et al. (2018) conceptualize the FEW nexus in terms of the interactions between resource impacts of products across different points along their supply chains. This perspective underscores the importance of tracing resource use and dependencies beyond the boundaries of production sites to account for the full life-cycle impacts embedded in global trade networks.

Relevant insights can be drawn from frameworks such as Total Site Integration (TSI), proposed by Varbanov (2014), which shares conceptual affinities with nexus thinking. TSI emphasizes optimizing the interactions not only among resources (e.g., water, energy) but also among facilities (e.g., industrial plants) and regions, in pursuit of both economic and environmental benefits. In doing so, it highlights the potential for designing interconnected systems that improve efficiency and reduce trade-offs at multiple levels of operation.

Despite the long-standing recognition of interaction-based systems thinking, nexus research still faces limitations. Many existing studies fall short in capturing the complex interactions among driving factors, such as climate change, technological development, and policy interventions, particularly across different temporal and spatial scales (Tan et al. 2025). There is absence of standardized methodologies for studying nexus interactions.

Within urban contexts, the urban nexus highlights the interaction between WEF provisioning systems, embedded in both material and social flows. Through this analysis, Covarrubias et al. (2019) identified two distinct interactional mechanisms:

- *Material-driven nexus*—where physical resource flows (e.g., energy transmission, water supply, food distribution) create linkages across sectors, often leading to unintended synergies or trade-offs.
- *Socially-driven nexus*—where institutional, political, or cultural interventions—such as new urban energy policies or land-use plans—reshape infrastructure and resource provisioning, thereby altering material flows.

These interactive dynamics are especially relevant in the context of urban sustainability. This implies that urban nexus interactions are situated within nested, multi-scale frameworks that span neighborhood, city, regional, and global levels. Bellezoni et al. (2021) identified the nexus relationships of FWEN (food–water–energy nexus) with urban green and blue infrastructure (GBI) within cities. GBI systems—such as parks, wetlands, and permeable surfaces—can enhance resilience and optimize the provisioning of food, water, and energy, thus acting as mediators of FWEN dynamics.

Therefore, nexus thinking provides a lend to see the dynamic aspects of interlinkages among nodes. Understanding how these interactions scale up or cascade down is vital for formulating coherent and context-sensitive governance strategies that align with broader sustainability goals.

2.4 Nexus as Networks

Beyond the technical discourses of interlinkage, integration, and interaction, nexus thinking has evolved to encompass a network-based perspective, particularly in the context of urban governance. To analyze the urban WEF nexus as a network, it is essential to examine not only resource flows but also the structure, connectivity, and function of the social and material components involved. Networks are composed of nodes (e.g., actors, infrastructures, institutions), flows (e.g., energy, nutrients, information), and linkages (relationships that connect nodes). These elements together define how resources are distributed, governed, and transformed within the urban fabric.

Covarrubias et al. (2019) advance this conceptualization by framing the urban nexus as a dynamic socio-material network. In their view, cities function as experimental and innovative spaces where water, energy, and food (WEF) systems are shaped by interacting flows, actors, goals, and governance mechanisms. A network's structure reflects its density, robustness, and spatial–temporal dynamics—how frequently and effectively nodes are connected. The scope of a network refers to its spatial and sectoral reach, while the function indicates the collective goals and values that shape the behavior and development of the network. In the urban WEF context, nodes such as water utilities, food distributors, or energy grids are connected through flows such as wastewater, heat, or agricultural nutrients (e.g., nitrogen, phosphorus, potassium). These flows link biophysical systems to social institutions and highlight the multifunctionality of urban infrastructure.

On above the physical dimensions, Covarrubias et al. (2019) further defined the governance powers within the nexus networks. The formation and evolution of WEF networks are influenced by four interrelated forms of power:

- *Networked power*—the concentration of decision-making authority within specific nodes or actors, such as dominant utilities or regulatory agencies. For example, the liberalization of utility systems in Europe has diminished state monopolies, redistributing control across public and private actors.

- *Network power*—the ability to set norms, standards, and protocols that determine inclusion within or exclusion from networks. Compatibility of goals and values—such as shared commitments to sustainability—can enable tighter, more coherent nexus networks.
- *Networking power*—the capacity of specific actors to include or exclude others based on alignment with existing network values, scope, or operational objectives. This gatekeeping function significantly affects who participates in WEF governance.
- *Network-making power*—the strategic capability to shape network structures and their functions, including redefining relationships, connecting new nodes, and reconfiguring shared values or goals. Actors with this power, often termed "switchers" or "programmers," can influence system transitions—such as guiding food systems toward decarbonization despite fossil fuel dominance.

Ultimately, the network view of the nexus emphasizes that the reconfiguration of urban WEF systems depends not only on physical interdependencies but also on the social structures, institutional capacities, and power dynamics embedded within urban systems. Effective network integration requires shared visions, enabling conditions for collaboration, and deliberate efforts to rewire both infrastructure and governance systems toward sustainable, resilient outcomes. The application of a network lens reveals both opportunities and constraints in urban WEF governance. For instance, initiatives that seek to connect food, energy, and water systems more sustainably often remain at the experimental or start-up stage (Covarrubias et al. 2019). This is due to persistent barriers, including misaligned values, institutional incompatibility, technical standards, and regulatory hurdles (e.g., health and safety concerns).

2.5 Moveable Nexus

Although the interpretations of nexus look different—qualitative interaction versus quantitative analysis—they are not mutually exclusive in fact and can be synthesized into a unified framework. The diversity of definitions and interests surrounding the nexus reflects both its interdisciplinary richness and its conceptual ambiguity. While this pluralism presents challenges for standardization, it also allows the nexus to serve as a flexible framework for addressing complex, multi-scalar problems in an increasingly interconnected world.

Figure 1 indicates that nexus is not only interlinkages of one or two nodes, not only the integrations of factors, but the dynamic interactions in the network of food–energy–water. Considering the complexity and dynamics of the dimensions, nexus thinking is in fact the governance of nexus web of nature, economy, and politics such as SDGs. Concepts have been proposed to handle this complexity. Kurian (2017) highlighted how the nexus framework can be instrumental in understanding vulnerabilities and interdependencies that influence systemic resilience and security. In this way, the nexus is not merely a conceptual lens for integration but also a platform for addressing emerging socio-ecological risks and guiding adaptive governance. Similarly, the WEF nexus has been variously narrowed to explore specific interactions (e.g., the energy implications of water pumping or the water requirements of agricultural production) (Valek et al. 2017; Grubert and Sanders 2018), and broadened to include new dimensions and sectors such as human development, equitable resource access (Sušnik and Zaag 2017), mineral extraction, and social structures (Schlör et al. 2018).

Yan and Roggema has introduced the concept of the Moveable Nexus (Yan and Roggema 2019; Roggema and Yan 2021) through the M-NEX Project, one of the 15 international initiatives funded under the Sustainable Urbanisation Global Initiative (SUGI)-NEXUS program from 2018 to 2022.

Rather than treating the WEF nexus as a static structure, the Moveable Nexus conceptualizes it as a living system that can reconfigure its components, relationships, and capabilities in response to evolving environmental, technological, and societal conditions. This adaptive quality makes the Moveable Nexus particularly relevant for multi-scalar urban planning—spanning buildings, neighborhoods, cities, and regions—and inclusive of diverse stakeholder engagements. As a philosophical and practical framework built on the practical M-NEX Project, the Moveable Nexus essentially means knowledge, technologies, and strategies developed in one locality can be transferred and adapted to another, fostering cross-pollination of ideas and solutions. This mobility invites fluidity, experimentation, and transdisciplinary explanations including,

- *Moveable geographies*—acknowledging that nexus solutions must traverse spatial scales and boundaries.
- *Moveable people*—recognizing migration, mobility, and demographic changes as integral to nexus planning.
- *Moveable thinking*—advocating for epistemological flexibility in how we conceptualize and govern resources.
- *Moveable flows*—supporting the dynamic management of resource cycles and infrastructure systems.
- *Moveable knowledge*—encouraging the exchange of ideas across disciplinary and cultural borders.
- *Moveable inventions*—promoting adaptable, modular technologies that can be scaled or relocated.
- *Moveable platforms*—creating institutional and digital infrastructures for collaboration and data sharing.

In short, the Moveable Nexus repositions the nexus not as a fixed analytical framework but as an adaptive strategy for sustainable urban futures. It emphasizes fluidity over rigidity, co-creation over top-down control, and systemic learning over static modeling—making it a forward-thinking paradigm well-suited to the complex, uncertain challenges from sustainable development goals even further in the Anthropocene.

3 Nexus Principles

3.1 Native Nexus Principles

The foundational objective of nexus thinking is to promote sustainable and efficient resource use in a manner that extends the availability of existing resources to meet the needs of growing populations while preserving ecosystem integrity (McGrane et al. 2018). This objective was embodied as three guiding principles at Bonn Conference (Hoff 2011; Weitz et al. 2014):

1. investing on nature to sustain ecosystem services,
2. doing more with less,
3. ensuring equitable access to resources, particularly for vulnerable and marginalized populations

The first principle underlying nexus thinking draw heavily on—and evolve from—other holistic approaches to environmental governance, such as Integrated Natural Resource Management (INRM), Integrated Water Resources Management (IWRM), and the broader agenda of sustainable development (Leck et al. 2015). A well-recognized critique of Integrated Water Resources Management is its insufficient attention to administrative and jurisdictional boundaries. This limitation is not exclusive to IWRM—it also affects other frameworks such as basin-based management, where natural geophysical boundaries often do not align with existing institutional or administrative units.

The second principle, resource efficiency, is variably defined in the literature. For example, Dawkins et al. (2010) describe resource efficiency from a supply-side perspective, focusing on reducing inefficiencies in consumption. The European Union's more institutional definition frames resource efficiency as the ability of the economy to "create more with less"—producing greater value with fewer inputs, while minimizing environmental impacts. Both definitions converge on a shared goal: maintaining ecosystem productivity and resilience (Koutra et al. 2022).

Regarding the third principle, literature criticize that the WEF nexus must address infrastructure planning across water supply, wastewater treatment, and energy systems, recognizing that the integration of these sectors is highly context-specific. Planning and implementation differ significantly across low-, middle-, and high-income regions, each with unique challenges, constraints, and priorities (Zarei et al. 2021).

Therefore, as an interdisciplinary concept, the nexus thinking integrates not only environmental and technical concerns but also social, political, and economic dimensions. This integration is essential for developing robust policies and interventions capable of navigating the complexity of resource systems in a rapidly changing world.

3.2 Implementing Nexus Principles

Over the past decade, extensive research has sought to develop effective governance models for the food–energy–water (FEW) nexus, emphasizing holistic and integrated approaches. While there is no universal blueprint or one-size-fits-all solution, various governance frameworks and principle-based approaches have been proposed by national governments, international organizations, and academic institutions to guide nexus implementation in practice.

Melo et al. (2020) proposed a hybrid framework that incorporates forest ecosystems into the FEW nexus, highlighting their role in securing food, water, and energy, and in advancing the Sustainable Development Goals (SDGs). They propose three key principles—mainstreaming forest restoration, empowering local communities, and implementing nature-based solutions—which align closely with Hoff's (2011) native guiding principles.

Yuan and Lo (2022) conducted a systematic review of governance strategies and highlighted how nexus thinking has evolved to respond to emerging sustainability challenges. Building upon Lockwood's (2010) foundational governance principles—legitimacy, transparency, accountability, inclusiveness, fairness, connectivity, and resilience—their research proposed a nine-principle framework tailored to assess and guide FEW nexus governance. These nine principles include: connectivity—fostering linkages between sectors and actors; innovation—encouraging novel solutions and adaptive management; equitability—ensuring fair access and benefit-sharing; participation—engaging stakeholders at all levels; coordination—harmonizing sectoral policies and institutional actions; sharing—promoting data, knowledge, and resource exchange; legitimacy—building trust and credibility in decision-making; empowerment—enabling communities and institutions to take action; and strategy—maintaining a long-term, goal-oriented planning process.

3.2.1 Nexus Implementation to Sustain Ecosystems

The application of nexus principles has been also extended to fragile ecosystems, such as the Himalayan region, where integrated resource governance is vital for sustainability. Rezaei Kalvani et al. (2024) developed a FEW nexus model for the Himalayan ecosystem, identifying key implementation principles to enhance resource security and ecosystem resilience. These include restoring water resource capacity through the conservation of soil, wetlands, glaciers, lakes, and aquifers; developing climate-smart agriculture and enhancing environmental and social infrastructure for effective water use; reducing downstream flood risks through ecosystem-based interventions; investing in biodiversity conservation and water resource management; and engaging mountain communities as stewards of the ecosystem. This mountain-specific nexus approach highlights the importance of context-sensitive governance and the inclusion of ecological dimensions in implementation strategies.

3.2.2 Nexus Implementation to Improve Resource Use Efficiency

Urban Nexus expands beyond the traditional WEF sectors with the main goal of improving resource efficiency and promoting integrated development in both urban and peri-urban areas to incorporate land use, social inclusion, waste management, and transportation (Vogt and Brekke 2014). Its overarching objective of Urban Nexus is to enhance access to basic services, improve service quality, and elevate quality of life, all within the limits of environmental boundaries (Vogt and Brekke 2014). As an operational methodology, the Urban Nexus promotes vertical and horizontal integration in governance, reinforcing existing urban development strategies and fostering institutional transformation. By fostering synergies across sectors, administrative jurisdictions, and technical domains, the nexus approach aims to boost institutional performance, optimize resource use, and advance service delivery. In doing so, the Urban Nexus counters the inefficiencies and redundancies of traditional sectoral planning, which often results in fragmented investments, elevated costs, and underutilized infrastructure (Vogt and Brekke 2014).

3.2.3 Nexus Implementation to Engage Community

A key theme across successful nexus implementation efforts is the active engagement of local communities. In the WEFF (water–energy–food–forest) nexus framework, community participation is central to restoring degraded ecosystems and enhancing regional resource security. Forest ecosystems, in particular, are recognized as foundational to sustainable livelihoods and long-term WEF security. This framework connects forest restoration with SDG acceleration, emphasizing the multi-scalar importance of landscape-based governance.

In summary, the implementation of nexus principles must move beyond abstract theory into adaptive, place-based strategies. Whether applied in urban contexts, national governance systems, or fragile mountain ecosystems, the effective realization of nexus governance depends on clearly defined principles, inclusive and equitable stakeholder engagement, integration of ecological knowledge, long-term strategic planning, and institutional and financial support. These principles serve not only as normative anchors but also as practical guides for advancing sustainable development through the operationalization of the nexus approach.

3.3 *Extended Principles*

As the scope of nexus thinking has expanded, so too have its guiding principles, evolving to integrate broader considerations such as ecosystem services (ES), nature-based solutions (NBS), circular economy (CE), and social equity. These extensions are essential for enhancing cross-sectoral and cross-scalar coherence between land use, spatial planning, and governance, particularly in the face of complex socio-ecological trade-offs (Fürst et al. 2017; Salam et al. 2017).

3.3.1 Nature-Based Solutions and Systemic Integration

Recent scholarship on urban NBS has introduced a set of five interrelated principles to guide design, planning, and governance (Kabisch et al. 2022). These principles are particularly relevant to nexus thinking, given their emphasis on integration and sustainability:

- *Systemic understanding*—the foundation principle, emphasizing the need to conceptualize NBS within interconnected systems of people, nature, and infrastructure.
- *Benefiting people and biodiversity*—prioritizing co-benefits for both humans and ecosystems.
- *Inclusive solutions for the long-term*—ensuring that NBS are durable, equitable, and responsive to diverse stakeholder needs.
- *Context consideration*—tailoring NBS to local ecological, cultural, and institutional conditions.
- *Communication and learning*—facilitating adaptive governance through participatory engagement and knowledge exchange.

These principles are interconnected. For instance, systemic understanding (P1) underpins all others, while context sensitivity (P4) and inclusive governance (P3) reinforce the need for communication and co-learning (P5).

3.3.2 Extended Nexus Dimensions and Complexity

Wang et al. (2021) expanded the conventional WEF nexus to include land use, GHG emissions, waste, and pollution, reflecting the growing recognition that resource systems are deeply embedded within complex sustainability challenges. However, implementing such extended nexuses is inherently difficult due to the multidimensionality of the sectors, the need for broad stakeholder engagement, and the financial costs of infrastructure transformation (Salam et al. 2017).

A further issue arises with the proliferation of "nexus" extensions: conceptual ambiguity. As the WEF nexus expands to WEL (water–energy–land) and beyond, the boundaries between domains become blurred. Wang et al. (2021) emphasize that nexuses must be treated as complex adaptive systems, rather than mere extensions of climate impact models. Case studies, therefore, require principles that embrace complexity and account for feedback, thresholds, and nonlinear dynamics.

3.3.3 Ecosystem Services and Urban Systems

Efforts to integrate urban context (UC), ecosystem services (ES), and nature-based solutions (NBS) have led to the identification of qualitative and causal linkages within the nexus (Babí Almenar et al. 2021). Zhou et al. (2025), for example, establishes ecological boundaries for ES provisioning in urban areas, distinguishing between supply-side flows (e.g., water, land) and regulatory capacities (e.g., climate regulation, pollution mitigation).

3.3.4 Circular Economy and Nexus Reduction Strategies

The circular economy (CE) provides an increasingly relevant perspective for nexus governance. Demir and Alp (2025) proposed a CE-based framework using the Circular Agriculture Priority Assessment (CAPA) index to guide policy, investment, and research aimed at securing water, energy, and food while maintaining ecosystem health in Türkiye. Despite the richness of CE literature, there is a notable gap in its application to nexus reduction strategies (Del Borghi et al. 2020). A life cycle perspective, supported by CE principles, can illuminate nexus interconnections across entire supply chains and support the transition toward sustainability.

3.3.5 Social Equity and Human-Centered Principles

Salam et al. (2017) proposed three additional nexus principles centered on human well-being and justice, critical for ensuring sustainable, long-term development:

- Placing people and their basic human rights at the center of the nexus;
- Fostering public awareness and political will to support nexus-informed decision-making;
- Involving local communities in planning and implementation to cultivate ownership and long-term sustainability.

These human-centered principles are particularly important for bridging the implementation gap—the divide between nexus thinking and real-world governance and behavior.

Therefore, the extended principles of nexus thinking signal a transition from purely technocratic models to multidimensional, adaptive, and inclusive frameworks. They highlight the need to engage ecosystem services, foster circularity, incorporate social equity, and address complexity at multiple scales. These emerging principles serve not only to guide nexus analysis but also to reshape urban planning, infrastructure, and governance toward a more resilient and sustainable future.

3.4 Moveable Nexus Principles

The Moveable Nexus is anchored in three guiding principles upon the native nexus principles (Roggema and Yan 2021). The first principle, Integration, seeks to mobilize resources in cities by reframing FEW systems as interconnected networks of stocks and flows. By mapping flows, assessing vulnerabilities, and replicating successful practices across scales, the Moveable Nexus promotes scalable and resilient solutions. The second principle, Interlinkage, seeks to connect FEW systems with knowledge and technology across multiple spatial and temporal levels. It identifies points of overlap—such as local food system in suburban area—the approach seeks to close resource loops and promote circularity and efficiency (Yan and Nakayama 2021). The third principle, Collation, seeks to merge the traditionally separated processes of design, participation, and evaluation into an integrated, iterative workflow. Each participatory workshop contributes insights that inform both the design and evaluation processes, enhancing coherence and adaptability.

The Moveable Nexus is particularly adept at addressing the dynamic challenges posed by climate change, urbanization, and shifting urban boundaries for resources. By visualizing solutions from the scale of rooftop gardens to regional infrastructures, it captures the inherent complexity of FEW systems. Importantly, the moveable nexus principles act as a bridge between human and natural systems, motivating stakeholders—including citizens, policymakers, researchers, and industry—to co-create and co-implement actionable strategies.

4 Nexus Approaches

As urban systems become more complex and interconnected, nexus-based approach is deemed essential for addressing unsustainable development trajectories and looming resource constraints. New forms of urban nexus science and decision-support tools can support more secure and equitable access to essential services by recognizing the interconnections between water, energy, and food, and by incorporating considerations of trade, investment, and climate policy (Sperling and Berke 2017). This section reviews how nexus principles are facilitated for resource management and strategic planning with multi-actor cooperation.

4.1 Nexus Framework

The Bonn Conference 2011 laid the foundation for subsequent models and applications. The development of nexus conceptual frameworks has played a crucial role in promoting integrated thinking in the management of food, energy, and water (FEW) systems. Early nexus research tends to cluster around three primary domains:

conceptual development, which clarifies the theoretical underpinnings of the FEW nexus; modeling approaches, which assess trade-offs and synergies and internalize socio-environmental impacts; and empirical applications, which investigate real-world drivers of FEW dynamics while promoting stakeholder collaboration (McGrane et al. 2018). For instance, FAO's 2014 Water–Energy–Food Nexus Framework embedded the nexus perspective within the broader goals of sustainable development (FAO 2014). FAO's framework emphasizes the importance of stakeholder dialogue, supported by scenarios, evidence, and policy options, as a driver for decision-making as its mission of eradicating hunger, reducing poverty, and managing natural resources sustainably.

Then, Ramaswami et al. (2017) proposed an urban system framework that contextualizes FEW linkages within city environments. This model captures both in-boundary and trans-boundary interactions, quantifies environmental impacts of community-scale FEW provisions, and visualizes vulnerabilities within urban supply chains. A case study in Delhi, for instance, integrated data on household food consumption by socioeconomic strata, along with visitor and industrial demands, thereby revealing the interdependencies within urban systems.

Despite this momentum, many of nexus frameworks remain largely theoretical and demand extensive, often inaccessible, data—particularly at local levels—making the operationalization of nexus thinking difficult (McGrane et al. 2018). Shannak et al. (2018) conducted a comprehensive review of WEF nexus models and exposed several critical gaps. Key limitations include the absence of truly integrated models that combine all three elements of the FEW nexus using transdisciplinary, multi-method approaches. Many models lack spatial variables such as land use and cropping patterns, limiting their utility in agricultural planning. There is also an overreliance on macro-level analyses that often overlook local-scale complexities shaped by geography and climate. Furthermore, insufficient treatment of spatial–temporal dynamics hampers the accurate modeling of natural resource variability. These shortcomings highlight the need for more comprehensive frameworks that incorporate political, economic, and social dimensions—particularly when designing tools for practical application.

Ghodsvali et al. (2019) emphasized the rising importance of transdisciplinary nexus methods, especially within governance strategies targeting urban climate neutrality. Their research underscores how aligning WEF system flows with shared sustainability goals—such as reducing CO_2 emissions—requires an understanding of the nexus not only as a technical system but as a socio-technical configuration. This broader framing accounts for the dynamic interactions between institutions, infrastructure, and social norms, reflecting the complex and coevolving nature of urban systems.

Higgins and Najm (2020) introduced a ladder-based conceptual and mathematical model to represent the urban FEW nexus. Their framework organizes resource flows into four hierarchical stages: Stage 0 (Biophysical base), Stage 1 (Natural resources), Stage 2 (Improved resource quality and availability), and Stage 3 (End-use efficiency). Progress along these stages reflects increasing levels of resource optimization, with transitions between stages requiring specific interventions. These stages are influenced by four crosscutting factors: material cycles, climate and ecosystem functions, policy and management, and population and economic dynamics. The model also maps out the feedback loops and interconnections among food, energy, water, carbon, and nutrients, illustrating how nexus effects manifest across different systems and scales.

Therefore, the evolution of nexus frameworks reveals a shift from siloed, sector-specific analyses to integrated, multi-scalar, and transdisciplinary approaches. Despite significant conceptual and methodological progress, practical implementation continues to face challenges such as limited data, insufficient spatial detail, and inadequate stakeholder involvement. Future work should focus on the development of adaptive, context-sensitive, and data-informed frameworks that can effectively support decision-making for sustainable and resilient FEW systems in both urban and regional contexts.

4.2 Nexus Assessment

The assessment of the water–energy–food (WEF) nexus is fundamental to understanding the interdependencies and trade-offs among resource systems. At its core, the nexus represents a systematic evaluation of the linkages between water, energy, and food, with the goal of improving resource-use efficiency and guiding sustainable management practices (Rezaei Kalvani et al. 2024). According to a comprehensive review by Albrecht et al. (2018), the application of WEF nexus in rigorous, policy-relevant assessments remains limited. Persistent gaps exist in how nexus principles are employed. Fewer than one-third of studies use specific and reproducible nexuses, and most fail to adequately capture cross-sectoral interactions. Over 70% rely exclusively on quantitative interlinkages, while only a quarter in corporate governance of social factors. Moreover, only a minority adopt mixed nexuses or integrate multiple disciplinary perspectives. This disconnection between the conceptual promise and its methodological execution has led to calls for transdisciplinary, mixed approaches that better reflect the complexity of socio-environmental systems. Additionally, few nexus assessments incorporate dimensions of urban and regional planning or account for transboundary resource flows—both essential for practical implementation and policy relevance.

Efforts to expand the scope of nexus assessments have pushed beyond the traditional WEF triad. Wang et al. (2021) proposed extending the nexus framework to include greenhouse gas emissions, pollution, land use, and waste, thereby offering a more comprehensive view of nexus-related impacts. However, such expansions risk conceptual dilution. In some cases, "nexus" has become a catch-all term that substitutes for existing analytical approaches such as multi-criteria optimization, input–output analysis (IOA), or material flow analysis—often without adding conceptual clarity. Caputo et al. (2021) proposed a conceptual framework to assess urban agriculture within the food–water–energy–people (FWEP) nexus. This approach evaluates not just resource efficiency and food production but also health impacts and motivational factors of people, reflecting a shift toward incorporating social dimensions and recognizing the multifunctionality of urban systems.

To strengthen analytical rigor, established tools such as Environmentally Extended Input–Output (EEIO) analysis and Life Cycle Assessment (LCA) have been adopted in nexus studies. EEIO allows researchers to trace embodied environmental impacts from trade, assess economic–environmental linkages, and understand intersectoral dependencies. However, it faces limitations, including assumptions of homogeneity for grouped products, low sectoral resolution, and exclusion of informal labor, as well as significant data gaps and reporting delays. LCA is gaining traction, particularly from a food systems perspective, but remains largely sector-specific, limiting its ability to assess cross-sectoral interactions. Furthermore, fragmented sectoral data and the lack of integrated institutional infrastructure hinder comprehensive assessments at relevant geographic scales.

To address the limitations of traditional tools, Galychyn et al. (2022) proposed a multi-criteria assessment framework that combines energy accounting with a hybrid-unit input–output model. Applied to Vienna and Austria, this approach evaluates the environmental support for urban socioeconomic systems and maps resource flows between urban industries and surrounding ecosystems. The framework allows for identifying efficiency improvements through both supply-side strategies, such as increasing renewable energy, and demand-side measures like consumption reduction. This integrative approach represents a significant step toward operationalizing nexus assessments within urban socio-ecological contexts.

Despite its conceptual appeal, the operationalization of the WEF nexus remains methodologically fragmented and data constrained. Subnational and local-level assessments are particularly constrained by limited data granularity. Data are often scattered across agencies or not systematically collected, posing major challenges for policy-relevant analysis. Developing robust and holistic nexus assessment tools is essential for translating theory into actionable insights and informing evidence-based policy to navigate the complex interdependencies of food, water, and energy systems.

Advancing nexus assessment requires greater use of reproducible, transdisciplinary methods; integration of qualitative and participatory approaches alongside quantitative tools; incorporation of spatial and temporal dynamics; improved data accessibility and interoperability across sectors; and the adoption of multi-criteria frameworks such as energy and hybrid IO models. Technological advances such as smart meters and digital sensors are beginning to support FEW data collection at both institutional and household levels, opening new possibilities for bottom-up monitoring of nexus dynamics.

4.3 Nexus Modeling

As cities confront growing climate and development pressures, integrative nexus models will be indispensable in designing evidence-based, cross-sectoral strategies for sustainable futures. Modeling plays a critical role in the operationalization of the food–energy–water (FEW) nexus by enabling the analysis of interdependencies, identification of synergies and trade-offs, and the design of integrated management strategies. The primary goal of nexus modeling is to overcome siloed approaches by revealing how decisions in one sector affect outcomes in others, thereby improving institutional coordination, preventing unintended consequences, and promoting co-benefits in resource governance (McGrane et al. 2018).

Zhang et al. (2018) identify two dominant conceptual approaches in nexus modeling. The first is *interaction-focused models*, which examine dynamic relationships between subsystems—such as water supply, agricultural production, and electricity generation—within the broader nexus system. The second is *linkage quantification models*, which assess the magnitude and direction of flows between nexus nodes using data-driven methods. Both approaches are scalable and have been applied across various spatial contexts, from individual households to transboundary regions.

The application of nexus models across different spatial scales reflects a diversity of priorities and challenges. At the regional level, such models can be integrated with urban growth scenarios to evaluate land-use change, infrastructure development, and trade-offs in cost, benefits, and risk. For instance, Haase et al. (2018) explored how FEW integration contrasts with urban shrinkage dynamics, underscoring the value of multidimensional optimization. McGrane et al. (2018) emphasize identifying effective entry points—specific locations or scenarios—where data and stakeholder input can guide better resource management. Their UK-based case study demonstrated how participatory workshops, combined with regional data analysis, uncovered localized nexus dynamics and highlighted issues related to data availability, scale, and institutional fragmentation.

At smaller scales such as communities or neighborhoods, modeling is often challenged by low data resolution. Here, the nexus often appears as a "black box," where resource inputs and outputs are measurable, but internal linkages and feedback are obscure. Nevertheless, Ma et al. (2023) proposed practical programs for applying the nexus at the community level, such as urban farming, stormwater reuse, and decentralized renewable energy. These interventions show how trade-offs can still be evaluated in data-limited settings.

Advancing nexus thinking requires the development of integrated modeling tools. Miralles-Wilhelm (2016), building on earlier work by Bazilian et al. (2011) and Bizikova et al. (2013), proposed a research agenda centered on integrated modeling for FEW nexus planning. These models aim to simulate multi-sectoral interactions, support the generation of scenarios, and inform long-term strategic planning under uncertainty. However, such models face ongoing challenges, including integrating cross-disciplinary data from hydrology, agronomy, energy systems, economics, and urban planning, as well as harmonizing scales and translating technical outputs into policy-relevant insights.

In summary, nexus modeling offers a powerful but complex set of tools for exploring the interconnections within FEW systems. To fully realize its potential, future research and practice should focus on addressing data gaps—particularly at the local level—promoting participatory modeling that integrates stakeholder insights, developing multi-scalar frameworks that connect spatial planning with resource flows, building scenario-based tools to assess long-term risks and resilience, and ensuring that models are flexible, adaptive, and translatable across varied urban and regional contexts.

4.4 Nexus Governance

Effective governance of the water–energy–food (WEF) nexus demands participatory, place-based, and integrative strategies that empower local stakeholders to collaboratively develop sustainable solutions. One of the most promising approaches to nexus governance is participatory modeling. This method brings together a diverse set of actors—policymakers, scientists, civil society, and citizens—to co-create shared representations of WEF systems, identify interdependencies, and evaluate potential trade-offs (Jones and White 2022). By blending qualitative insights with quantitative tools, participatory modeling not only promotes mutual learning but also builds a common understanding of complex challenges and generates broader consensus (McGrane et al. 2018).

Participatory processes help shift stakeholder perspectives, highlight the depth of systemic interlinkages, and build lasting collaborative networks. A case in point is the participatory assessment of the WEF nexus in a catchment in Plymouth and the Tamar Valley, UK, where stakeholders identified critical factors influencing sustainable livelihoods. The process yielded practical tools such as trade-off matrices and conceptual diagrams, along with intangible outcomes such as strengthened governance relationships and a shared understanding of local challenges (McGrane et al. 2018).

However, implementing nexus governance in practice comes with notable challenges. For instance, while water and energy usage are frequently monitored through household metering (especially in countries such as the UK), food consumption and waste data are rarely collected in a systematic manner. This uneven data landscape restricts cities from designing informed and effective policies for sectors such as restaurants and food processing. Additionally, data sharing remains a bottleneck due to institutional silos, privacy issues, and limited cross-sectoral collaboration—even when advanced infrastructure such as smart meters is available.

Regarding to these issues, Halbe et al. (2015) proposed an iterative, multilevel governance framework involving individual, organizational, and governmental actors. In this model, stakeholders first develop conceptual diagrams based on their own perspectives, which then guide the creation of quantitative models. These models are used to explore scenarios and assess strategies such as aquaponics, graywater reuse, and decentralized resource systems. The process of iterating between modeling and deliberation strengthens adaptive learning, enhances stakeholder ownership, and improves the implementation of nexus-based policies. Moreover, it supports both horizontal (across sectors) and vertical (across scales) integration in governance.

Ultimately, effective nexus governance must move beyond traditional sectoral silos. It should embrace participatory, transdisciplinary, and adaptive approaches that combine multiple knowledge systems and engage stakeholders across scales. Core elements of successful WEF governance include stakeholder coproduction of strategies, integration of diverse data sources, acknowledgment of spatial and informational asymmetries, iterative design for continuous learning, and alignment between local initiatives and higher-level sustainability goals. By applying these principles, the nexus framework can become a practical governance tool for managing urban and regional resource systems in a more integrated, resilient, and equitable manner.

4.5 Nexus Design

Cities, as centers of population and economic activity, have historically prioritized economic efficiency over resource sustainability and quality of life. This development trajectory has transformed surrounding lands—once critical for food, energy, and water (FEW) provisioning—into zones dominated by industrial production and urban infrastructure. Consequently, modern cities have become resource-intensive, environmentally constrained, and increasingly vulnerable. Economic growth has not necessarily translated into improved urban living; many cities are now grappling with congestion, stress, poverty, and inequality. The vital interconnections among FEW systems, which are essential for urban resilience and human well-being, remain largely unrecognized and poorly integrated into city planning.

The Moveable Nexus, developed under the M-NEX project as part of the SUGI-NEXUS program, was an evolution upon the nexus approaches aforementioned (Roggema and Yan 2021). The M-NEX project extends the principles articulated at the Bonn Nexus Conference (Hoff 2011)—including resource mobilization, technological integration, and accessibility enhancement—into a moveable urban design platform. This platform incorporates design methods, evaluation tools, and participatory mechanisms to embed FEW considerations into tangible urban interventions. These interventions range from innovative policy strategies and building systems to public spaces, commercial infrastructure, and community-based programs.

The M-NEX project contains three core objectives: First, it seeks to translate existing FEW research into urban practice by employing design-based interventions. Second, it consolidates a variety of analytical tools into a unified decision-support platform tailored for FEW integrations. Third, it applies these tools in participatory, design-led workshops embedded within urban living labs. To support these goals, the project has developed three interconnected knowledge platforms: a Design Platform to guide co-creative design processes, an Evaluation Platform to integrate metrics and feedback for assessing FEW performances, and a Participation Platform to ensure inclusive stakeholder engagement through public workshops and living labs. With its open-access nature, the Moveable Nexus ensures that design outcomes and lessons are publicly shared, making the framework transferable across different bioregions. Ultimately, it represents a scalable, replicable, and participatory model for advancing FEW-integrated urban development. Through the integration of knowledge, the dismantling of design silos, and the embrace of systemic thinking, M-NEX demonstrates how design-led research can effectively operationalize nexus principles in practice.

- The M-NEX design methodology is grounded in codesign and reflexive practices, implemented through iterative stakeholder engagement. The process includes several key steps:
- *Inventory and spatial assessment*: Identify existing or potential FEW-related resources, such as rooftops, vacant buildings, and underutilized land suitable for urban agriculture and other sustainable uses.
- *Solution design*: Create interventions that improve land-use efficiency, support local food production, and deliver ecosystem services, while minimizing energy and water consumption using integrated technologies and systems thinking.
- *Nexus matrix composition*: Map and activate material flows and intersectoral connections, developing cross-disciplinary strategies tailored to the city's socio-ecological context.
- *Evaluation of solutions*: Assess the environmental costs and co-benefits of proposed interventions, analyzing their spatial, temporal, and systemic effects across ecological and social systems.
- *Iteration and delivery*: Present alternative solutions to stakeholders, refining them through continuous engagement, evaluation, and scenario testing.

This coproduction process draws together natural, social, industrial, and financial dimensions into a shared design framework. Resource mobilization, in this context, means activating and linking untapped capital—whether physical, human, or institutional—across disciplines and administrative boundaries to enable new flows of materials, services, and innovations in governance.

By embedding FEW systems into urban design and planning, the Nexus Design approach offers a transformative framework for developing resilient, adaptive, and inclusive urban environments. Instead of viewing FEW systems as external infrastructure concerns, this approach integrates them into the core of urban form, function, and governance. It shifts the role of cities from passive consumers to active co-creators of sustainable futures, thereby mainstreaming the nexus into the built environment and opening new pathways for systemic innovation.

5 Conclusions

Globally, FEW resources are fundamental for human survival and require coordinated, efficient management across urban–rural and local–global scales. Yet, in most cities, these resource flows are divided into isolated sectors—water services, power grids, and food supply chains—concealing their systemic interdependencies. For urban residents, FEW systems directly affect their daily lives and shape their relationships with public and private service providers (Artioli et al. 2017). These complexities require a systems-thinking approach capable of capturing cross-sectoral interactions and feedback loops (Bazilian et al. 2011). The Urban Nexus approach presents strategic potential for optimizing urban systems in alignment with local, national, and global sustainability goals. It can address pressing issues such as resource scarcity, pollution, and ecosystem degradation while enabling greater resilience, efficiency, and inclusivity in urban development (Vogt and Brekke 2014).

Despite this potential, the realization of integrated frameworks continues to be hindered by gaps in data, information, and knowledge that limit our capacity to understand FEW interconnections. Although research on the nexus has expanded, there remains a lack of systematic tools and platforms that can fully account for the complexity, trade-offs, and synergies inherent in FEW systems. Future research should prioritize enhanced data integration and sharing, the development of interoperable software platforms for nexus modeling and analysis, and the strengthening of transdisciplinary collaboration among scientists, policymakers, and local communities (Liu et al. 2017).

In conclusion, unlocking the transformative potential of the FEW nexus in urban contexts will require a combination of conceptual innovation, participatory governance, technological integration, and sustained cross-sector collaboration.

Acknowledgments This chapter is a comprehensive review on theory and practice of the M-NEX project granted by the SUGI-Nexus (Sustainable Urbanisation Global Initiative/Food–Water–Energy Nexus) by Belmont Forum (No. 11314551).

References

Abdul Salam P (2017) Water-Energy-Food (WEF) Nexus and sustainable development. In Salam P, Pandey V, Shrestha S, Anal AK (eds) Water-energy-food nexus: principles and practice. John Wiley Sons, p 223. https://doi.org/10.1038/s41893-020-00608-z

Albrecht TR, Crootof A, Scott CA (2018) The water-energy-food nexus: a systematic review of methods for nexus assessment. Environ Res Lett 13(4):043002

Alcamo J (2017) Systems thinking for advancing a nexus approach to water, soil and waste. J Environ Econ Manag 67(2):189–208

Allan T, Keulertz M, Woertz E (2015) The water–food–energy nexus: an introduction to nexus concepts and some conceptual and operational problems. Int J Water Resour Dev 31(3):301–311

Allouche J, Middleton C, Gyawali D (2014) Nexus nirvana or nexus nullity? https://scholar.google.com/citations?user=gwVPXkoAAAAJ&hl=en&oi=sra

Al-Saidi M, Elagib NA (2017) Towards understanding the integrative approach of the water, energy and food Nexus. Sci Total Environ 574:1131–1139

Artioli F, Acuto M, McArthur J (2017) The water-energy-food nexus: an integration agenda and implications for urban governance. Polit Geogr 61:215–223

Avellán T et al (2017) Making the water–soil–waste nexus work. Sustainability 9:1–17

Babí Almenar J et al (2021) Nexus between nature-based solutions, ecosystem services and urban challenges. Land Use Policy 100:104898

Bazilian M, Rogner H, Howells M, Hermann S, Arent D, Gielen D, Steduto P, Mueller A, Komor P, Tol RSJ, Yumkella KK (2011) Considering the energy, water and food nexus: towards an integrated modelling approach. Energy Policy 39(12):7896–7906. https://doi.org/10.1016/j.enpol.2011.09.039
Bellezoni RA et al (2021) Understanding and conceptualizing how urban green and blue infrastructure affects the food, water, and energy nexus. J Clean Prod 289:125825
Bizikova L et al (2013) Nexus: towards a practical planning and decision support framework. IISD
Cairns R, Krzywoszynska A (2016) Anatomy of a buzzword. Environ Sci Pol 64:164–170
Caputo S et al (2021) Applying the food-energy-water nexus approach to urban agriculture. Urban For Urban Green 58:126934
Covarrubias M (2019) The nexus between water, energy and food in cities. Sustain Sci 14(2):277–287
Covarrubias M, Spaargaren G, Boas I (2019) Network governance and the urban nexus of water, energy, and food. Energy Sustain Soc 9(1):1–11
Dawkins E, Roelich K, Barrett J, Baiocchi G (2010) Securing the future: the role of resource efficiency. https://ueaeprints.uea.ac.uk/id/eprint/29502/
Del Borghi A et al (2020) Circular economy approach to reduce water–energy–food nexus. Curr Opin Environ Sci Health 13:23–28
Demir E, Alp E (2025) A framework for assessing the circular economy potential in the water and agriculture sectors in Türkiye. Sustain Prod Consum 54:335–347
Endo A et al (2017) A review of the current state of research on the water, energy, and food nexus. J Hydrol Reg Stud 11:20–30
FAO (2014) The water-energy-food nexus – a new approach in support of food security and sustainable agriculture. Food and Agriculture Organization of the United Nations, pp 1–11
Francisco ÉC et al (2023) Food-energy-water (FEW) nexus: sustainable food production governance through system dynamics modeling. J Clean Prod 386:135825
Fürst C, Luque S, Geneletti D (2017) Nexus thinking–how ecosystem services can contribute to enhancing the cross-scale and cross-sectoral coherence between land use, spatial planning and policy-making. Int J Biodivers Sci Ecosyst Serv Manag 13(1):412–421
Galychyn O et al (2022) A multi-criteria framework for assessing urban socio-ecological systems. Cleaner Environ Syst 5:100080
Ghodsvali M et al (2019) Review of transdisciplinary approaches to food-water-energy nexus. Environ Sci Pol 101:266–278
Glibo I, Koenigstorfer J (2023) Understanding the nexus of sustainable development and sport: the systemsthinking perspective. Sustain: Sci Pract Policy 19(1). https://doi.org/10.1080/15487733.2023.2240664
Grubert E, Sanders KT (2018) Water use in the United States energy system. Environ Sci Technol 52(11):6695–6703
Haase A, Wolff M, Rink D (2018) From shrinkage to regrowth: the nexus between urban dynamics, land usechange and ecosystem service provision. In Future City. Springer International Publishing, pp 197–219. https://doi.org/10.1007/978-3-319-59324-1_11
Halbe J et al (2015) Governance of transitions towards sustainable development. Water Int 40(5–6):877–894
Higgins CW, Najm MA (2020) An organizing principle for the water-energy-food nexus. Sustain Sci Pract Policy 12(19):8135
Hoff H (2011) Understanding the nexus. Background Paper for the Bonn2011 Nexus Conference
Hoff H et al (2019) A nexus approach for the MENA region—from concept to knowledge to action. Front Environ Sci 7:48
Jones JL, White DD (2022) Understanding barriers to collaborative governance. Environ Sci Pol 127:111–119
Kabisch N, Frantzeskaki N, Hansen R (2022) Principles for urban nature-based solutions. Ambio 51(6):1388–1401. https://doi.org/10.1007/s13280-021-01685-w
Koutra S, Balsells Mondejar M, Becue V (2022) The nexus of 'urban resilience' and 'energy efficiency' incities. Curr Res Environ Sustain 4(100118):100118. https://doi.org/10.1016/j.crsust.2021.100118

Kurian M (2017) The water-energy-food nexus: trade-offs, thresholds and transdisciplinary approaches tosustainable development. Environ Sci Policy 68:97–106. https://doi.org/10.1016/j.envsci.2016.11.006

Laspidou CS, Mellios NK, Spyropoulou AE, Kofinas DT, Papadopoulou MP (2020) Systems thinkingon the resource nexus: modeling and visualisation tools to identify critical interlinkages for resilient and sustainablesocieties and institutions. Sci Tot Environ 717(137264):137264. https://doi.org/10.1016/j.scitotenv.2020.137264

Lazaro LLB et al (2022) Ten years of research on the water-energy-food Nexus: an analysis of topics evolution. Front Water 4:859891

Leck H, Conway D, Bradshaw M, Rees J (2015) Tracing the water–energy–food nexus: description, theory and practice. Geogr Compass 9(8):445–460

Liu J et al (2017) Challenges in operationalizing the water–energy–food nexus. Hydrol Sci J 62(11):1714–1720

Lockwood M (2010) Good governance for terrestrial protected areas. J Environ Manag 91(3):754–766

Ma N et al (2023) Comprehensive review of food-energy-water nexus at the community scale. J Clean Prod 420:138311

Machell J et al (2015) The water energy food nexus-challenges and emerging solutions. Environ Sci Water Res Technol 1(1):15–16

Mcgrane SJ, Acuto M, Artioli F, Chen PY, Comber R, Cottee J, Farr-Wharton G, Green N, Helfgott A,Larcom S, Mccann JA, O'Reilly P, Salmoral G, Scott M, Todman LC, van Gevelt T, Yan X (2018)Scaling the nexus: Towards integrated frameworks for analysing water, energy and food. Geogr J. https://doi.org/10.1111/geoj.12256

Melo FPL, Parry L, Brancalion PHS, Pinto SRR, Freitas J, Manhães AP, Meli P, Ganade G, Chazdon RL (2020) Adding forests to the water–energy–food nexus. Nat Sustain 4(2):85–92. https://doi.org/10.1038/s41893-020-00608-z

Micheli P et al (2019) Doing design thinking. J Prod Innov Manag 36(2):124–148

Miralles-Wilhelm F (2016) Development and application of integrative modeling tools in support of food-energy-water nexus planning. J Environ Stud Sci 6(1):3–10

Nature (2016) Buzzword off. 'Nexus' is enjoying new-found popularity. But what does it actually mean? Nature 538:140. https://doi.org/10.1038/538140b

Newell JP, Goldstein B, Foster A (2019) A 40-year review of food–energy–water nexus literature and itsapplication to the urban scale. Environmental Research Letters: ERL [Web Site] 14(7):073003. https://doi.org/10.1088/1748-9326/ab0767

Owen A et al (2018) Identifying critical supply chains and final products. Appl Energy 210:632–642

Paavola M (2020) Transport and mobility in an urban circular economy. https://lutpub.lut.fi/handle/10024/160766

Purwanto A et al (2021) Water-energy-food nexus: critical review. Sustainability 13(4):1919

Ramaswami A, Boyer D, Nagpure AS, Fang A, Bogra S, Rao-Ghorpade BBEC (2017) An urban-systems framework to assess the trans-boundary food-energy-water nexus: implementation in Delhi, India. Environmental Research Letters: ERL [Web Site], 12, online

Rezaei Kalvani S et al (2024) Analysis of pros and cons in using the water–energy–food Nexus approach. Sustainability (Switzerland) 16(7):2605

Roggema R, Yan W (2021) The moveable nexus, transforming thinking on cities. In R. Roggema (ed), TransFEWmation: Towards design-led food-energy-water systems for future urbanization. Springer, Cham, pp 3–8

Romero-Lankao P et al (2017) The food-energy-water nexus and urban complexity. Nat Clim Chang 7(4):233–235

Schlör H et al (2018) The FEW-nexus city index—measuring urban resilience. Appl Energy 210:382–392

Shannak S, Mabrey D, Vittorio M (2018) Moving from theory to practice in the water–energy–food nexus: Anevaluation of existing models and frameworks. Water-Energy Nexus 1(1):17–25. https://doi.org/10.1016/j.wen.2018.04.001

Sperling JB, Berke PR (2017) Urban nexus science for future cities. Curr Sustain Renew Energy Rep 4(3):173–179

Sušnik J, Staddon C (2022) Evaluation of water-energy-food (WEF) nexus research. J Am Water Resour Assoc 58(6):1189–1198

Sušnik J, van der ZP (2017) Correlation and causation between the UN Human Development Index andnational and personal wealth and resource exploitation. Econ Res-Ekon Istraživanja 30(1):1705–1723. https://hrcak.srce.hr/file/285109

Tan Z et al (2025) Synergistic optimization and interaction evaluation of water-energy-food-ecology nexus under uncertainty. Sustain Cities Soc 124:106291

Tye MR, Wilhelmi OV, Pierce AL, Sharma S, Nichersu I, Wróblewski M, Goszczyński W, Wendel J, Laborgne P, Heyder M, Nichersu I (2022) The food water energy nexus in an urban context: connectingtheory and practice for nexus governance. Earth Syst Gov 12:100143. https://doi.org/10.1016/j.esg.2022.100143

United Nations (2012) The future we want. https://sustainabledevelopment.un.org/content/documents/733FutureWeWant.pdf

United Nations (2015) Transforming our world: the 2030 agenda for sustainable development. https://app.paperpile.com/view/?id=595545fd-9eb6-4b8a-9cb0-f499f0bfb264

Urban Nexus (2013a) Health and quality of life in urban areas. Urban nexus WP3 synthesis report. European Union's Seventh Framework Programme [FP7/2007-2013]

Urban Nexus (2013b) Nexus synthesis report: competing for urban land. European Union's Seventh Framework Programme [FP7/2007-2013]. http://www.urban-nexus.org.eu

Urban Nexus (2013c) Synthesis report: urban climate resilience. European Union's Seventh Framework Programme [FP7/2007-2013]

Valek AM et al (2017) Quantification of the urban water-energy nexus in México city. Sci Total Environ 590–591:258–268

Varbanov PS (2014) Energy and water interactions: implications for industry. Curr Opin Chem Eng 5:15–21

Vogt C, Zimmermann M, Brekke K (2014) Operationalizing the urban NEXUS. ICLEI–GIZ. http://resilient-cities.iclei.org/fileadmin/sites/resilient-cities/files/Full_papers/Urban_NEXUS_Publication_ICLEI-GIZ_2014_web.pdf

Wang X-C et al (2021) Extended water-energy nexus contribution to environmentally-related sustainable development goals. Renew Sust Energ Rev 150:111485

Weitz N, Nilsson M, Davis M (2014) A nexus approach to the post-2015 agenda. SAIS Rev Int Aff 34(2):37–50

White DJ et al (2018) The water-energy-food nexus in East Asia. Appl Energy 210:550–567

Yan W, Roggema R (2019) Developing a design-led approach for the food-energy-water nexus in cities. Urban Planning 4(1):123–138. https://doi.org/10.17645/up.v4i1.1739

Yan W, Murakami Y, Yasuda A, Makihara T, Fujimoto R, Nakayama S (2021) Developing the eXtended tangible user interface as an experimental platform for Geo CPS. In Sakurai M, Shaw R (eds) Emerging technologies for disaster resilience. Springer: Singapore, 1st ed., pp 11–32. https://doi.org/10.1007/978-981-16-0360-0_2

Yuan M-H, Lo S-L (2022) Principles of food-energy-water nexus governance. Renew Sust Energ Rev 155:111937

Zarei S et al (2021) Environmental sustainability: a review of the water–energy–food nexus. Aqua 70(2):138–154

Zhang C, Chen X, Li Y, Ding W, Fu G (2018) Water-energy-food nexus: concepts, questions and-methodologies. J Clean Prod 195:625–639. https://doi.org/10.1016/j.jclepro.2018.05.194

Zhang P et al (2019) Food-energy-water (FEW) nexus for urban sustainability. Resour Conserv Recycl 142:215–224

Zhou Q et al (2025) Translating planetary boundaries into city systems. Sustain Prod Consum 56:66–79

Chapter 2
Investing on Nature for Sustainable Ecosystem Services

Shuang Gan and Wanglin Yan

Abstract Sustainable urban development depends on natural resources, especially ecosystem services. However, as global urbanization accelerates, the conflict between urban expansion and natural resource conservation has become more apparent. With population trends shifting from growth to decline, this conflict may evolve in new ways. Balancing urban expansion and natural resource protection in this context is now a key challenge for sustainable development. Farmland, as one of the most impacted resources by urban expansion, deserves particular attention. This chapter focuses on how the nexus principles of investing on nature contributes to improve conservation of farmland ecosystems during the transition from population increase to decline. It highlights the importance of farmland in periphery of metropolitan regions, examines how population changes affect urban expansion and farmland use, and explores how urban expansion leads to farmland loss and reduced ecosystem services. Strategies for balancing urban development and farmland protection are also discussed. By investing in natural resources, promoting compact urban planning, and optimizing resource distribution, it is possible to use land more efficiently, protect ecosystems, and foster urban–rural coordination. These approaches can ease the pressure of urban expansion on farmland while achieving a balance between ecological, social, and economic benefits. This study provides practical insights and guidance for future urban planning and farmland conservation.

Keywords Urban sustainability · Urban expansion · Farmland preservation · Population growth and decline · Ecosystem services

S. Gan (✉)
Graduate School of Media and Governance, Keio University, Fujisawa, Kanagawa, Japan
e-mail: ganshuang3104@keio.jp

W. Yan
Faculty of Environment and Information Studies/Graduate School of Media and Governance, Keio University, Fujisawa, Kanagawa, Japan

W. Yan (ed.), *Urban Nexus Approach for Sustainable Cities*,
https://doi.org/10.1007/978-981-95-6156-8_2

1 Introduction

Natural resources, especially ecosystem services, provide essential support for human survival and urban development. There is a strong nexus between ecosystem services and urban development, with urban expansion being one of the primary manifestations of this relationship. However, urban expansion often involves the conversion of natural land, leading to the loss of ecosystem services and creating conflicts with natural resource conservation. Therefore, finding a balance between urban expansion and natural resource protection has become a critical challenge for achieving sustainable urban development.

Forests, grasslands, farmlands, wetlands, and rivers are vital natural resources that underpin ecosystem services. Among them, forests, wetlands, and farmlands are the most sensitive to urban expansion, with farmlands being particularly vulnerable (Ma et al. 2024). Unlike forests, farmlands are typically located on plains, making them a preferred target for urban development (Wang et al. 2022). This chapter focuses on farmlands and their ecosystem services. Farmland ecosystems, especially paddy fields, act as artificial wetlands and provide both ecological and environmental benefits. They not only supply food but also deliver key ecosystem services such as flood control, climate regulation, and biodiversity support to surrounding areas (Matsuno et al. 2006; Yoon 2009). However, the loss and fragmentation of farmlands can significantly degrade these services (Osawa et al. 2020; Katayama et al. 2015; Masumoto et al. 2003; Jiang et al. 2020).

Population is one of the main factors affecting land use and ecosystem changes (Li et al. 2015). As the population grows, the demand for land increases, leading to urban expansion. This often results in the conversion of natural land into built-up areas (Ramankutty et al. 2002), causing ecosystem damage and the loss of ecosystem services. Although global population growth is still the dominant trend, countries such as Japan and Germany, with low birth rates and aging populations (Hospers and Reverda 2015), have seen a decline since the late twentieth century. This trend is expected to spread to more countries, becoming a new global pattern. Population decline may present new challenges for sustainable urban development. On one hand, it reduces the demand for land, easing pressure on natural areas (Brosse 1983). On the other hand, it may cause problems such as vacant housing, inefficient land use, and a loss of biodiversity (Kobayashi et al. 2020).

Given these challenges, effective strategies are needed to balance urban expansion and farmland protection while adapting to population dynamics. In this process, natural investment, land-use efficiency, and service allocation are key nexus principles of sustainable land management, playing a vital role.

The principle of natural investment emphasizes preserving and enhancing ecosystem services by investing in natural capital, such as farmland, forests, and wetlands. It can be categorized into direct investment, physical capital investment, and human capital investment (Ruggeri 2009). In global farmland protection, current efforts focus on three key trends: (1) conserving agricultural landscapes to promote biodiversity, such as establishing protected zones; (2) advancing agricultural technology to improve resource efficiency; and (3) integrating ecological compensation with carbon markets to incentivize sustainable agriculture. These measures not only boost agricultural productivity but also strengthen urban ecological security and sustainability.

The efficiency principle focuses on optimizing resource allocation and land use to balance urban development with natural resource conservation. This study examines how to optimize urban layouts within limited land resources to minimize the impact of expansion on natural areas, particularly farmland. Globally, key strategies for improving land-use efficiency include: (1) spatial planning and land management policies, such as urban growth boundaries, land consolidation, and transit-oriented development; (2) economic incentives and land market mechanisms, such as land value and vacant land taxes; and (3) smart city innovations, including rooftop and urban farming. These approaches not only enhance land-use efficiency but also reduce the ecological impact of urban expansion.

The principle of service allocation emphasizes the fair distribution of resources between urban and rural areas to promote balanced development, ensuring that urban expansion does not come at the expense of rural communities and natural resources. Rural depopulation is a key driver of this imbalance (Pan et al. 2022), as population decline and aging lead to labor shortages, accelerating farmland abandonment and conversion. Additionally, disparities in infrastructure, such as education and healthcare, along with economic gaps, further widen the urban–rural divide. Globally, key strategies to enhance service allocation include: (1) improving infrastructure to ensure equitable access to public services; (2) fostering urban–rural integration to promote industrial linkages and balanced population distribution; and (3) implementing rural revitalization initiatives to strengthen agricultural resilience and boost rural economies.

It is essential to recognize that natural investment, efficiency, and service allocation are interconnected and mutually reinforcing. Natural investment establishes the ecological foundation for urban development, efficiency optimizes land use, and service allocation ensures the fair distribution of ecological and social benefits. Integrating these principles enables a dynamic balance between urban expansion and farmland protection, driving sustainable development.

The relationship between urban expansion and farmland reduction has been widely studied, focusing on factors such as the mechanisms of impact, spatial distribution and patterns, environmental and ecological effects, social and economic consequences, and models for prediction. However, most of these studies focus on population growth, with little attention given to how urban expansion and farmland changes are affected by population decline. To fill this gap, this study will explore the connection between population dynamics, urban expansion, and farmland change. It will examine how the relationship between urban expansion and farmland protection changes as the population shifts from growth to decline, and how to balance these two factors in the context of declining population, ultimately achieving sustainable urban development.

2 The Importance of Farmland Ecosystem Services

2.1 Definition and Classification of Farmland Ecosystem Services

The natural resources and ecosystem services provided by farmlands play an irreplaceable role in maintaining human societal development and ecological balance. These services include provisioning, regulating, cultural, and supporting functions.

1. *Provisioning services*: Farmlands are the primary production base for major global food crops, such as rice, wheat, and corn, directly meeting the basic survival needs of humans. In addition, farmlands can provide economic crops and fibers, such as cotton and flax, which are used as industrial raw materials, as well as energy crops such as sugarcane and rapeseed.
2. *Regulating services*: Paddy fields play an active role in flood control, groundwater recharge, climate regulation (carbon sequestration), water purification, and soil conservation. For example, studies show paddy fields help manage water resources and reduce flood risks through water retention, especially in urbanizing areas (Hao et al. 2015).
3. *Supporting services*: These services include nutrient cycling, soil fertility maintenance, and providing habitats for diverse species. Paddy fields maintain biodiversity by supporting a variety of flora and fauna, which is particularly relevant in traditional agricultural landscapes (Yan et al. 2019).
4. *Cultural services*: These fields hold sociocultural importance, especially in regions such as Asia, where rice farming is deeply integrated into cultural traditions and rural landscapes, contributing to the cultural identity and heritage of communities (Tran Thi et al. 2023).

Furthermore, agricultural ecosystems do not exist in isolation but rather form a complex landscape mosaic consisting of forests, rivers, and farmland, with paddy fields playing a particularly significant role. In Japan, this interconnected landscape is known as Satoyama (Natuhara 2013). Satoyama acts as a "backyard" for paddy fields, providing a more diverse ecological environment that supports a wider range of biodiversity and serves as a vital habitat for various species. For centuries, Satoyama has played an important role in shaping Japan's agricultural practices, ecological balance, and cultural heritage.

2.2 Trends in Farmland Changes and Their Impacts

Globally, farmland landscapes are undergoing significant and complex changes, mainly driven by urbanization, industrialization, and socioeconomic shifts. In rapidly urbanizing areas, large amounts of arable land are being converted into urban areas, leading to a sharp decline in farmland. For example, from 1990 to 2020,

Japan's arable land decreased by 16.98% (Ministry of Agriculture, Forestry and Fisheries), and from 2001 to 2015, China lost 738,200 hectares of arable land. At the same time, urbanization has changed the spatial structure of regional landscapes, causing farmland to become fragmented into scattered patches of various sizes and shapes. In some mountainous and remote areas, traditional farmland such as terraced fields are being abandoned due to difficult farming conditions and the loss of agricultural labor, further worsening farmland abandonment.

This trend of change has multiple effects on both the environment and social well-being. The loss and fragmentation of farmland reduce agricultural productivity, threaten food security, and hinder agricultural modernization. At the same time, the degradation of traditional agricultural landscapes weakens ecosystem services. Once paddy fields are converted into urban land, these lost services are rarely reversible (Lee et al. 2015). The most concerning functions affected by the reduction of paddy fields include biodiversity maintenance, flood regulation, and climate regulation. Research by Katayama et al. (2015) shows that the loss of paddy fields negatively impacts biodiversity by destroying habitats for wetland species and causing habitat fragmentation. Osawa et al. (2020) found that converting paddy fields into urban areas also reduces the region's ability to retain floodwater. Lee et al. (2015) showed that paddy fields, as urban green infrastructure, help reduce the urban heat island effect, and fewer paddy fields reduce this cooling effect. Additionally, as farmlands are abandoned, the surrounding environment may suffer, with risks such as increased fire hazards, loss of traditional landscape diversity, and further decline in agricultural ecosystem services. Therefore, in the face of rapid urbanization and economic development, finding ways to protect farmland while optimizing its ecosystem services has become a key issue in global sustainable development.

3 The Nexus Between Population Change, Urban Expansion, and Farmland Protection

3.1 The Relationship Between Urban Expansion and Farmland Loss During Population Growth

During the population growth phase, urban land expands into suburban areas, leading to a rapid increase in urban and residential land, which results in farmland being occupied and fragmented. Population growth also drives the demand for living and working spaces, further accelerating the conversion of farmland to urban areas (Zheng et al. 2022). Urbanization not only leads to the conversion of farmland but also contributes to its abandonment. As rural populations migrate to cities (Kamada and Nakagoshi 1997) and many farmers shift to nonagricultural work for higher incomes, the farming population declines, labor shortages occur, and farmland abandonment increases (Grădinaru et al. 2015). Rapid urbanization also affects food production and consumption systems (Godfray et al. 2010), which further promotes farmland conversion (Davis et al. 2016). Government policies play a key role in this process. For example, in Vietnam, the government has used land expropriation policies to promote the conversion of farmland to urban land (Nguyen and Kim 2020). In addition to external factors such as urbanization, population growth, and policies, studies show that internal factors such as elevation, slope, soil quality, and precipitation can either promote or slow down farmland conversion (Bürgi et al. 2004).

3.2 The Relationship Between Urban Expansion and Farmland Loss During Population Decline

During the population decline phase, aging becomes more severe (Kabisch 2005), and with significant population loss, the agricultural labor force further decreases, leading to more farmland conversion. In theory, population decline should reduce the demand for housing and infrastructure, causing built-up areas to shrink. However, in many shrinking cities, land expansion and urban shrinkage often occur simultaneously (Großmann et al. 2013; Slach et al. 2019). Haase et al. (2014) found that even during urban shrinkage, a decrease in population and increased aging may still create a demand for more housing, leading to further farmland conversion. However, this conversion of farmland to urban land is less than in the population growth phase (Lauf et al. 2016). On the other hand, urban shrinkage can also have positive effects on farmland. It increases the availability of open space, creating opportunities for farmland expansion, though this effect is delayed (Haase et al. 2012).

3.3 Population Change, Urban Expansion, and Farmland Protection in the Tokyo Metropolitan Area

The Tokyo Metropolitan Area is one of the most densely populated urban regions in the world. Although the overall population has continued to grow, this growth has been unevenly distributed—concentrated mainly in areas designated for urban development, while some peripheral zones, particularly peri-urban areas under urbanization control, have experienced population decline since 2015 (Table 2.1). These peri-urban areas contain extensive farmland that provides crucial ecosystem services to the metropolitan region (Fig. 2.1). However, they are increasingly challenged by the conflicting pressures of urban expansion and farmland preservation. As population growth in the urban core continues to drive demand for new development, urban expansion has pushed outward, encroaching on suburban farmland. From 2000 to 2020, the built-up land area in the peri-urban areas of the Tokyo Metropolitan Area increased from 273,150 hectares to 308,584 hectares, while the total farmland area declined from 1.70×10^7 hectares to 1.56×10^7 hectares, reflecting a clear trend of urban sprawl and agricultural land loss.

Table 2.1 Trends in population, built-up land, and farmland change in the peri-urban areas of Tokyo Metropolitan Area

Year		2000	2005	2010	2015	2020
Population (persons)	Total	24,100,295	24,713,313	25,614,035	25,989,154	26,305,422
	UPA	17,994,255	18,472,540	19,259,137	19,622,574	19,966,128
	UCA	5,869,598	5,992,268	6,096,046	6,097,812	6,087,686
Built-up land area (ha)		273,151	286,300	295,987	302,740	308,584
Farmland area (ha)		178,066	167,654	161,720	157,181	156,390
Forest area (ha)		172,478	163,017	160,328	159,911	154,954
Waterbody area (ha)		12,821	13,257	13,956	12,173	10,519
Others area (ha)		46,251	52,539	50,776	50,762	52,320
Total land area (ha)		682,767	682,767	682,767	682,767	682,767

UPA urbanization promotion area, *UCA* urbanization control area in zoning

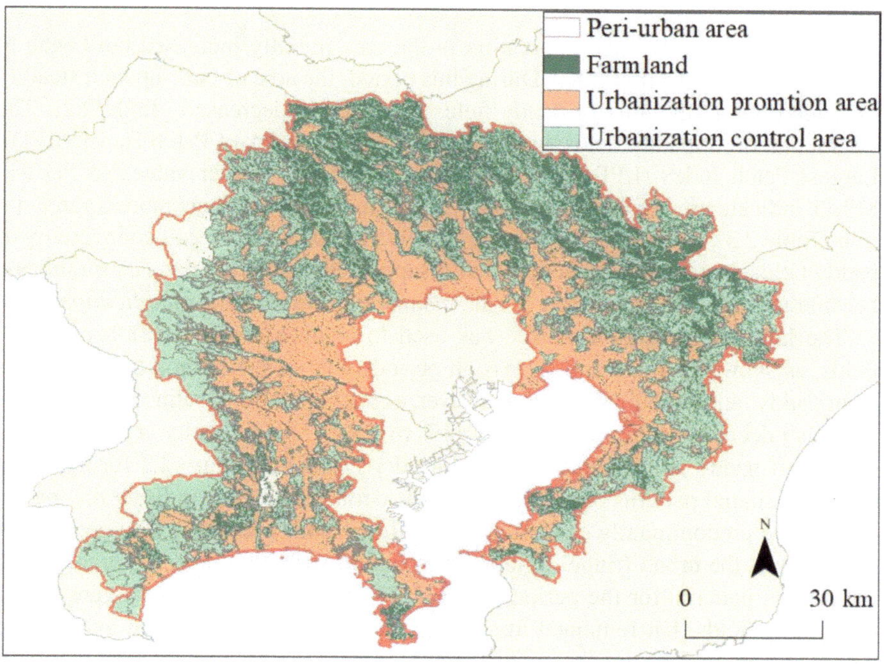

Fig. 2.1 Farmland distribution in the peri-urban areas of Tokyo Metropolitan Area

3.4 *Case Study Analysis*

The Odawara area is located about 80 km southwest of Tokyo, forming a small urban zone (Fig. 2.2). The Sakawagawa River flows from northeast to southwest through the region, with both the city and farmland surrounding the river.

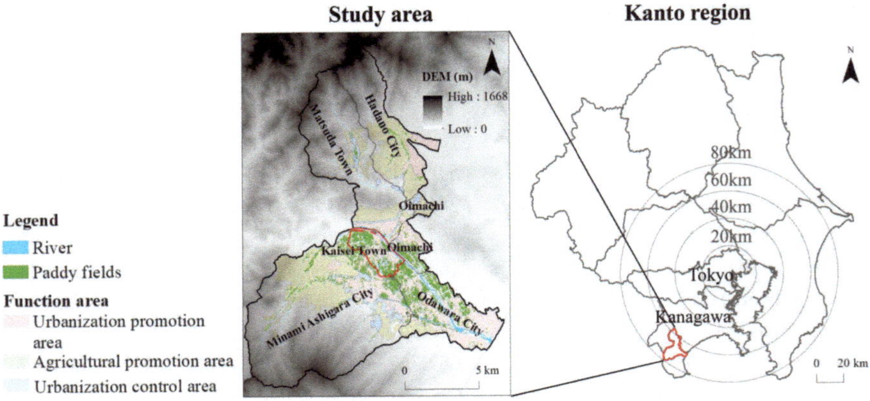

Fig. 2.2 Location and basic information of the study area

From 1995 to 2015, the population in the area initially increased but began to decline significantly after 2010. During this period, the area of built-up land steadily expanded, while the area of paddy fields consistently decreased (Table 2.2). The fragmentation of paddy fields was analyzed using the index of Patch Density (PD), Largest Patch Index (LPI), and Split Indices (SPLIT). Higher values of PD and SPLIT indicate greater fragmentation, while a smaller LPI reflects more fragmentation (Table 2.3). The data shows that from 1995 to 2015, the overall connectivity of paddy fields decreased. Smaller patches disappeared, and larger patches shrank and became more fragmented. (Gan & Yan, *manuscript submitted for publication*).

The land use transition matrix was used to analyze the conversion of paddy fields, as shown in Table 2.4. Across all periods, the most common transition was from paddy fields to built-up land. However, as the population declined, this conversion also decreased, dropping from 4.44% to 2.68%. Additionally, a considerable portion of paddy fields was converted into dry farmland. Figure 2.3 further illustrates the spatial patterns of paddy field conversion. The results indicate that paddy fields were predominantly converted into built-up land, with these changes mainly occurring at the urban fringe. In terms of spatial distribution, Fig. 2.3 presents the conversion patterns for the periods 2005–2010 and 2010–2015. The figure shows that paddy fields that remained unchanged were primarily located in agricultural promotion areas, whereas the conversion of paddy fields to built-up land was concentrated in urban promotion areas, particularly along the boundary between urban and agricultural regions.

Based on an analysis of 128 census blocks within the study area, multiple linear regression was used to identify the key driving factors of paddy field loss during periods of population growth and decline. As shown in Table 2.5, during periods of population growth, the expansion of built-up land was the most significant driver of

Table 2.2 Changes in population, built-up land area, and paddy field area in the study area from 2005 to 2015

Factors	Year 2005	2010	2015
Population (1000 person)	195.41	197.60	193.25
Built-up land area (ha)	3702.72	3900.29	4012.73
Paddy field area (ha)	837.87	752.79	726.60

Table 2.3 Changes in the fragmentation of paddy fields

Index	Year 2005	2010	2015
PD	6.25	6.51	6.02
LPI	0.67	0.42	0.41
SPLIT	11,925.09	26,564.04	27,196.42

PD Patch Density, *LPI* Largest Patch Index, *SPLIT* Split Index

Table 2.4 Land use transfer matrix in 2005–2010 and 2010–2015 (%)

	PF	DL	UL	AL	BL	Total
2005–2010						
PF	90.58	3.30	0.69	0.99	4.44	100.00
DL	0.64	95.22	0.60	0.92	2.62	100.00
UL	0.39	1.97	92.19	3.96	1.49	100.00
AL	2.13	30.43	2.92	59.05	5.48	100.00
BL	0.17	0.69	0.52	0.06	98.56	100.00
2010–2015						
PF	96.18	0.64	0.22	0.28	2.68	100.00
DL	0.05	96.60	0.94	0.13	2.28	100.00
UL	0.28	0.60	94.38	0.01	4.73	100.00
AL	1.51	8.99	1.39	83.48	4.64	100.00
BL	0.04	0.36	0.31	0.00	99.28	100.00

Note: Paddy field (PF); dry land (DL); unused land (UL); abandoned land (AL); built-up land (BL)

paddy field loss, with larger expansions leading to greater losses. Additionally, the rate of aging played an important role, exerting a negative influence on paddy field loss. In contrast, during periods of population decline, paddy field loss was no longer solely driven by built-up land expansion. Instead, it was influenced by a combination of factors, including paddy field fragmentation, built-up land expansion, changes in household numbers, and the rate of farmer population change. Moreover, during this period, the influence of agricultural protection zones became more pronounced, indicating an improvement in their effectiveness in safeguarding farmland.

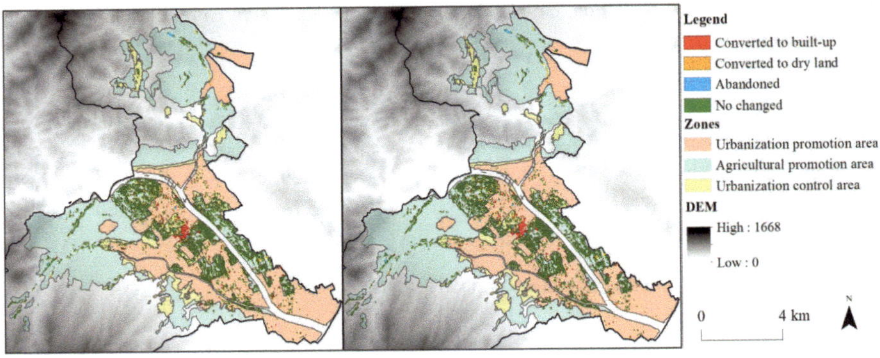

Fig. 2.3 Spatial distribution of paddy field conversion

Table 2.5 Results of multiple linear regression

Factors	Population growth	Population decline
PCR	0.16	−0.09
FCR	0.02	−0.08*
ACR	−0.23*	−0.01
HCR	−0.10	0.14
BIR	1.03***	0.22***
PD	−0.07	0.22***
AA	0.10	−0.13*
Adjusted R^2	0.84	0.49
p-value	<0.01	<0.01

FCR farmer change rate, *ACR* aging change rate, *HCR* household change rate, *PD* paddy density, *AA* area of agricultural promotion areas
* and *** represent 0.05 and 0.001 significance levels, respectively

4 Efforts to Balance Urban Expansion and Farmland Protection

4.1 Ecological Protection and Urban Expansion Balancing Strategy Driven by Natural Investment

Natural investment provides an important pathway to balance urban expansion and paddy field protection, with its core focus on achieving sustainable development through the protection, restoration, and optimization of natural resources. Land-use zoning, as a key tool, can effectively guide the implementation of natural investment. By scientifically delineating land functional zones, it enables the coordination of ecological and urban functions.

The successful practices of various countries demonstrate the important role of natural investment in balancing urban and agricultural land resources. For example, Japan has effectively limited urban sprawl and protected agricultural land through compact city development and land-use zoning policies, particularly by designating agricultural promotion areas, which have helped create a stable environment for agricultural communities. In the Netherlands, the "Green Heart" protection area uses land-use zoning to clearly define the boundaries between agricultural and urban land, ensuring the long-term sustainable use of paddy fields. In recent years, China's establishment of permanent basic farmland and the ecological protection red line system have also integrated land-use zoning with natural investment, prioritizing the protection of ecosystem functions and high-value agricultural resources amid urban expansion.

With the help of land-use zoning, local governments can incorporate green infrastructure investments into urban planning, such as ecological corridors, wetland restoration, and agricultural modernization support projects. These measures not only reduce the damage caused by urban expansion to natural resources but also enhance the productivity and ecological value of farmland by improving the ecological service functions of natural resources, thus achieving a win-win situation for both urban expansion and farmland protection.

4.2 Compact City and Efficient Land Use Path for Farmland Protection

The efficiency principle of nexus emphasizes the optimal allocation and utilization of resources, with the core idea of maximizing land-use benefits, providing strong support for balancing urban expansion and farmland protection. Within this framework, the compact city model becomes an important approach to achieving efficient land use. By increasing development density within the city and optimizing spatial layout, it reduces the encroachment on farmland resources.

Japan's compact development integrates transit-oriented development (TOD), which focuses on high-density residential and commercial areas around transit stations. By concentrating urban functions and transportation networks around rail stations, Tokyo effectively reduces the encroachment on surrounding farmland. Singapore maximizes land-use efficiency through high-density development and a well-established public transportation network, while preserving key natural resources. In China, the "Land Reclamation and Conversion" policy and intensive land use have optimized urban–rural land allocation, protecting a large amount of farmland. France also promotes compact development through "Urban Planning Contracts," prioritizing the use of existing urban space, reducing urban sprawl, and effectively maintaining the balance between agricultural ecology and urban functions through farmland protection policies. The Netherlands has effectively limited urban sprawl by promoting high-density development within cities. Its urban planning emphasizes optimizing public transportation and bicycle networks to reduce the encroachment on agricultural land while enhancing the efficiency of urban space usage. Similarly, Curitiba in Brazil has focused urban functions along bus rapid transit (BRT) corridors through high-density development. By adopting mixed land-use planning, the city has successfully reduced urban sprawl and improved land-use efficiency.

The compact city model not only improves land-use efficiency but also optimizes infrastructure layout. By promoting urban renewal and the redevelopment of old industrial areas instead of new developments, it further alleviates the pressure of urban expansion on farmland and achieves coordinated development between urban areas and agriculture.

4.3 Service Distribution-Oriented Urban–Rural Coordinated Development and Farmland Protection Strategy

The service distribution principle emphasizes the equitable allocation of resources and services, particularly in the process of urban expansion and farmland protection. By balancing the development needs of urban and rural areas, it aims to achieve the synergy of social, economic, and ecological benefits. Effective service distribution can reduce the one-way migration of populations to urban areas caused by uneven resource allocation, fundamentally alleviating the pressure of urban expansion on farmland resources.

Germany's urban–rural integration development strategy provides valuable experience. By extending public services and infrastructure to rural areas, Germany has not only improved the quality of life for rural residents but also reduced the overreliance on urbanization. The Netherlands' "Balanced Development of Towns and Villages" policy enhances the attractiveness of rural areas by improving education, healthcare, and transportation services, thereby protecting farmland resources while promoting urban expansion. China's recent rural revitalization strategy also incorporates the service distribution principle. By channeling resources and funds to rural areas, it has boosted economic vitality and promoted agricultural modernization and farmland protection. Korea's "Saemaul Undong" (New Village Movement) aimed to narrow the urban–rural gap by improving rural infrastructure, enhancing agricultural production conditions, and increasing farmers' incomes. By reducing excessive rural-to-urban migration, the movement helped alleviate the pressure of urban expansion on farmland to some extent, promoting balanced urban–rural development.

The urban–rural collaborative development guided by the service distribution principle not only effectively alleviates the pressure of urban expansion on natural resources but also creates more opportunities in rural areas, promoting social equity. This strategy, while achieving a balance between urban development and agriculture, also fosters sustainable development between urban and rural areas.

5 Conclusion

This study explores the dynamic changes in the relationship between urban expansion and farmland protection during periods of population growth and decline. The findings suggest that urban expansion typically leads to a reduction in farmland and loss of ecosystem services, with population growth intensifying this trend. However, population decline may slow down urban expansion, though it can also result in inefficient land use.

Through nexus strategies such as natural investment, efficient land use, and coordinated urban–rural development, farmland can be protected and ecosystem service functions enhanced while accommodating urban expansion. Successful cases, such as Japan's compact city development, the Netherlands' Green Heart protection area, and China's farmland protection policies, provide valuable experiences for addressing the conflict between urban expansion and farmland preservation. In the future, urban planning should focus on ecological protection and sustainable development, promoting urban–rural coordination and ecological balance.

Acknowledgments This work was jointly supported by Taikichiro Mori Memorial Research Grants, 2023.

References

Brosse U (1983) Population decline and the environment. Raumforsch Raumordn 41(5–6):189–195

Bürgi M, Hersperger AM, Schneeberger N (2004) Driving forces of landscape change—current and new directions. Landsc Ecol 19:857–868

Davis KF, Gephart JA, Emery KA, Leac AM, Galloway JN, D'Odorico P (2016) Meeting future food demand with current agricultural resources. Glob Environ Chang 39:125e132

Godfray HC, Beddington JR, Crut IR, Haddad L, Lawrence D, Muir JF, Pretty J, Robinson S, Thomas SM, Toulmin C (2010) Food security: the challenge of feeding 9 billion people. Science 327:812e818

Grădinaru SR, Iojă CI, Onose DA, Gavrilidis AA, Pătru-Stupariu I, Kienast F, Hersperger AM (2015) Land abandonment as a precursor of built-up development at the sprawling periphery of former socialist cities. Ecol Indic 57:305–313

Großmann K, Bontje M, Haase A, Mykhnenko V (2013) Shrinking cities: notes for the further research agenda. Cities 35:221–225

Haase A, Rink D, Großmann K (2012) Urban shrinkage as a shock? Challenges for modelling human environmental interaction. International Congress on Environmental Modelling and Software, p 101

Haase D, Haase A, Rink D (2014) Conceptualizing the nexus between urban shrinkage and ecosystem services. Landsc Urban Plan 132:159–169

Hao L, Sun G, Liu Y, Wan J, Qin M, Qian H, Liu C, Zheng J, John R, Fan P, Chen J (2015) Urbanization dramatically altered the water balances of a paddy field dominated basin in southern China. Hydrol Earth Syst Sci 19:3319–3331

Hospers G, Reverda N (2015) Introduction: population decline. Springer, pp 1–5

Jiang P, Li M, Cheng L (2020) Dynamic response of agricultural productivity to landscape structure changes and its policy implications of Chinese farmland conservation. Resour Conserv Recycl 156:104724

Kabisch S (2005) Empirical analysis on housing vacancy and urban shrinkage. In: Vestbro DU, Hürol Y, Wilkinson N (eds) Methodologies in housing research, Gateshead. Urban International Press, Newcastle upon Tyne, pp 188–205

Kamada M, Nakagoshi N (1997) Influence of cultural factors on landscapes of mountainous farm villages in western Japan. Landsc Urban Plan 37(1/2):85–90
Katayama N, Baba YG, Kusumoto Y, Tanaka K (2015) A review of post-war changes in rice farming and biodiversity in Japan. Agric Syst 132:73–84
Kobayashi Y, Higa M, Higashiyama K, Nakamura F (2020) Drivers of land-use changes in societies with decreasing populations: a comparison of the factors affecting farmland abandonment in a food production area in Japan. PLoS One 15(7):e0235846
Lauf S, Haase D, Kleinschmit B (2016) The effects of growth, shrinkage, population aging and preference shifts on urban development—a spatial scenario analysis of Berlin, Germany. Land Use Policy 52:240–254
Lee YC, Ahern J, Yeh CT (2015) Ecosystem services in peri-urban landscapes: the effects of agricultural landscape change on ecosystem services in Taiwan's western coastal plain. Landsc Urban Plan 139:137–148
Li F, Zhang SW, Bu K, Yang JC, Wang Q, Chang LP (2015) The relationships between land use change and demographic dynamics in western Jilin province. J Geogr Sci 25(5):617–636
Ma S, Deng G, Wang L, Hu H, Fang X, Jiang J (2024) Telecoupling between urban expansion and forest ecosystem service loss through cultivated land displacement: a case study of Zhejiang Province, China. J Environ Manage 357:120695
Masumoto T, Nozoe M, Yoshimura A, Matsuda S (2003) A paddy runoff model in hilly rural areas for evaluating runoff change due to abandoned rice cultivation. Trans Jpn Soc Irrig Drain Rural Eng 71(2):35–44. (in Japanese with English abstract)
Matsuno Y, Nakamura K, Masumoto T, Matsui H, Kato T, Sato Y (2006) Prospects for multifunctionality of paddy rice cultivation in Japan and other countries in monsoon Asia. Paddy Water Environ 4:189–197
Natuhara Y (2013) Ecosystem services by paddy fields as substitutes of natural wetlands in Japan. Ecol Eng 56:97–106
Nguyen Q, Kim DC (2020) Reconsidering rural land use and livelihood transition under the pressure of urbanization in Vietnam: a case study of Hanoi. Land Use Policy 99:104896
Osawa T, Nishida T, Oka T (2020) High tolerance land use against flood disasters: how paddy fields as previously natural wetland inhibit the occurrence of floods. Ecol Indic 114:106306
Pan M, Huang Y, Qin Y, Li X, Lang W (2022) Problems and strategies of allocating public service resources in rural areas in the context of county urbanization. Int J Environ Res Public Health 19(21):14596. https://doi.org/10.3390/ijerph192114596
Ramankutty N, Foley JA, Olejniczak NJ (2002) People on the land: changes in global population and croplands during the 20th century. Ambio 31(3):251–257
Ruggeri J (2009) Government investment in natural capital. Ecol Econ 8:1723–1739
Slach O, Bosak V, Krticka L, Novacek A, Rumpe P (2019) Urban shrinkage and sustainability: assessing the nexus between population density, urban structures and urban sustainability. Sustainability 11(15):4142
Tran Thi N, Trink Phuong N, Phung Thanh T (2023) Assessment paddy rice ecosystem services on awareness and satisfaction level of local people in Hongthai, Vietnam. Int J Sci Res Manag 11(1):237–243. https://doi.org/10.18535/ijsrm/v11i01.fe01
Wang Y, Dong P, Liao S, Zhu Y, Zhang D, Yin N (2022) Urban expansion monitoring based on the digital surface model—a case study of the Beijing–Tianjin–Hebei plain. Appl Sci 12:5312
Yan F, Zhang S, Su F (2019) Variations in ecosystem services in response to paddy expansion in the Sanjiang plain, Northeast China. Int J Agric Sustain 17:158–171
Yoon CG (2009) Wise use of paddy fields to partially compensate for the loss of natural wetlands. Paddy Water Environ 7:357–366
Zheng YH, Long HL, Chen KQ (2022) Spatio-temporal patterns and driving mechanism of farmland fragmentation in the Huang-Huai-Hai plain. J Geogr Sci 32(6):1020–1038

Part II
Urban Nexus Methods

Chapter 3
Nexus Approaches and Methods for Urban Sustainability

Wanglin Yan

Abstract Food, energy, and water (FEW) are not only the foundation of survival but also represent material wealth and spiritual fulfillment. Traditionally, urban planning has focused on land, buildings, and infrastructure, often neglecting the sustainability of resource supply, quality of life, and environmental impact. The urban FEW nexus perspective addresses these shortcomings. Approaching FEW from an urban design perspective can visualize scientific research outcomes and bring urban issues into reality. The author advocates for the moveable nexus method, a design-led Urban Nexus research approach. It consists of three components: (1) capturing urban systems at multiple scales based on FEW, (2) visually evaluating the current state of urban decarbonization and the effectiveness of initiatives using the environmental load indicator FEWprint, and (3) visualizing the connections among diverse stakeholders through actor networks to promote cocreation. This chapter reviews Urban Nexus research methods, explains the M-NEX approach, and introduces its application.

Keywords Urban Nexus · Design-led Urban Nexus · Environmental load · Actor-network · CO_2 emission

1 Introduction

Urbanization has expanded huge area of built-up development and pushed vast amount of people working and living in urban and suburban cities in last century. The sustainability of cities is questioned in resource management, quality of life, and environment impacts. Cities are sustained by food, energy, and water (FEW) transported from distant locations, imposing significant environmental burdens.

W. Yan (✉)
Faculty of Environment and Information Studies/Graduate School of Media and Governance, Keio University, Fujisawa, Kanagawa, Japan
e-mail: w_yan@keio.jp

W. Yan (ed.), *Urban Nexus Approach for Sustainable Cities*,
https://doi.org/10.1007/978-981-95-6156-8_3

Consequently, cities consume 70% of energy and account for 86% of CO_2 emissions (Yan and Roggema 2019). Water consumes significant energy from its source to households. For urban residents, food, energy, and water (FEW) are the foundation of life, provided as services by governments and businesses, and its sufficiency is essential for maintaining life and quality of living. Thus, FEW is not only a foundation for survival but also a measure of material wealth and spiritual fulfillment. Traditional urban planning has focused on land, buildings, and infrastructure, often neglecting the sustainability of resource supply, consumption life, and environmental impact.

Urban FEW nexuses must consider the following three dimensions: (1) stable supply of FEW resources (resilience), (2) improvement in quality of life (QoL), and (3) reduction of environmental load (CO_2). The first dimension focuses on production and supply, the second on consumption and lifestyle, and the third on the global environment. Urban sustainability requires rethinking the relationships among FEW (Yan and Roggema 2019). Three dimensions correspond to the complex relationships in cities are as follows:

- *Relationship between FEW production and consumption sites (spatial proximity, urban, and regional scales)*: While FEW is closely interconnected at production sites, it is often segregated by sector in cities. The challenge is to determine the extent of local production and consumption and how to shorten the distance between production and consumption sites.
- *Relationship between cost and benefit (value judgments by businesses and individuals)*: The production and provision of FEW services come with environmental costs. Sustainable development requires mechanisms that ensure personal benefits and corporate profits without increasing environmental burdens, creating marginal effects.
- *Relationship between workplaces and living spaces (lifestyle and community roles)*: Urban life has traditionally symbolized material wealth, with FEW seen as something that can be obtained with money. However, cities are not just places of work but also spaces for living and fulfillment. It is time to reconsider how we work and live in cities, as long commutes and suburban living may not be desirable lifestyles.

Researchers have managed to investigate, analyze, and model the relationships qualitatively and quantitatively. No simple solutions can sort the intertwined relationships easily. Urban Nexus addresses the problem by focusing on urban dynamic systems with nodes, interlinkages, and interactions for integrative effects. Since the nexus concept has reemerged in 2011, it gained widespread attention. Substantial research has been published by using systems thinking or resilience thinking methods to investigate nexus effects. This has led to criticism that nexus approaches are conceptually driven and lack novelty. The borrowed methods and tools are not specifically designed to evaluate nexus interactions and feedback.

This chapter first reviews nexus methods, then introduces nexus design approach to model the three relationships. Stemmed from the native nexus guiding principles, the nexus design method consists of three components: (1) capturing urban systems at multiple scales based on FEW, (2) visually evaluating the current state of urban decarbonization and the effectiveness of initiatives using the environmental load indicator FEWprint, and (3) visualizing the connections among diverse stakeholders through actor networks to promote cocreation.

2 Methods and Tools for Nexus Approach

2.1 Outline of Methods for Nexus Approach

As an emerging academic field, borrowing traditional methods is not inherently bad, and leveraging them effectively is a starting point.

Zhang et al. (2019) has summarized systems nexus methods concisely in space of the types of methods and research objectives as illustrated in Fig. 3.1. The methods were categorized into environmental load assessment, system analysis and simulation, and management optimization. Research objectives include understanding and quantifying nexus effects, evaluation and prediction, and integration and optimization. Some methods and objectives overlap, making it difficult to separate them strictly.

Environmental load assessment includes input–output analysis and life cycle assessment. Evaluation, prediction, and simulation include network analysis, system dynamics modeling, and evaluation indicators. Integration and optimization

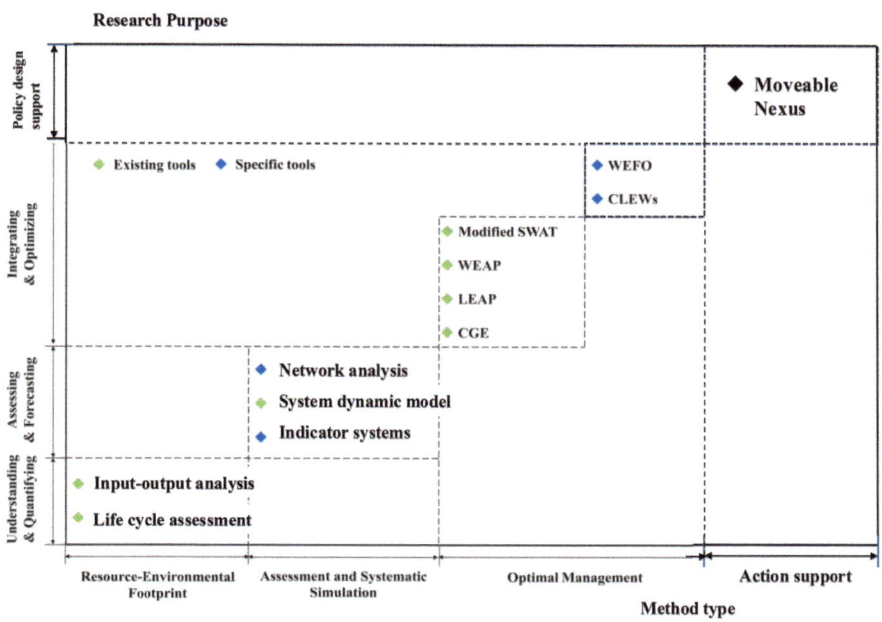

Fig. 3.1 Landscape of nexus approaches. (Extended by author from Zhang et al. 2019)

include General Equilibrium Models (CGE), Long-range Energy Alternatives Planning (LEAP), Water Evaluation and Planning (WEAP) tool, and Soil and Water Assessment Tool (SWAT) (Howells et al. 2013). These methods are used to incorporate multiple sectors into traditional models, reflecting the new perspective of nexus as a system analysis tool. Other traditional methods such as LCA, MFA, and IAM are also used, often focusing on energy, water, and food from a governance perspective, with most studies conducted at national or regional scales. However, as indicated in Fig. 3.1, this review uses nexus approach as a collection of analytic method, lacking perspectives to support policymaking and actions in practice.

The engagement of multiple stakeholders in nexus research and practice is also considered important as shown in Fig. 3.2 by Ghodsvali et al. (2019). The authors highlighted the perspective of social participation by using similar two-dimensional space similar to Zhang et al. (2019), with research methods in horizontal axis and research purposes in vertical axis.

The participatory nexus methods are also often categorized as "analytical" or "dialogical." Analytical methods involve engaging specific stakeholders in specific shared activities, while dialogical methods engage stakeholders in the implementation process to elicit diverse values and commitments. Techniques such as living labs, collaborative design, and online communication are used for vision-building and social learning. Workshops and participatory scenario analysis are commonly used. The functionality of these methods depends on the extent and purpose of stakeholder involvement.

The systems approach is quantitative while the participatory approach is qualitative. The former aims to clarify the hidden interlinkages and nexus effects of sectors

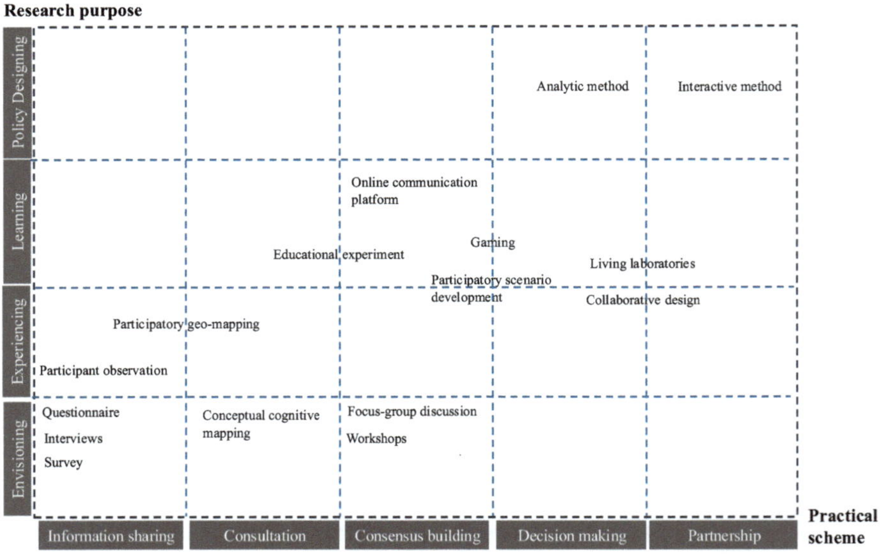

Fig. 3.2 Multiple stakeholder engagement within nexus research. (Extended by author from Ghodsvali et al. 2019)

for improving cooperations with scientific evidence, while the latter directly advocates the connectivity to stakeholders with intensive communications. There are always challenges in practice to integrate the two approaches in limitation of resources.

2.2 Input–Output Analysis

The primary benefit of the nexus approach is tracing connections across different industries and sectors, revealing trade-offs and synergies. Input–output tables are created at national, regional, and municipal levels. Cross-sector relationships are well represented in economic input–output tables. Converting monetary flows to material flows allows for the evaluation and analysis of resource-use efficiency, consequently interlinkages of sectors (Munksgaard et al. 2005; White et al. 2018). This is known as Input–Output Analysis (IOA). Using IOA in nexus analysis reveals the structure of production and consumption and their material flows, moving beyond the black box of input–output relationships. Extending this to estimate CO_2 emissions and waste is called Environmental IOA (EIOA) or Environmentally Extended IOA (EEIO) (Schaffartzik et al. 2014).

Input–output tables have limitations, such as coarse industry classifications, uniform price conversions for monetary and material flows, exclusion of unpaid labor, and data availability issues at national, regional, and urban levels (Kitzes 2013; Schaffartzik et al. 2014). Comparing datasets across countries often relies on assumptions, failing to account for diversity within parameters. Environmental accounting is useful for annual national data reconciliation but often cannot be used for smaller spatial and temporal scales. For example, average energy intensity comparisons are often constructed for cross-country comparisons but cannot account for differences in national energy systems. Additionally, information is often fragmented and domain-specific, making it difficult to provide insights into cross-sector dependencies (Mcgrane et al. 2018).

2.3 Life Cycle Assessment

Life Cycle Assessment (LCA) is a widely accepted environmental management tool for quantifying resource consumption and environmental and health impacts over the entire life cycle of a process, product, or activity, consequently cross sectors. LCA consists of four main stages: (1) goal and scope definition, (2) life cycle inventory analysis, (3) life cycle impact assessment, and (4) interpretation of results. LCA uses existing software (e.g., SimaPro) (Herrmann and Moltesen 2015) and inventory databases (e.g., Ecoinvent) (Pfister et al. 2016) to identify physical connections, flows, and environmental impacts of defined systems, making it a useful tool for tracing nexus from a physical perspective.

LCA is widely used in food production and consumption assessments and water–energy nexus studies, but how it can be used to analyze nexus without focusing on individual sectors remains unclear. Additionally, LCA is challenging for full-domain nexus analysis as it requires considering all sectors and activities within a region. New spatiotemporal dynamic LCA methods contrast with static, nonspatial traditional LCA methods, offering potential for exploring nexus in geographic contexts and combining with future scenarios (Maier et al. 2017; Raschio et al. 2017). However, building spatially and temporally explicit inventory databases requires significant data, including process and element flow locations, times, and background environmental characteristics (Mcgrane et al. 2018). This requirement limits the application of LCA in a specific industrial scope.

2.4 System Dynamics

Dynamics emphasize interactions of elements in a system. System dynamics (SD) is an approach to understanding and modeling the interactions in complex natural and social systems (Wolstenholme 1982; Meadows 2009). The Earth's environment, from global to micro scales, continuously flows materials and energy from inputs to outputs, providing resource accumulation (stocks) and services (flows) (Meadows 2016). SD quantitatively describes the connections within such a system and dynamically simulates interactions between elements, making it widely used as a technical method aligned with nexus thinking (Chhipi-Shrestha et al. 2017; Cai et al. 2019). Various tools have been developed for SD (Bala et al. 2017; Sterman 2001), with Vensim being a representative example (Größler et al. 2003; Kirkwood 1998; García 2020).

Building the systems dynamics model requires full understanding of the quantitative interlinkages of nexus nodes. The functionality of the system dynamics model will depend on the coverage of elements, the data availability of the related factors with acceptable spatial and temporal accuracies.

2.5 Sustainability Indicators

Indicator development is a popular approach to evaluate integrative effects. This is the same in nexus research. The success of an indicator depends on development objectives and data availability (Mcgrane et al. 2018). The application of indicators in nexus research is beginning to emerge but is progressing slowly (Willis et al. 2016). Existing evaluation indicators often have numerous subindicators and require large amounts of input data from diverse sources. Many indicators are designed for specific applications at diverse levels, lacking scheme to integrate and compare comprehensively.

Ecological footprint (EF) is a unique indicator by using unit of global land area to interactively evaluates ecosystem, water, land, carbon, energy, and material use efficiency for exploring sustainable boundaries (Santamouris et al. 2007; Mcmanus and Haughton 2006). National EF allocates environmental loads to final consumers of products and services, showing differences in global environmental loads among countries (WWF). The main difference between footprints and traditional environmental accounting is the consideration of direct and indirect resource consumption (upstream consumption). Footprint studies apply various approaches corresponding to the analysis object and scale (e.g., from products to national scales) (Fang and Heijungs 2015; Fang et al. 2016; Fang et al. 2015; Vanham et al. 2019). Mabhaudhi et al. (2019) proposed Nexus City Index but a collection of individual indicators.

2.6 Participatory Workshops

Participatory workshops, based on solid social learning theory, bring stakeholders together to integrate knowledge and perspectives, generating qualitative and quantitative insights. Workshops ground knowledge in the field, providing participants with perspectives to support innovative and appropriate policy and action design (Midgley and Richardson 2007). Participatory workshops effectively integrate knowledge from multiple stakeholders, sectors, and disciplines. Nexus problem workshops not only produce new outputs such as qualitative data and conceptual models that better capture interdependencies and trade-offs but also foster perspective shifts, knowledge enhancement, new insights, and connections among participants. This allows for the discovery of nexus interrelationships through discussion that may not be visible in quantitative data analysis.

Organizing, implementing, and sustaining workshops requires a solid activity platform and time to build trust among actors. The timing and media representation of quantitative information and qualitative discussions can significantly impact workshop progress and outcomes. Effective workshop practices that exclude such sensitive aspects are not yet established and are often passed down through the experience of workshop facilitators. Living lab gains attentions as a support platform for networking participants and disseminate knowledge. Desing Thinking as another support took is often applied in participatory workshop for integrate knowledge in practice.

2.7 Nexus Tools

In research, the term "tool" can be broadly understood as multifunctional, corresponding to many processes such as empirical research, indicator development, modeling, system tools, and decision support, or narrowly understood as enabling quantitative calculations and evaluations. In sense of broad understanding, Albrecht et al. (2018) systematically reviewed nexus tools and methods in 73 papers. Kaddoura et al. (2017) similarly reviewed existing nexus modeling tools used for integrated policy formulation, aiming to help policymakers, practitioners, and institutions identify the most suitable tools for their modeling needs. As conclusions of the review works, most of the nexus methods are not clearly correlated with nexus conceptualizations. In other words, at the tool level, nexus approaches often use existing tools, and their usage is largely unrelated to nexus thinking. This reflects the difficulty of addressing complex FEW issues by handling three or four sectors simultaneously. The main functions of nexus tools are understanding nexus complexity, considering financial elements in tools, recognizing the importance of multiple nexus approach directions, incorporating different time scales, and enhancing tool accessibility. These tools require vast amounts of data, and coordination among tools for individual nexus sectors is insufficient (Kaddoura and El Khatib 2017).

In sense of narrow interpretation, nexus tools are understood alongside nexus modeling and nexus analysis. Only a very few examples exist, such as "Nexus Tool 2.0." This tool provides actual calculation and evaluation tools at national scale (Daher and Mohtar 2015). These kinds of tools can only be developed as the having matured methodologies and enough data support at operational stage.

Liu et al. (2024) modeled the system dynamics of household food, water, and energy nexus effects and provided scenario analysis tool. By applying the system dynamics (SD) methodology, the FEW consumptions and generation originating from home appliances were modeled and simulated. The research clearly revealed nexus effects that water-related and food-related energy use accounted for approximately 55% of the total energy use. In addition, water-related energy use showed high uncertainty, suggesting a management potential of approximately 24% for reduction, and was significantly correlated with household carbon emissions. The model requires support of Home Energy Management System (HEMS).

3 Nexus Design Approach

Nexus methods are mostly borrowed from systems thinking or design thinking arenas so far. They are not clearly correlated with nexus conceptualizations (Albrecht et al. 2018). The tools are not specifically designed to evaluate nexus interactions and feedback. As critized by Galaitsi et al. (2018), using these methods, the empirical FWE nexus research has not produced a discernible intellectual toolkit, nor has it validated claims that nexus approaches can improve resource management and governance outcomes.

3.1 Concept

Nexus thinking offers three guiding principles for implementation: investing in the natural environment, improving resource-use efficiency, and delivering services without leaving anyone behind. Previous approaches have largely focused on conceptual proposals and system analysis, lacking direct solutions for urban design and management. "Few studies incorporate practices that specifically address policy uptake and on-the-ground implementation (e.g. participatory stakeholder engagement; collaborative research; relevant scales of analysis)." (Galaitsi et al. 2018). The M-NEX project has proposed the moveable nexus philosophy and build a series of procedures to implement the concept. Nexus Matrix is a new development of the moveable nexus family for model the nexus effects (Yan et al. 2021).

Indeed, food, energy, and water, which form the basis of urban life, account for about 70% of global CO_2 emissions. Furthermore, among urban infrastructure systems, food, energy, water, mobility, buildings, waste, and green/public spaces account for about 87% of CO_2 emissions (Ramaswami et al. 2016).

Due to high human activity density, urban areas often exceed natural supply capacity, leading to overshoot its biocapacity. These cities rely on resource imports transboundary and consequently become main emitter of greenhouse gases. It is estimated that CO_2 emissions associated with the production, transportation, consumption, and disposal of FEW account for 70% of total urban emissions. Therefore, decarbonization and sustainability hinge on increasing local supply and reducing external imports. This is considered the essence of nexus thinking. It can be illustrated as Fig. 3.3.

Food, energy, and water are inherently interconnected in nature, providing resources and services to each other. Traditionally, cities have been viewed as hubs

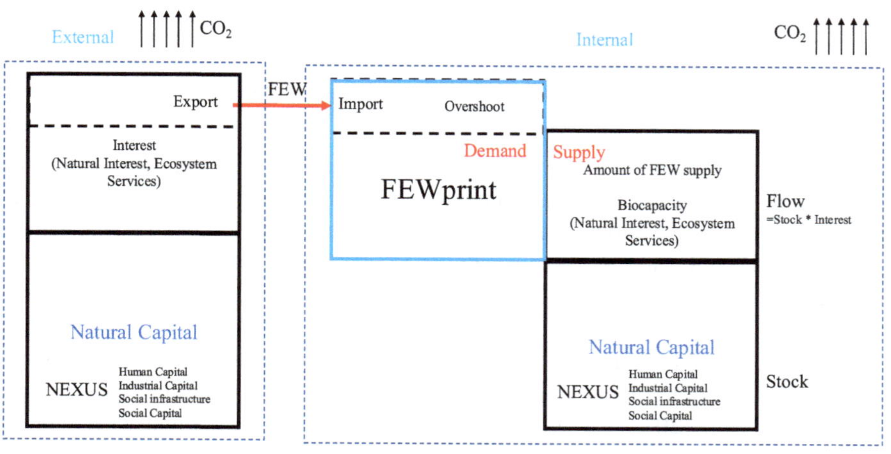

Fig. 3.3 Mechanism of sustainable urbanization. (By author)

of production and consumption, with little consideration for ecological production of food, energy, and water. In cities, food, energy, and water are often managed by separate departments. Governments strive to ensure stable supplies, while businesses focus on providing quality services. Residents assume that as long as they pay, they can freely use these resources. However, this assumption is being challenged by climate change, extreme weather, and population decline. Achieving 46% decarbonization by 2030 and carbon neutrality by 2050 will require a thorough review of the current urban systems, finding potentials of resources for increasing local production and consumption, enhancing resource-use efficiency by collaboration and cooperation across sectors and projects.

The issue is not a lack of resources but their underutilization. Cities are often seen as consumption spaces, with limited options beyond energy savings. Urban development in conventional ways consumes natural resources and brings negative impacts to environment. In fact, cities are treasure troves of nexus, with diverse land, buildings, technologies, and sectors in close proximity. Urban development changes physical conditions and bring anthropogenic environment under human control. Taking this advantage, cities possess tremendous potentials to be developed. The possibilities embedded in food, energy, water, and land development are inventoried in Table 3.1.

For example, installing solar panels on rooftops enables local energy production and consumption, and adding batteries can create an independent energy system. In suburban areas, converting unused farmland into community gardens for local food production is common. These practices leverage the geographical proximity of producers and consumers. Promoting environmental symbiosis using region-specific resources can bring new opportunities to rural areas. In other words, reverse urbanization can generate new services and values from food, energy, and water. Realizing these potentials requires a shift in mindset and supportive nexus methods. Nexus design approach is developed as an alternative to this challenge.

3.2 Urban Nexus Matrix

Urban infrastructures have been developed and operated for economic efficiency and human activities. The infrastructures are defined broadly as the systems that provide water, energy, food, shelter, transportation and communication, waste management, and public spaces (Ramaswami et al. 2012), caused heavy CO_2 emissions counted for 86.6% globally (IPCC, 2014). Urban services are provided on the infrastructure in- and trans-city boundaries.

Because of rapid industrialization and urban expansion, urban infrastructures become highly complex, lacking effective description and quantification methods. Ramaswami et al. (2017) captures both in-boundary and trans-boundary FEW interactions, quantifies environmental impacts of community-scale FEW provisions, and visualizes vulnerabilities within urban supply chains in Delhi, for instance.

Table 3.1 Transformative FEW nexuses view on influences of urbanization

Category	Development style	Negative view on urban development (−)	Future-oriented view on urban redevelopment (+)
Food	Open field cultivation	F: Decrease in productivity, abandonment of cultivation E: Worker phenomenon, increased machine utilization W: Soil contamination, soil erosion	F: Increasing demand for organic E: Promoting the use of biomass resources W: Promoting rainwater infiltration function
	Protected cultivation	F: Single breed E: Consumption of chemical fertilizers and fossil fuels W: Reduced rainwater absorption	F: High productivity E: Solar Sharing W: Effective use of rainwater and groundwater
	Urban cultivation	F: Single breed E: Energy consumption W: Drainage	F: High productivity E: Renewable energy available W: Water recycling available
	Organic cultivation	F: High cost E: Intensive manpower	F: Increasing demand for health foods E: Renewable energy available W: Rainwater harvesting
	Food manufacturing	F: Increasing food waste E: Increased energy for processing W: Increasing wastewater	F: Development of the food processing industry E: Distribution base close to consumers W: Adoption of water recycling technologies
	Food transport	F: Food loss E: Long-distance transportation, CO_2 emissions W: Virtual water	F: Cool delivery, maintaining freshness E: Electric car, transportation by freight train
	Food sales	F: Food waste E: Energy for transportation	F: Deliver to everyone E: Optimal store placement W: Water recycling in stores
	Food waste	F: Food loss E: Increased processing costs W: Wastewater treatment	E: Biomass utilization
	Food environment	E: Accessibility dependency	F: Convenient access to ingredients
	Dietary habit(home-cooking)	F: Access to ingredients Others: Reduced home-cooking	F: Cooking tools developed
	Dietary habit(ready-made meals)	E: Access to the shop F: Food loss	Others: Saving cooking time
	Dietary habit(eat-out)	E: Access to the store F: Food loss	Other: Communication

(continued)

Table 3.1 (continued)

Category	Development style	Negative view on urban development (−)	Future-oriented view on urban redevelopment (+)
Energy	Solar farm (ground)	F: Decrease in food production W: Decrease in greenery, decrease in water recharge	E: Increase in renewable energy
	Solar farm (rooftop)	F: Competition with rooftop vegetable gardens	E: Increase in renewable energy
	Wind farm	Noise concerns	E: Increase in renewable energy
	Bio farm	W: Water circulation treatment	F: New agriculture, new ingredients E: Increase in renewable energy
	Hydropower	W: Stabilization of incoming water	E: Increase in renewable energy
	Solar heating	E: Competition with PV Others: Rooftop view	E: Increasing use of renewable energy W: Warm water is provided
Water	Dam	F: Decline in agricultural land W: Deterioration of water quality Others: Changes in natural landscapes	F: Fishing and aquaculture W: Stable supply, flood control E: Hydroelectric power generation
	Underground Water-take	Others: The ground subsided	E: Heat pump W: Local production and local consumption of water resources
	Rainwater harvesting	E: Cleaning the aquarium	F: Irrigation of the vegetable garden E: Water saving W: Flood control
	Forest for water conservation	E: CO_2 emissions from forestry machinery	F: Fruit E: Biomass, CO_2 absorption W: Rainwater infiltration, flood control

(continued)

Table 3.1 (continued)

Category	Development style	Negative view on urban development (−)	Future-oriented view on urban redevelopment (+)
Combined	Residential land development	F: Farmland decreases and farmers decrease E: Urban heat island W: Decreased water absorption and water permeability	F: Concentrated and expanding consumer demand E: Concentrated and expanding consumer demand E: Increased centralized supply efficiency W: Concentrated and expanding consumer demand
	Commercial land development	E: Urban heat island E: Increased mobility, concentration of consumption W: Decreased water absorption and water permeability	F: Increasing demand for food among workers E: Increase efficiency with centralized supply W: Concentrated and expanding consumer demand
	Industrial land development	F: Decrease in farmland and decrease in the number of farmers E: Urban heat island W: Industrial water, increased wastewater	F: Increasing demand for food among workers E: Increased movement and transportation E: Increase efficiency with centralized supply
	Compact city	F: Widening access gaps E: Bias toward centralized supply W: Increased water pressure due to higher floors	F: Concentration of consumer demand E: Increase efficiency with centralized supply W: Increase efficiency with centralized supply
	Compact city	E: Maintenance cost of greenery W: Landscape tends to be a priority	F: Promotion of local production for local consumption E: Expansion of renewable energy, mitigation of UHI W: Conservation of water and greenery Others: Maintenance of natural landscapes
	Green infrastructure	Other: Increased maintenance costs	F: Promoting local production for local consumption E: Heat island mitigation W: Increased water absorption and permeability

Physically, cities are filled with buildings and structures, creating numerous surfaces both inside and outside. The seven items—food, energy, water, buildings, mobility, green/public spaces, and waste disposal—are ubiquitous. Some are highly utilized, while others remain underutilized due to siloed systems or poor economic viability. In advanced cities, the seven urban infrastructures rely heavily on centralized external resource supply. Oppositely, centralized infrastructures in developing cities or some areas are weak where nature-based small systems are alive. These implies the urban infrastructure could be a mix of large and small scales. This combination could happen not only in local regions but also in modern cities. For instance, rooftop solar power generation and greening are prime examples. Installing solar panels on vacant land, rooftops, and roadside belts, combined with core electrical infrastructure, forms a centralized-distributed energy system.

Similar ideas and technologies apply to local food production and green infrastructure, complementing energy, improving water cycles, and enriching urban life. Cristiano et al. (2021) proposed a conceptual framework that evaluates the role of green and blue infrastructure (GBI) in urban FEW systems. Their approaches demonstrate how GBI interventions—such as green roofs, wetlands, and rain gardens—affect ecosystem services linked to food, water, and energy. The model distinguishes between direct (in-boundary) and indirect (trans-boundary) environmental effects while incorporating external drivers, including socioeconomic, biophysical, and socio-ecological factors.

Compared to massive urban infrastructure, these efforts are small-scale and decentralized but can complement large systems, building sustainable decentralized cooperative systems. While these technologies are not entirely carbon-zero, innovation is rapidly advancing. The day when cities become centers of food, energy, and water production, circulation, and regeneration is not far off. However, this is not a task for a single sector but requires the entire city to function as a single system, emphasizing local production and consumption, ecosystem services, and resource circulation for decarbonized, sustainable urban development.

Based on the in- and trans-boundary observations, we composite a "Nexus Matrix" to describe the nexus relationships within urban systems. Figure 3.4 illustrates the concept of the Nexus Matrix. The rows represent services provided by large-scale trans-boundary infrastructures noted as food (F), energy (E), water (W), buildings (B), mobility (M), green spaces (G), and waste disposal (D) while the columns represent services produced within boundary as f, e, w, b, m, g, and d.

The cells of the matrix connect external provision (rows) with internal efforts (columns) to generate nexus effects. Urban decarbonization must promote local production and consumption, ecosystem services, and resource circulation within the city, reducing dependence on external sources.

The implications of Nexus Matrix may vary by context. In low-income or large cities, especially in Africa and Asia, urban agriculture and food security are often prioritized. In contrast, middle- and high-income cities tend to focus on food and energy services such as cooling from green spaces. Nexus Matrix can reflect these variations for different urban priorities, governance capacities, and levels of investment.

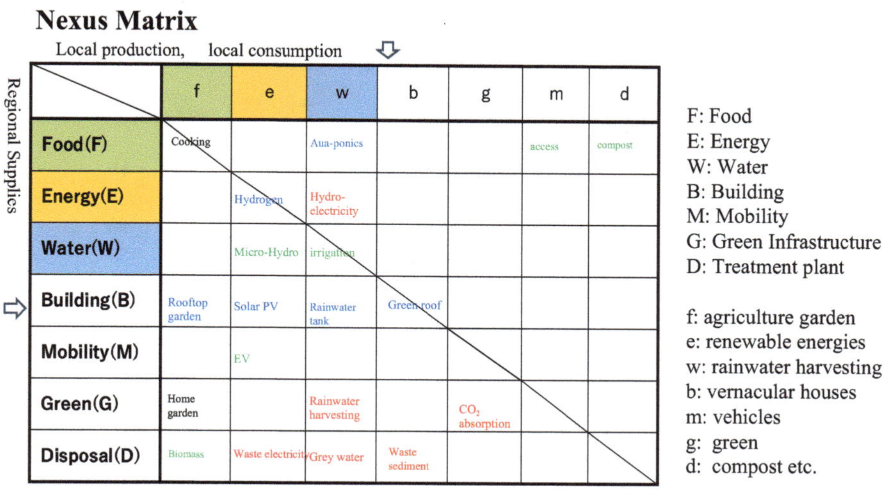

Fig. 3.4 Nexus matrix

3.3 *FEWprint*

To improve resource-use efficiency, it is crucial to comprehensively consider urban infrastructures beyond single sectors or spaces and quantitatively evaluate the environmental inputs and socioeconomic performance of decarbonization projects. Previously, carbon footprint (CF) and ecological footprint (EF) have been developed as indicators for decarbonization, but they evaluate sectors separately, overlooking nexus. Urban planning and design focus on spaces, using metrics such as floor area ratio, building coverage ratio, and setback ratio, all based on spatial area. Recognizing this, the M-NEX project adapted the EF concept for urban planning, summing the urban space area occupied by resource consumption from seven sectors—food, energy, water, etc.—and the forest area needed to absorb CO_2 emissions, creating FEWprint (Yan and Nakayama 2021; ten Caat 2022). The mechanism is illustrated in Fig. 3.5. In other words, the current environmental load of the city (1) serves as the baseline, and urban renewal scenarios (2) and the creation and use of new spaces (3) are converted into spatial area and CO_2-absorbing forest area to support comparative analysis of decarbonization effects.

3.4 *Nexus Platform*

With the above concepts and evaluation indicators, the moveable platform was developed by the M-NEX project as a support tool for decarbonized urban development. Figure 3.6 shows its functional groups.

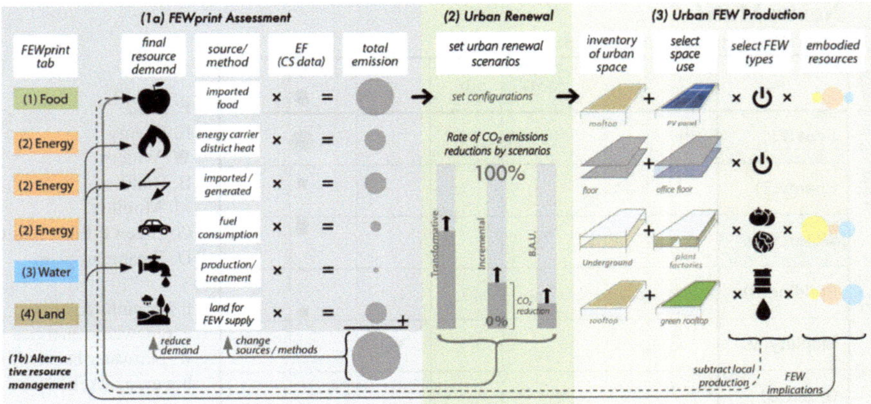

Fig. 3.5 Evaluation tool FEWprint for urban design support. (ten Caat 2022)

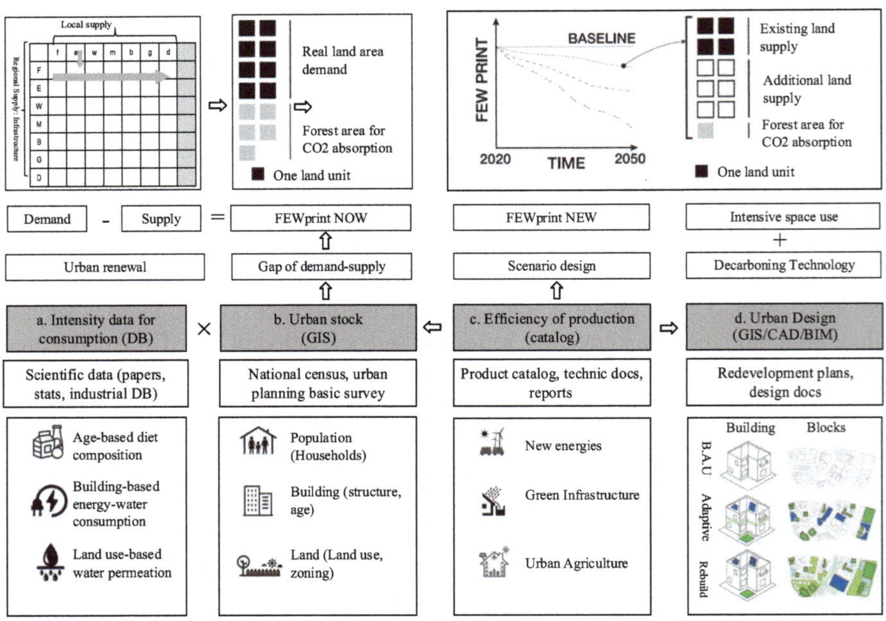

Fig. 3.6 Functions of the M-NEX platform

The moveable platform involves (1) capturing urban resource consumption demands for food, energy, and water from population, buildings, and land use, and collecting intensity unit values from research papers, statistics, and industry databases. (2) For a target city, GIS (Geographic Information Systems) and statistics are used to investigate population, land, and building stocks. These stocks are input into the Nexus Matrix, multiplied by unit values to calculate FEWprint NOW, representing the current spatial use and environmental load of the city. The left side of Fig. 3.6 illustrates this current evaluation function and mechanism.

The right side of Fig. 3.6 shows the function and mechanism for supporting urban development toward decarbonization. Urban decarbonization effects are significantly influenced by construction, management, and operation technologies, so current evaluations and future plans must consider technological innovation. To support this, a technology catalog (3) is prepared, providing information on current technologies and anticipated innovations (new energy, infrastructure, industry, and business) for current evaluations and urban development scenario design (d).

Cities frequently change in both soft and hard aspects due to population shifts, building updates, infrastructure modifications, and lifestyle changes. Soft changes such as population mobility and lifestyle are learned from statistical and life surveys and reflected in unit values and input data. Hard changes such as building updates and infrastructure modifications depend on urban development plans and projects. These can be categorized into three scenarios: maintenance (no active involvement), guided promotion (actively utilizing urban development policies and systems), and creative development (large-scale redevelopment projects such as land readjustment). Finally, the environmental load FEWprint NEW for each scenario is estimated, calculating existing supply land, new supply land, and CO_2 absorption land.

Thus, M-NEX evaluates the current state of a city's population, buildings, and land use, predicts the impact of change scenarios on food, energy, and water production and consumption, visually represents spatial creation and decarbonization effects, and supports decarbonized, sustainable urban development with a collection of knowledge and tools.

3.5 Applications

In the metropolitan area, urbanization accelerated after the Meiji era, with urban areas expanding significantly during the high economic growth period of the 1960s–1980s. Planned developments with land readjustment projects coexisted with unplanned urban sprawl. Previous research and methods often focused on differences in living environments from perspectives such as convenience and safety, but methods for comparing and evaluating sustainability from a global perspective and supporting data-driven decarbonized sustainable urban development were not established.

Cities, with dense populations and buildings, are highly consumptive of food, energy, and water, but compact spatial use has also improved resource efficiency. In other words, resource efficiency and environmental load vary with a city's population composition, land, and building use. Future decarbonized urban development requires tailored considerations, and M-NEX is a powerful support tool for this. To demonstrate the platform's effectiveness, four districts from the city center to the suburbs were selected, as shown in Fig. 3.6. These four districts are all existing urban areas, representing typical urban development scenarios: (I) densely packed wooden housing (no land readjustment), (II) suburban residential development (with farmland consolidation), (III) suburban low-rise housing (with land readjustment), and (IV) suburban housing complex (with land readjustment), with expected differences in FEWprint.

For the calculations, unit values for FEW production and consumption and CO_2 emissions were collected, as shown in Table 3.2. These values were gathered from various sources. Table 3.3 shows the FEWprint calculation process and results. The four districts differ in land area and building floor area, so ⑭–㉔ are all converted to hectares (Fig. 3.7).

Table 3.2 Unit values for FEWprint calculation (corresponding to Fig. 3.6a)

Item	CO_2 emission	Unit	Source	Required land area	Unit	Source
① Vegetable production/kg	0.60	g-CO_2	a	0.48	m^2	f
② Electricity supply/kWh	0.34	kg-CO_2	b	0.24	10^{-3} m^2	g
③ City gas supply/kWh	3.27	kg-CO_2	c	0.15	10^{-3} m^2	h
④ Water supply/m^3	0.34	kg-CO_2	d	1.13	10^{-3} m^2	i
⑤ CO_2 absorption forest/kg-CO_2				1.14	m^2	j
⑥ PV annual generation (/m^2)	61.58	kWh	e			

Note: Unit values are cited from the following sources:

a: Yoshikawa et al.'s paper (below) https://www.jstage.jst.go.jp/article/proer2000/34/0/34_0_245/_article/-char/ja/

b: Tokyo Electric Power Company's integrated report (2019) https://www.tepco.co.jp/about/ir/library/annual_report/pdf/201910tougou-j.pdf

c: Tokyo Gas website https://support.tokyo-gas.co.jp/faq/show/8954?category_id=839&site_domain=eee

d: Yokohama Waterworks Bureau https://www.city.yokohama.lg.jp/kurashi/sumai-kurashi/suido-gesui/suido/torikumi/kankyohozen/kankyo-houkokusyo.files/0032_20210330.pdf Tokyo Metropolitan Government Bureau of Waterworks https://www.waterworks.metro.tokyo.lg.jp/files/items/18996/File/h30_01.pdf

e: *FY2014 Tokyo Household Energy Consumption Trends Survey Report. Same as E*

f: Ministry of Agriculture, Forestry and Fisheries FY2017 Crop Statistics Survey

g: http://www.aec.go.jp/jicst/NC/senmon/vision/siryo/vision06/siryo2.pdf https://www.env.go.jp/earth/ondanka/ghg-mrv/emissions/results/material/yoin_2015_2_3a.pdf http://www.kyuden.co.jp/effort_renewable-energy_photovoltaic_faq.html, https://eneken.ieej.or.jp/data/5459.pdf, https://www.chuden.co.jp/energy/renew/biomass/bio_chuden/

h: https://www.tokyo-gas.co.jp/IR/library/pdf/anual/21japanese.pdf, https://www.tokyo-gas.co.jp/Press/20070718-02.html, https://www.keihin.ne.jp/pdf/surveillance_map_front_04.pdf, https://www.tokyo-gas.co.jp/news/press/20210311-01.html, https://www.tokyo-gas.co.jp/Press/20081218-01.html

i: Yokohama City (only Yokohama City data is listed in Table 3.1), https://www.city.yokohama.lg.jp/kurashi/sumai-kurashi/suido-gesui/suido/suishitsu/josuijo/nishiya/gaiyou.html, https://www.city.yokohama.lg.jp/kurashi/sumai-kurashi/suido-gesui/suido/suishitsu/josuijo/kosuzume/gaiyou.html, https://www.city.yokohama.lg.jp/kurashi/sumai-kurashi/suido-gesui/suido/suishitsu/josuijo/kawai/gaiyou.html. Tokyo (applied Tokyo data to I, II), Tokyo Metropolitan Government Bureau of Waterworks FY2020 Business Annual Report https://www.waterworks.metro.tokyo.lg.jp/suido-jigyo/gaiyou/nenpou_r02.html

j: Forestry Agency, https://www.rinya.maff.go.jp/j/sin_riyou/ondanka/20141113_topics2_2.html

Table 3.3 Evaluation of urban environmental load FEWprint in M-NEX

District	I	II	III	IV	Source	Formula
Urban stock						
⑦ Total Population (2015)	282	61	24	128	A	
⑧ Area (ha)	3.75	0.60	0.31	0.37	B	
⑨ Population Density (ha)	80	102	76	357	A/B	
⑩ Number of Buildings	21	21	7	27	C	
⑪ Total Floor Area (ha)	4.15	0.42	0.15	0.46	C	
⑫ Distance to Nearest Station (m)	60	1000	1350	150	GIS	
⑬ Distance to City Center (km)	22	22	15	8	GIS	
Demand						
⑭ Annual Vegetable Demand (t/ha)	6.75	9.76	7.08	32.74	D	
⑮ Annual Electricity Demand (kWh/ha)	89,768.00	133,240.00	73,954.84	800,270.27	E	
⑯ Annual Gas Demand (m^3/ha)	10,454.40	15,470.00	8187.00	85,929.73	E	
⑰ Annual Water Demand (m^3/ha)	5179.84	7674.00	3956.13	41,724.32	F	
Supply						
⑱ Farmland/Garden Area (m^2)	0	0	1463	0	B	
⑲ PV Installed Area (m^2)	0	102.80	91.90	102.50	G	
FEW Actual Area						
⑳ Required Land Area, F (m^2/ha)	3235.54	4681.90	1953.03	15,705.53		⑭ ÷ ① – ⑱

(continued)

Table 3.3 (continued)

District	I	II	III	IV	Source	Formula
㉑ Required Land Area, E (m^2/ha)	23.11	32.78	17.62	203.44	(⑮ − ⑲ ÷ ⑥) ÷ ② + ⑯ ÷ ③	
㉒ Required Land Area, W (m^2/ha)	5.85	10.51	5.42	57.16	⑰ ÷ ④	
Subtotal (m^2/ha)	3264.51	4725.19	1958.06	15,966.13	⑳ + ㉑ + ㉒	
(A1) Required Actual Area (m^2)	12,241.90	2835.12	1958.06	15,966.13	Subtotal × ⑧	
(A2) FEW Occupied Area (m^2/person)	43.41	46.48	25.29	46.15	A1/⑦	
(A3) Residential Occupied Land Area (m^2/person)	132.98	98.36	129.17	28.91	⑧/⑦	
(A4) Residential Occupied Floor Area (m^2/person)	147.16	68.85	62.50	35.94	⑪/⑦	
Virtual Area						
㉓ CO_2 Absorption Forest Area (m^2/ha)	79,724.2	115,562.1	62,548.1	655,168.9		(⑭ × ① + (⑮ − ⑲ ÷ ⑥) × ② + ⑯ × ③ + ⑰ × ④) × ⑤
(A5) CO_2 (kg/ha)	69,933.50	101,370.24	54,787.82	574,709.57	㉓/j	
(A6) CO_2 (kg)	262,250.63	60,822.15	16,984.22	212,642.54	A5 × ⑧	
(A7) CO_2 (kg/person)	929.97	997.08	707.68	1661.27	A6/⑦	
(A8) CO_2 (kg/unit floor m^2)	6.32	14.48	11.32	46.23	A6/⑪	
㉔ *FEWprint (ha/ha)*	8.30	12.03	6.44	67.11	⑳ + ㉑ + ㉒ + ㉓	

Note: A: 2015 National Census, B: Urban Planning Basic Survey, C: ZENRIN Building Point Data, D: National Health and Nutrition Survey (2019), E: FY2014 Tokyo Household Energy Consumption Trends Survey Report, F: Tokyo Metropolitan Government Bureau of Waterworks, G: Google Map visual inspection

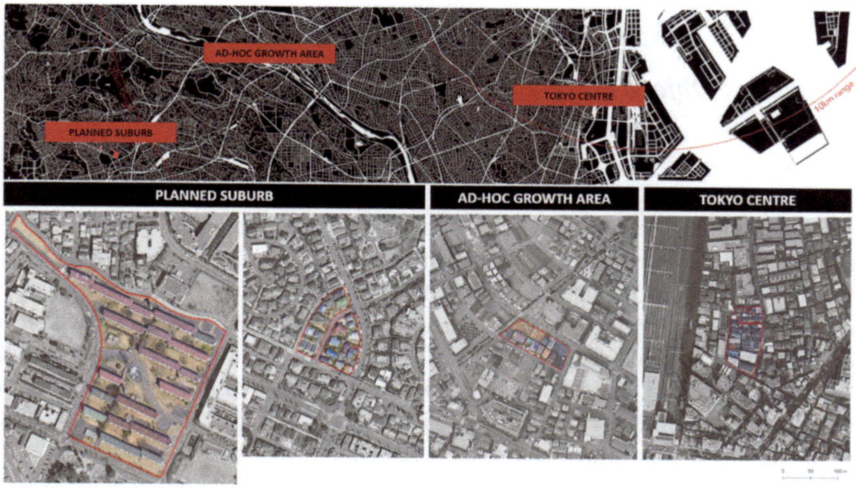

(I) Suburban condominum (II) Suburbban detached houses (III)Suburban in developing (IV) Densely developed central area

Fig. 3.7 Location and urban form of target districts for case calculations

The results show that the suburban housing complex (I), a postal housing area, had reduced occupancy due to planned redevelopment, resulting in larger per capita land and floor area. The suburban residential development (II) had a small population and few buildings, with remaining farmland, resulting in the smallest FEWprint. The suburban low-rise housing (III) had the second-highest environmental load. The densely packed wooden housing (IV) in the city center had the largest FEWprint, occupying 67.11 times the actual land area. Its proximity to the city center and high proportion of single-person households led to smaller per capita land and floor area but higher FEW consumption. Thus, the FEWprint indicator and M-NEX platform approach FEW supply and demand, supply potential, and CO_2 emissions from the perspective of nexus, quantitatively evaluating differences in environmental load due to development methods based on population, household, dietary composition, and land and building use.

Food, energy, and water are delivered to consumers through long supply chains from upstream to downstream. Traditionally, decarbonization was seen as an industry issue, with few methods to make consumers feel its impact. Using M-NEX, individuals can review their consumption structures, compare urban initiatives, and quantitatively analyze the differences and effects of development methods. Linking these results to lifestyle and business practices can strongly support decarbonized urban development.

4 Discussions

4.1 Nexus Object/Trigger

The FEW nexus, as well as its expansion to FEW-X nexus, becomes increasingly complex as the number of objects grows, making it difficult to handle the entire system. In applying nexus approaches, system analysis is often conducted, focusing on key elements and driving forces, known as nexus triggers (Fig. 3.3). Triggers are elements that bring about changes in the entire nexus. For example, expanding algae biofuel production requires inputs of land, water, fertilizer, and energy, which in turn affect regional food production capacity and energy and water demand (Smith 2014). Water, energy, and food supply can all be considered triggers for the FEW nexus. Often, multiple triggers exist, but identifying a single trigger first can simplify modeling or conceptualization (Hoff 2011).

The nexus is defined as connections, cores, or centers, but its actual nature varies depending on the problem at hand. To shift from thinking about things to thinking about concerns, gathering stakeholders is essential. Newell et al. propose using urban metabolism as a boundary object in cities, attracting diverse scholars and practitioners to more fully capture the scale, complexity, and interactions of a particular system. "Framing the urban FEW metabolism as a boundary object can attract a diverse group of scholars and practitioners to more fully capture the scale, complexity, and interactions of a particular system." (Newell et al. 2019). FEW systems (e.g., food supply chains, bioenergy production, wastewater treatment) could similarly serve as empirical boundary objects to collaboratively develop integrative approaches and responses to sustainability and resilience challenges. Here, boundary objects refer to research investigation objects. Papadopoulou et al. (2020) adopted this idea and applied it to low-carbon economy policy analysis. "Problem and related research questions define the boundaries of investigation and the nexus components involved. For example, if the problem is water allocation in a region where there is a power plant, irrigated farmland, and populated areas, the nexus object of investigation is that of water-energy-agriculture." (Papadopoulou et al. 2020). Therefore, when attempting a nexus approach, it is helpful to select what and what to use as nexus objects. Specifically, a piece of land, a building, a small shop, or a large commercial facility can be considered. Focusing on these and examining the connections, interactions, and integration effects within their boundary areas.

4.2 System Scale and Boundary

Since the nexus aims to elucidate trade-offs and synergies in resource use, defining the system boundary for performance evaluation is crucial. "It is of great importance to define the system boundary when conducting nexus analysis, as different system boundary definitions result in different results." (Zhang et al. 2018a, b). System boundary setting depends on objectives and data availability. A broad system boundary allows for comprehensive consideration of diverse nexuses but requires more data and makes hypothesis and model construction more challenging. A narrow system boundary allows for more detailed examination of specific nexus relationships but limits conclusions to that context.

Nexus thinking seeks to address global issues (climate change, finite resource bases, population growth, and migration), but nexus practices often emerge at local scales (household > region > country, etc.). Closely related to spatial scale is administrative jurisdiction scale, where political units such as cities, regions/states, and countries implement policies and governance (Bhaduri et al. 2015). Temporal scale is also important for the nexus. Scenario planning is often conducted as a tool for interpreting climate change and population predictions in the future. Due to the multifaceted nature of scale in the nexus, it relates to various issues in nexus research (Cash et al. 2006). Cash et al. identified three problems related to scale: ignorance, mismatch, and plurality. Mismatches in nexus scale and level are easily exemplified by the inconsistency between human actions and the environment in which physical processes operate. For example, knowledge of global climate change has advanced to better understand uncertainties surrounding future global resource security but is often of little use in solving more regional challenges such as river basin or farm management. Pluralists often (mistakenly) assume that there is one perfect solution to a given problem, denying the different needs of all stakeholders involved in the process and the multiple different outcomes that nonlinear relationships can produce (Mcgrane et al. 2018).

4.3 Data Availability

Closely related to scale are issues of data availability and accessibility. McCarl et al. (2017), as well as recent papers by Ramaswami et al. (2017), Newell and Ramaswami (2020), and Ramaswami et al. (2016), emphasize the importance of data accessibility in WEF nexus analysis and discuss the increasing problem of availability at smaller granularities. At national and cross-border levels, open-access international databases are used. At lower levels, much of the data relevant to nexus evaluation has been managed within sector-specific institutions for many years. Furthermore, due to cost and data confidentiality, researchers often cannot easily access existing critical data from other sources.

High-precision nexus analysis requires data at the household or building level, or even individual movement level. Smart meters, HEMS, and BEMS are becoming more widespread, and big data accumulation is progressing.

Tokyo Electric Power Company, in collaboration with NTT Data, has launched the Grid Data Bank. However, this data is only available to specific companies for specific purposes and is not publicly available for research. For supermarkets to function well, they rely on large infrastructure and logistics, serving as hubs for selling products produced locally, domestically, and overseas. At the same time, many supermarkets are implementing proactive programs to reduce environmental impact. For example, Tesco operates a 24-hour monitoring service, collecting energy and water consumption data from all facilities and transmitting it to a monitoring facility in Bengaluru (The Guardian 2013). Such monitoring is a step toward supermarkets assessing their role in the broader nexus and addressing issues of sustainability and environmental accountability in their practices, but the data is rarely made public due to ownership issues (McCarl et al. 2017) and is not available for research.

5 Conclusions

Urban development has traditionally been planned and designed based on vast national censuses and urban planning basic surveys, creating livable spaces. Under the global goal of decarbonization, it is necessary to reconsider planning objectives and support evidence-based planning, design, and consensus-building. To this end, it is essential to view cities holistically, considering both hard and soft aspects as well as external and internal factors. This chapter introduced the M-NEX platform, which uses the Nexus Matrix to describe decarbonized, centralized-distributed cooperative urban systems and FEWprint to present a method for decarbonized urban development centered on food, energy, and water (FEW). The platform can be applied to various urban areas, visually showing differences in environmental load based on population composition and urban form. The tool was applied to a large-scale creative development project in a central business district, confirming significant decarbonization effects.

This data-driven support was made possible by the accumulation of vast amounts of data through previous informatization efforts, as well as the increased accessibility of nonstandard research results and technical information. The former has already accumulated a vast amount of data through previous informatization efforts, and it is time to utilize it highly. The latter is scattered throughout the internet in various forms. The spread of open data policies has increased accessibility, but collection and organization remain challenging. M-NEX has created a framework for organizing unit value databases and technology catalogs, accumulating case-based data. With the advancement of DX, further enrichment is expected.

Acknowledgments This chapter is developed at the project "Moveable NEXUS: Design-Led Urban Food, Water and Energy Management Innovation in New Boundary Conditions of Change" (No. 11314551), granted by Belmont Forum under the Collaborative Research Area on Sustainable Urbanization Global Initiative (SUGI)/Food–Water–Energy Nexus (2018–2021).

References

Albrecht TR, Crootof A, Scott CA (2018) The water-energy-food nexus: a systematic review of methods for nexus assessment. Environ Res Lett 13(4):043002

Bala BK, Arshad FM, Noh KM (2017) System dynamics. Springer Nature Singapore

Bhaduri A, Ringler C, Dombrowski I, Mohtar R, Scheumann W (2015) Sustainability in the water–energy–food nexus. Water Int 40(5–6):723–732

Cai Y, Cai J, Xu L, Tan Q, Xu Q (2019) Integrated risk analysis of water-energy nexus systems based on systems dynamics, orthogonal design and copula analysis. Renew Sust Energ Rev 99(August 2018):125–137

Cash DW, Adger WN, Berkes F, Garden P, Lebel L, Olsson P, Pritchard L, Young O (2006) Scale and cross-scale dynamics: governance and information in a multilevel world. Ecol Soc 11(2):12. https://www.jstor.org/stable/26265993

Chhipi-Shrestha G, Hewage K, Sadiq R (2017) Water–energy–carbon nexus modeling for urban water systems: system dynamics approach. J Water Resour Plan Manag 143(6):04017016

Cristiano E, Deidda R, Viola F (2021) The role of green roofs in urban water-energy-food-ecosystem nexus: a review. Sci Total Environ 756(February):143876

Daher BT, Mohtar RH (2015) Water–energy–food (WEF) nexus tool 2.0: guiding integrative resource planning and decision-making. Water Int 40(5–6):748–771

Fang K, Heijungs R (2015) Investigating the inventory and characterization aspects of footprinting methods: lessons for the classification and integration of footprints. J Clean Prod 108(December):1028–1036

Fang K, Heijungs R, De Snoo GR (2015) Understanding the complementary linkages between environmental footprints and planetary boundaries in a footprint–boundary environmental sustainability assessment framework. Ecol Econ J Int Soc Ecol Econ 114(June):218–226

Fang C, Liu H, Li G (2016) International progress and evaluation on interactive coupling effects between urbanization and the eco-environment. J Geogr Sci 26(8):1081–1116

Galaitsi S, Veysey J, Huber-lee A (2018) Where is the added value? A review of the water-energy-food nexus literature. SEI

García JM (2020) Theory and practical exercises of system dynamics: modeling and simulation with Vensim PLE. In: Preface John Sterman. Amazon Digital Services LLC

Ghodsvali M, Krishnamurthy S, de Vries B (2019) Review of transdisciplinary approaches to food-water-energy nexus: a guide towards sustainable development. Environ Sci Pol 101(April):266–278

Größler A, Stotz M, Schieritz N (2003) A software interface between system dynamics and agent-based simulations: linking Vensim® and RePast®. In: Proceedings of the 21st System Dynamics Society International Conference. Citeseer. http://citeseerx.ist.psu.edu/viewdoc/download?doi=10.1.1.452.6032&rep=rep1&type=pdf

Herrmann IT, Moltesen A (2015) Does it matter which life cycle assessment (LCA) tool you choose?—A comparative assessment of SimaPro and GaBi. J Clean Prod 86(January):163–169

Hoff H (2011) Understanding the nexus. background paper for the Bonn2011 nexus conference: Stockholm Environment Institute, no. November: 1–52

Howells M, Hermann S, Welsch M, Bazilian M, Segerström R, Alfstad T, Gielen D et al (2013) Integrated analysis of climate change, land-use, energy and water strategies. Nat Clim Chang 3(7):621–626

Kaddoura S, El Khatib S (2017) Review of water-energy-food Nexus tools to improve the Nexus modelling approach for integrated policy making. Environ Sci Policy 77:114–121

Kirkwood CW (1998) System dynamics methods. College of Business Arizona State University USA. https://nutritionmodels.com/papers/Kirkwood1998.pdf

Kitzes J (2013) An introduction to environmentally-extended input-output analysis. Resources 2(4):489–503

Liu H, Yan W, Kobayashi H (2024) Modelling the system dynamics of household food, water, and energy nexus effects. Heliyon 10:e25886

Mabhaudhi T, Nhamo L, Mpandeli S, Nhemachena C, Senzanje A, Sobratee N, Chivenge PP et al (2019) The water–energy–food nexus as a tool to transform rural livelihoods and Well-being in southern Africa. Int J Environ Res Public Health 16(16):2970
Maier M, Mueller M, Yan X (2017) Introduction of a spatiotemporal life cycle inventory method using a wind energy example. Energy Procedia 142(December):3035–3040
McCarl BA, Yang Y, Srinivasan R, Pistikopoulos EN, Mohtar RH (2017) Data for WEF nexus analysis: a review of issues. Curr Sustain Renew Energy Rep 4(3):137–143
Mcgrane SJ, Acuto M, Artioli F, Chen PY, Comber R, Cottee J, Farr-Wharton G et al (2018) Scaling the nexus: towards integrated frameworks for analysing water, energy and food. Geogr J 185(4):419–431. https://doi.org/10.1111/geoj.12256
Mcmanus P, Haughton G (2006) Planning with ecological footprints: a sympathetic critique of theory and practice. Environ Urban 18(1):113–127
Meadows D (2009) Thinking in systems: a primer, vol 89. earth
Meadows D (2016) Loss of biodiversity. In: People, planet, and progress beyond 2015. CABI Reviews
Midgley G, Richardson KA (2007) Systems thinking for community involvement in policy analysis. E:CO 9(1–2):167–183
Munksgaard J, Wier M, Lenzen M, Dey C (2005) Using input-output analysis to measure the environmental pressure of consumption at different spatial levels. J Ind Ecol 9(1–2):169–185
Newell JP, Ramaswami A (2020) Urban food-energy-water systems: past, current, and future research trajectories. Environ Res Lett 15(5):050201. https://doi.org/10.1088/1748-9326/ab7419
Newell JP, Goldstein B, Foster A (2019) A 40-year review of food–energy–water nexus literature and its application to the urban scale. Environ Res Lett 14(7):073003
Papadopoulou C-A, Papadopoulou MP, Laspidou C, Munaretto S, Brouwer F (2020) Towards a low-carbon economy: A nexus-oriented policy coherence analysis in Greece. Sustain Sci Pract Policy 12(1):373
Pfister S, Vionnet S, Levova T, Humbert S (2016) Ecoinvent 3: assessing water use in LCA and facilitating water footprinting. Int J Life Cycle Assess 21(9):1349–1360
Ramaswami A, Chavez A, Chertow M (2012) Carbon footprinting of cities and implications for analysis of urban material and energy flows. J Ind Ecol 16(6):783–785
Ramaswami A, Russell AG, Culligan PJ, Sharma KR, Kumar E (2016) Meta-principles for developing smart, sustainable, and healthy cities. Science 352(6288):940–943
Ramaswami A, Boyer D, Nagpure AS, Fang A, Bogra S, Rao-Ghorpade BBEC (2017) An urban systems framework to assess the trans-boundary food-energy-water nexus: implementation in Delhi, India. Environ Res Lett 12(2):025008. online
Raschio G, Smetana S, Contreras C, Mathys A, Heinz V (2017) Spatiotemporal tools for regional low-carbon development: linking LCA and GIS to assess clusters of GHG emissions from cocoa farming in Peru. In: Matsumoto M, Masui K, Fukushige S, Kondoh S (eds) Sustainability through innovation in product life cycle design. Springer Singapore, Singapore, pp 969–980
Santamouris M, Paraponiaris K, Mihalakakou G (2007) Estimating the ecological footprint of the heat Island effect over Athens, Greece. Clim Chang 80(3–4):265–276
Schaffartzik A, Sachs M, Wiedenhofer D, Eisenmenger N (2014) Environmentally extended input-output analysis. In: In social ecology working paper 154. Institute of Social Ecology, Vienna. s
Smith VH (2014) Progress in algae as a feedstock for bioproducts. Ind Biotechnol 10(3):159–161
Sterman JD (2001) System dynamics modeling: tools for learning in a complex world. Manag (Andover) 43(4):8
ten Caat PN, Tenpierik MJ, Sanyal T, Tillie NMJD, van den Dobbelsteen AAJF, Thün G, Cullen S, Nakayama S, Karanisa T, Monti S (2022) Towards fossil free cities—emission assessment of food and resources consumption with the FEWprint carbon accounting platform. Cleaner Environ Syst 4:100074

The Guardian (2013) Tesco cuts energy use with 24-hour monitoring system. https://www.theguardian.com/sustainable-business/tesco-cuts-energy-use-monitoring. Last accessed date: 026/01/30

Vanham D, Leip A, Galli A, Kastner T, Bruckner M, Uwizeye A, van Dijk K et al (2019) Environmental footprint family to address local to planetary sustainability and deliver on the SDGs. Sci Total Environ 693(November):133642

White DJ, Hubacek K, Feng K, Sun L, Meng B (2018) The water-energy-food nexus in East Asia: a tele-connected value chain analysis using inter-regional input-output analysis. Appl Energy 210:550–567

Willis HH, Groves DG, Ringel JS, Mao Z, Efron S, Abbott M (2016) Developing the Pardee RAND food-energy-water security index. rand.org. https://www.rand.org/content/dam/rand/pubs/tools/TL100/TL165/RAND_TL165.pdf

Wolstenholme EF (1982) System dynamics in perspective. J Oper Res Soc 33(6):547–556

Yan W, Nakayama S (2021) Redesigning the urban food life through the participatory living lab platform: practices in suburban areas of the Tokyo metropolitan region. In: Roggema R (ed) TransFEWmation: towards design-led food-energy-water systems for future urbanization. Springer, Cham, pp 209–234. https://doi.org/10.1007/978-3-030-61977-0

Yan W, Roggema R (2019) Developing a design-led approach for the food-energy-water nexus in cities. Urban Plan 4(1):123–138

Yan W, van Delbsteen A, Keeffe G, Thün G, and Roggema R (2021) Developing and implementing the design-led nexus approach for sustainable urbanisation. In CITIES 20.50--Creating habitats for the 3rd millennium: smart--sustainable--climate neutral. Proceedings of REAL CORP 2021, 26th international conference on urban development, regional planning and information society, 251–60. CORP--Competence Center of Urban and Regional Planning

Zhang C, Chen X, Li Y, Ding W, Fu G (2018a) Water-energy-food nexus: concepts, questions and methodologies. J Clean Prod 195:625–639

Zhang X, Chen N, Chen Z, Lixin W, Li X, Zhang L, Di L, Gong J, Li D (2018b) Earth-science reviews geospatial sensor web: a cyber-physical infrastructure for geoscience research and application. Earth Sci Rev 185(July):684–703

Zhang P, Zhang L, Chang Y, Xu M, Hao Y, Liang S, Liu G, Yang Z, Wang C (2019) Food-energy-water (FEW) nexus for urban sustainability: a comprehensive review. Resour Conserv Recycl 142(July 2018):215–224

Chapter 4
System Dynamics in the Urban Food–Water–Energy Nexus: A Systematic Review

Hanyu Liu and Wanglin Yan

Abstract The food–energy–water (FEW) nexus provides a critical framework for analyzing complex interdependencies within urban resource system, while the inherent complexity of urban system presents significant challenges for effective management and policy formulation. System dynamics (SD) modeling is widely applied to capture the complex interactions among urban systems. However, its application in urban FEW nexus research remains limited, with notable gaps such as the lack of post-modeling optimization analyses and insufficient integration of spatial dimensions. Therefore, this study reviews existing SD models applied to the urban FEW nexus, synthesizing their structures and applications and discussing future directions for model development. The systematic review of 179 journal articles reveals that although urban FEW management is the most frequently addressed theme, research in this field remains fragmented. Keywords clusters analysis highlights dominant research clusters focused on water management and energy–carbon modeling, while also exposing underexplored areas such as spatial modeling, household-scale analysis, and cross-sector integration. Scenario analysis emerges as the most widely adopted post-modeling approach, followed by sensitivity analysis. These findings inform the discussion on an ideal system dynamics modeling framework for the urban FEW nexus, emphasizing spatial–temporal perspectives, regional interdependencies, and post-modeling optimization. This study contributes to advancing FEW nexus understanding and supports the development of more effective tools for managing its complexities.

Keywords Urban food–energy–water nexus · System dynamics · Systematic review · FEW subsystems · Keywords clysters analysis

---

H. Liu (✉)
Graduate School of Media and Governance, Keio University, Fujisawa, Kanagawa, Japan
e-mail: liuhanyu@keio.jp

W. Yan
Faculty of Environment and Information Studies/Graduate School of Media and Governance, Keio University, Fujisawa, Kanagawa, Japan

W. Yan (ed.), *Urban Nexus Approach for Sustainable Cities*,
https://doi.org/10.1007/978-981-95-6156-8_4

1 Introduction

As urbanization accelerates globally, cities are increasingly confronted with complex challenges in managing essential resource, particularly within the interdependent sectors of water, energy, and food. These three systems are intrinsically linked: water is essential for energy production, energy is required for water treatment distribution, and both are critical inputs for food production. Global demand for all three resources is projected to rise significantly, an increase of nearly 50% by 2050, placing substantial pressure on urban infrastructures to meet these needs sustainably (United Nations 2018). For instance, in China, rapid urban expansion has led to substantial increases in both water and energy consumption, while agricultural productivity has struggled to match the escalating food demand, further highlighting the complexity of balancing FEW resources in urban contexts (Zhang et al. 2019).

The food–energy–water (FEW) nexus has emerged as a critical framework for understanding the synergies and trade-offs among these interdependent resource systems. By examining how water, energy, and food systems function and interact, the FEW nexus approach seeks to enhance resource-use efficiency, optimize system-wide synergies, and mitigate negative externalities, including environmental and social impacts (Kurian 2017). Effective management of these cross-sectoral interactions is essential for achieving sustainable development, as inefficiencies or mismanagement in one sector can significantly compromise outcomes in others. For instance, Singapore's pursuit of food security is intricately linked to its water and energy policies, with technologies such as vertical farming and desalination imposing substantial energy demands (Chan 2015). As urban populations continue to grow, a comprehensive understanding of the FEW nexus becomes increasingly vital to support integrated policy-making that prevents resource depletion, reduces systemic risks, and promotes long-term sustainability (Laspidou et al. 2020; Qi et al. 2022).

To address the complex interactions within the FEW nexus, system dynamics (SD) modeling has emerged as an effective tool for simulating the nonlinear relationships among FEW components and capturing their dynamic behaviors over time. A key strength of SD lies in its ability to evaluate how policy interventions, technological advancements, or external shocks influence the stability and resilience of urban FEW systems. This modeling approach is particularly valuable in accounting for external drivers such as climate change, which alters system dynamics through rising temperatures and shifting precipitation patterns that directly affect water availability, food production, and energy demand. For example, in Mexico City, SD models have been applied to assess the impacts of changing rainfall patterns on water reserves and food security, offering insights for long-term adaptation strategies (Brouwer et al. 2020).

However, many existing FEW nexus studies remain limited in scope, often focusing on bilateral interactions such as water–food or water–energy linkages, rather than capturing the full complexity of trisectoral interdependencies. Similarly, most urban FEW SD models emphasize sector-specific dynamics, such as water treatment systems or agricultural production, without incorporating cross-sectoral flows or system-wide integration. This lack of systemic modeling hinders the ability to coordinate policies across resource sectors and exacerbates unintended trade-offs (Smajgl et al. 2016). Furthermore, the formulation of effective cross-sectoral policies requires an understanding of the sociopolitical dimensions of each system, yet these factors are frequently overlooked in model design (Allouche et al. 2015; Foran 2015).

The FEW nexus framework itself is undergoing continual conceptual refinement, increasingly incorporating environmental, economic, social, and political dimensions. Recent literature emphasizes not only the physical linkages among systems but also the role of governance structures in shaping outcomes related to social equity, externalities, and socio-ecological resilience (Scott et al. 2015). Nevertheless, despite these conceptual advancements, corresponding progress in SD modeling has lagged behind. Many existing models remain confined to technical system interactions and fail to integrate broader governance, economic, or social dynamics. Moreover, while the FEW nexus emphasizes the optimization of synergies and the mitigation of negative impacts, most SD applications stop at quantifying interactions, with limited attention paid to post-modeling analyses such as optimization and decision support.

Although previous studies have explored a wide range of methodological approaches within the food–energy–water (FEW) nexus literature, few have systematically reviewed the application of system dynamics (SD) modeling in urban FEW contexts. Despite the increasing use of SD modeling in this field, a comprehensive assessment of its application in urban nexus studies remains lacking. There is a clear need for a structured and critical evaluation of existing urban FEW SD models to identify best practices, improve subsystem integration, and enhance post-modeling analyses, thereby facilitating the development of more robust and policy-relevant nexus assessments. To address this gap, this study systematically reviews existing SD models applied to the urban FEW nexus, synthesizing their research themes, subsystem structures, integrated analytical methods, and post-modeling applications, while also identifying future directions for model advancement. By providing a comprehensive overview, this study contributes to the refinement of SD-based urban FEW modeling and supports the formulation of more effective, integrated strategies for sustainable urban resource management.

2 Methods

2.1 Literature Collection and Analysis Methods

This systematic review was conducted using the Scopus database to identify relevant literature on the application of system dynamics within the context of the urban food–water–energy (FEW) nexus. A Boolean search strategy was employed using the following search terms: ((energy OR electricity OR gas) AND (water)) OR ((energy OR electricity OR gas) AND (food OR agriculture)) OR ((water) AND (food OR agriculture)) AND ("system dynamics") AND (urban OR city)). This approach aimed to capture studies that discussed the interactions between water, energy, and food, with a specific focus on system dynamics modeling in urban settings (Table 4.1).

To ensure relevance, inclusion and exclusion criteria were established. Studies were included if they applied system dynamics models to explore interactions within the FEW nexus at an urban or city scale. Only peer-reviewed journal articles were considered. Exclusion criteria were applied to studies that focused on resource interactions beyond the urban scale, did not address the interconnections between water, energy, and food, or did not employ system dynamics as a methodology.

The initial search yielded 179 records, from which 52 were removed due to irrelevance. This left 127 records that were screened based on title and abstract. Subsequently, 125 full-text reports were sought for retrieval, with no exclusions due to unavailability. However, after a detailed review, 95 reports were excluded, some of them only focused on one or two resources and their interactions, and some of these focused beyond the urban scale. Ultimately, 30 studies met the criteria for inclusion in this review, as shown in the flowchart (Fig. 4.1). For each study, data related to system dynamics modeling objectives, key variables, and policy implications were extracted. The findings were synthesized to evaluate how system dynamics models have been applied to the urban FEW nexus, and to identify common patterns, gaps in the literature, and areas for future research.

Table 4.1 Inclusion/exclusion criteria for literature

No.	Include criteria	Exclude criteria
1	Articles written in English	Articles written in languages other than English
2	Articles with full-text access available	Articles without full-text access
3	Journal articles	Manuscripts, conference papers, reports, etc.
4	Articles consisting of at least three pages	Short papers or summaries with fewer than three pages
5	Articles with unique titles	Articles with duplicate titles
6	Non-review articles	Review articles
7	Studies focusing on food, water, energy, and their interactions	Studies focusing solely on one or two resources or their interactions
8	Studies related to urban areas	Studies unrelated to urban areas

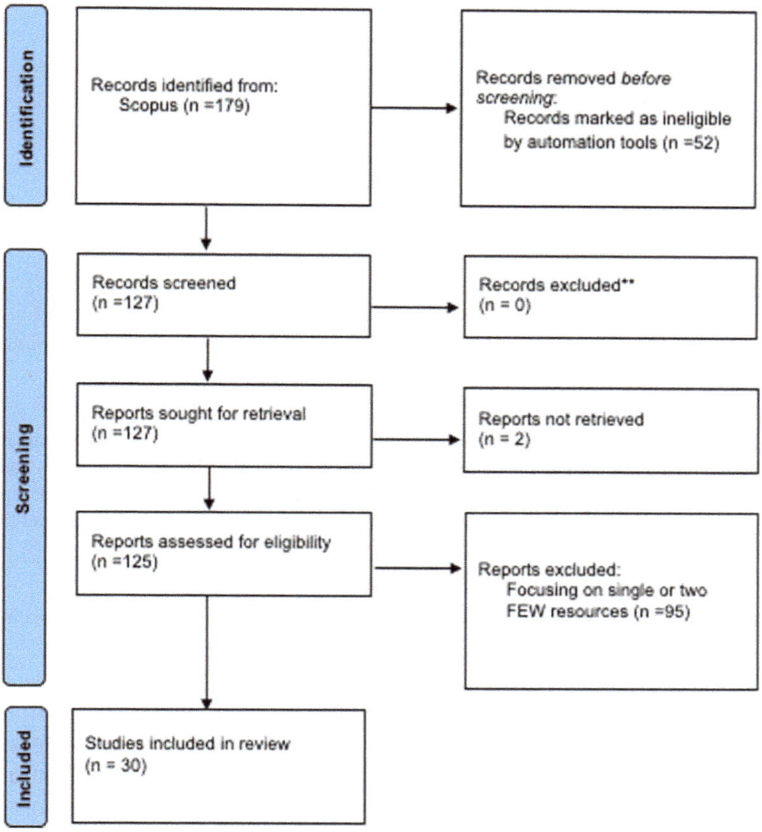

Fig. 4.1 Flowchart for the system dynamics in the urban food–water–energy nexus

3 Results

3.1 Temporal Trends in Urban FEW Nexus Using System Dynamics Modeling

Figure 4.2 presents a timeline visualization of keyword co-occurrence clusters in urban FEW nexus research (2013–2025). The literature is organized into major thematic clusters shown as horizontal tracks, with nodes indicating frequently co-occurring keywords. Node size reflects keyword frequency, node color represents the year of emergence, and red rings highlight burst keywords that gained rapid attention in a short period. It offers a dynamic and structured perspective on the intellectual development of the field between 2013 and 2025, illustrating the evolution of key research themes through keyword co-occurrence clusters. Additionally, Table 4.2 highlights the keywords with the strongest citation bursts, capturing short periods of intensified scholarly attention. Together, these visualizations provide a

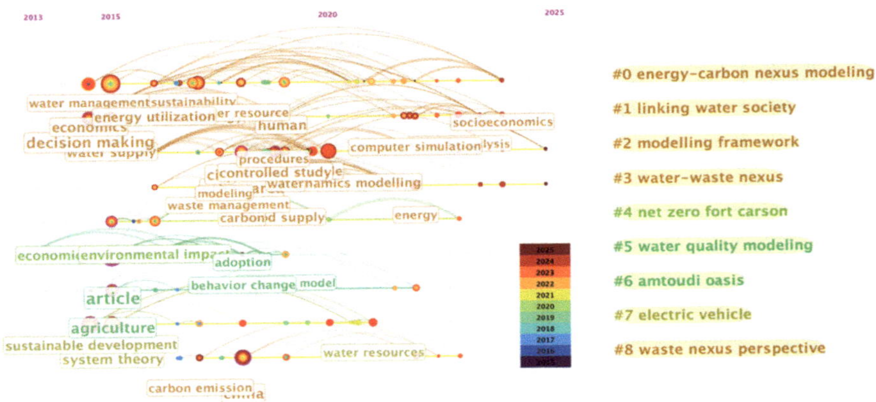

Fig. 4.2 Timeline visualization of keyword co-occurrence clusters in urban FEW nexus research (2013–2025). Each horizontal band represents a major thematic cluster, with nodes indicating frequently co-occurring keywords. Node size reflects keyword frequency, node color represents the year of emergence, and red rings highlight burst keywords that gained rapid attention in a short period

Table 4.2 Top 12 keywords with the strongest citation bursts in urban FEW nexus with system dynamics modeling (2015–2025)

Keywords	Year	Strength	Begin	End
Climate change	2015	2.60	2017	2018
Canada	2017	1.78	2017	2018
Environmental protection	2018	2.83	2018	2019
City	2018	1.76	2018	2020
Water management	2015	2.32	2019	2021
Water pollution	2019	2.26	2019	2020
Water demand	2020	2.19	2020	2021
Life cycle	2020	1.82	2020	2023
Water resources	2021	2.95	2021	2022
Energy	2022	2.83	2022	2023
Scenario analysis	2023	2.36	2023	2025
Carbon footprint	2023	2.16	2023	2025

temporal map of conceptual evolution, linking the structural patterns of keyword relationships with their sequential significance in the development of urban FEW nexus research using system dynamics modeling.

The largest and most foundational cluster identified is Cluster 0, centered on "energy–carbon nexus modeling." With 42 members and an average year of 2020, this cluster represents a core of system dynamics-based simulation efforts addressing carbon emissions, water resource planning, and energy–food synergies. The most cited concepts in this group include "water energy," and it features several high-impact studies such as Wen et al. (2022) and Liu et al. (2024a, b). Closely related is Cluster 1, focused on “linking water and society.” This group explores the socioeconomic and governance dimensions of nexus interactions, incorporating terms such as “decision-making” and “water supply.” The appearance of strong citation bursts such as “scenario analysis” and “water pollution” further indicates the rising interest in policy-relevant simulation. Cluster 2, labeled “modeling framework,” forms a methodological bridge in the network. With high citation counts and bursts centered on “system dynamics modeling” and “urban area,” this cluster underscores the technical backbone of the field, supporting scenario analysis and optimization strategies in urban and regional settings. Complementing this, Cluster 3 emphasizes the “water–waste nexus,” connecting circular economy approaches, decarbonization, and infrastructure modeling. The burst strength of terms such as “energy” and frequent references to waste management strategies point to an increasing interest in urban metabolism and resource recovery systems. Several other clusters highlight emerging and specialized areas of research. Cluster 4 focuses on the Net Zero Fort Carson initiative, offering real-world applications of nexus frameworks in military contexts. Cluster 5, defined by “water quality modeling,” emphasizes stakeholder engagement and basin-level simulation of nutrient flows and pollutant control. Cluster 6 addresses the Amtoudi Oasis case, while Cluster 7, labeled “electric vehicle,” expands nexus considerations to include sustainable transport systems. This latter cluster features key terms such as “system theory” and “sustainable development,” indicating a broadening conceptual scope. The “waste nexus perspective” (Cluster 8) focuses on urban carbon accounting, infrastructure management, and the intersection of waste and emissions. Meanwhile, the “urban water cycle” cluster (Cluster 9) aligns with emerging trends in circular economy modeling and water reuse. Both Clusters 8 and 9 exhibit temporal maturity and conceptual distinctiveness, suggesting that the FEW nexus field is moving toward more integrated, system-scale frameworks.

3.2 Research Subjects and Objectives of Urban FEW System Dynamics Modeling

The categorization of articles based on urban food–energy–water (FEW) nexus system dynamics (SD) modeling highlights the varying research focus across different subjects. The highest number of studies (6) fall under urban FEW management, indicating a strong emphasis on holistic system-level integration of food, energy, and water resources in urban environments. Urban water management and FEW uses of urban agriculture are also well-explored, with five articles each, reflecting the significance of water sustainability, resource-efficient farming practices, and agricultural production in urban settings. Similarly, urban land use (four articles) represents a key area of research, focusing on the spatial and infrastructural dimensions of FEW interactions (Fig. 4.3).

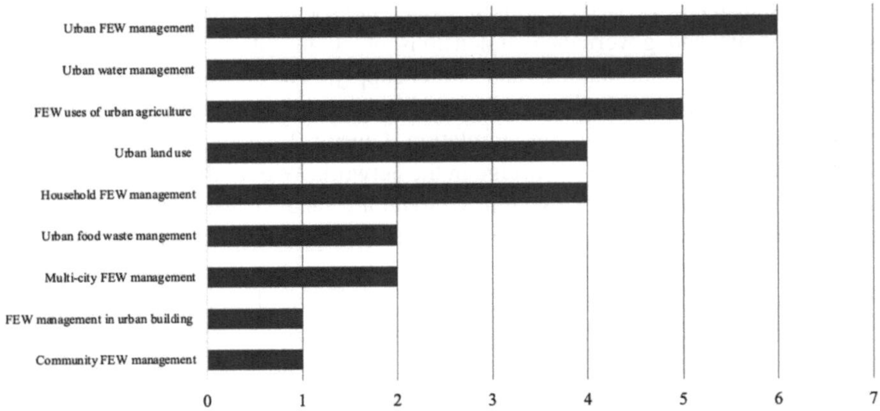

Fig. 4.3 Number of articles categorized under the subjects listed

At the household and community levels, household FEW management (three articles) investigates consumption patterns, waste management, and sustainability strategies at a microscale, while urban food waste management (two articles) and multicity FEW management (two articles) focus on specific challenges such as waste reduction, circular economy principles, and intercity resource coordination. Additionally, FEW management in urban buildings (one article) and community FEW management (one article) are less represented in the literature but offer valuable insights into localized and infrastructure-based FEW solutions (Table 4.3).

3.3 Subsystems and Integrated Methods in Urban FEW System Dynamics Modeling

Table 4.4 presents the integration of subsystems within system dynamics (SD) modeling across various research themes related to the food–energy–water (FEW) nexus in urban contexts. The findings illustrate how different urban FEW studies incorporate distinct but interconnected subsystems to examine resource management, policy implications, and sustainability transitions. Water-related subsystems, including water supply infrastructure, water treatment and recycling, and water demand, are predominantly addressed in studies on urban water management and multicity FEW management. These aspects highlight the critical role of water availability, treatment processes, and consumption patterns in shaping urban sustainability. Notably, water demand is considered across multiple research themes, reflecting its fundamental role in resource allocation and sustainability assessment. Similarly, energy-related subsystems, which encompass energy supply and energy consumption, are primarily integrated into studies on urban FEW management, multicity FEW management, and the application of FEW in urban agriculture. These investigations assess the reliance on both fossil fuels and renewable energy sources while

Table 4.3 Research objectives of system dynamics modeling in urban FEW nexus studies

Subject	Research objective	Reference
Urban water management	Analyze the interconnections between water, energy, and food (FEW) in urban water management, focusing on policy optimization, resource recovery, sustainability assessment, environmental impacts, and long-term planning for drought mitigation	Fagan et al. (2007), Das and Chirisa (2021), Wang et al. (2021), Shiu et al. (2023), and Egieya et al. (2024)
Urban land use	Assess the impacts of land-use changes on carbon emissions, resource allocation, and sustainability by integrating the food–energy–water (FEW) nexus. The models focus on urbanization, economic development, ecological balance, and policy-driven land optimization	Fu et al. (2017), Fang et al. (2019), Kumar et al. (2021), and Yu et al. (2021)
Urban food waste management	Assess the environmental, social, and economic impacts of urban food waste management, integrating the food–energy–water–climate (FEWC) nexus and circular economy principles to optimize resource efficiency and sustainability	Parsa et al. (2023, 2024)
Urban FEW management	Analyze the food–energy–water (FEW) nexus in urban areas, focusing on resource interactions, governance, sustainability transitions, policy impacts, and optimization strategies to enhance urban resilience and efficiency	Hu et al. (2019), Soares Dal Poz et al. (2022), Valencia et al. (2022a, b), and Wen et al. (2022), Li et al. (2022), Mirindi et al. (2024), and Zhou et al. (2024)
Multicity FEW management	Evaluate the coordination of food–energy–water (FEW) resources across multiple cities, addressing regional disparities, supply–demand dynamics, resource security, and policy recommendations for urban sustainability	Ouyang et al. (2021) and Li et al. (2024)
FEW management in household and community	Assess food–energy–water (FEW) interactions at household and community levels, focusing on consumption behavior, resource efficiency, net-zero strategies, and policy optimization for sustainability	Hussien et al. (2017), Du et al. (2022), Liu et al. (2024a, b), Procter et al. (2016), and Valencia et al. (2022a, b)
FEW uses of urban agriculture	Analyze the food–energy–water (FEW) interactions in urban agriculture, focusing on resource efficiency, sustainability, technological innovations, and carbon reduction strategies for optimal urban food production	Baganz et al. (2021), Song et al. (2022), Huang and Chang (2021), Wang et al. (2023), and Francisco et al. (2023)

evaluating energy consumption in residential, industrial, mobility, and water treatment sectors. The widespread consideration of energy consumption underscores its significant influence on urban metabolic processes and resource efficiency.

Food-related subsystems, including food production, food processing and distribution, and food waste management, are central to research on urban food waste management, the application of FEW in urban agriculture, and FEW management in household and community contexts. These studies explore sustainable

Table 4.4 Subsystems involved in urban FEW nexus system dynamics modeling

Subsystems	Subjects						
	Water management	Land use	Food waste management	Urban FEW management	Multicity FEW management	FEW management in household and community	FEW uses of urban agriculture
Water supply (infrastructure, groundwater, surface water)	✓			✓	✓		✓
Water treatment and recycling (wastewater treatment, water recycling, desalination)	✓			✓			
Water demand (domestic, industrial, agricultural)	✓	✓	✓	✓	✓	✓	✓
Energy supply (coal, oil, gas, renewables)				✓	✓		✓
Energy consumption (residential, industrial, mobility, water treatment)	✓	✓	✓	✓	✓	✓	✓
Food production (agriculture, aquaponics, vertical farming)		✓	✓	✓	✓		✓
Food processing and distribution (manufacturing, retail, transport)			✓			✓	
Food waste management (composting, anaerobic digestion, recycling)			✓			✓	
Land use (urban expansion, agricultural land, ecological zones)		✓					
Population	✓	✓		✓	✓		✓
Economic factors (GDP, investments, costs and benefits)	✓	✓		✓	✓		
Environmental impact (carbon emissions, pollution, climate adaptation)	✓		✓	✓			✓

agricultural practices, urban food supply chains, and waste reduction strategies, particularly within circular economy frameworks. Land use and population growth are predominantly analyzed within urban land use studies, given their importance in shaping spatial planning, ecological balance, and infrastructure expansion. Meanwhile, economic factors, such as policy regulations, trade mechanisms, and investment strategies, are commonly considered in studies on urban water management, urban land use, and multicity FEW management. These investigations assess the governance and financial implications associated with urban sustainability transitions.

Environmental impact, particularly in terms of carbon emissions, pollution, and climate adaptation, is a key focus in research on urban FEW management, urban food waste management, and the application of FEW in urban agriculture. The emphasis on environmental considerations highlights their importance in optimizing resource use and informing policy design. The distribution of subsystems across various research themes underscores the multidimensional nature of urban FEW dynamics, where interactions among water, energy, and food systems are closely interwoven with economic, environmental, and demographic factors. The integration of system dynamics modeling across these themes demonstrates a comprehensive and interdisciplinary approach to understanding urban resource sustainability challenges and opportunities.

3.4 Integrated Analytical Methods Supporting Urban FEW Nexus SD Modeling

Table 4.5 shows integrated methods applied in urban FEW system dynamics modeling. These methods have been applied across different urban FEW nexus subjects to enhance the understanding and management of resource interactions. The Urban Metabolism Analysis method was employed in community FEW management to evaluate resource flows, energy consumption, and waste production within urban environments, providing insights into sustainability and efficiency. Life Cycle Assessment (LCA) was widely used in FEW applications related to urban agriculture, water management, and buildings, enabling the evaluation of environmental impacts across the entire lifecycle of food, energy, and water resources. In multicity FEW management, Catastrophe Progression was applied to model sudden shifts and potential risks in resource systems, while Coupled Coordination Degree Analysis was used to assess the level of coordination among FEW subsystems, identifying inefficiencies and guiding improvements in multicity resource integration.

Within urban FEW management, a diverse set of methods was incorporated to improve decision-making and resource optimization. Artificial Intelligence (AI) was utilized for predictive modeling and data-driven optimization, while the Delphi Method facilitated expert-driven consensus in policy and strategy development.

Metabolic Analysis was applied to assess input–output dynamics within FEW systems, supporting resource efficiency and waste reduction. Multi-criteria Decision-Making was used to evaluate conflicting criteria in FEW management strategies, aiding urban planning and governance. Additionally, Optimization Tools were implemented to enhance FEW system efficiency by minimizing resource use and maximizing sustainability outcomes. A Cost–Benefit–Risk Tradeoff Analysis was also employed to evaluate the economic, environmental, and social trade-offs involved in FEW management, assisting in comprehensive decision-making. For urban land use studies, the CLUE-S model (Conversion of Land Use and its Effects) was used to simulate land-use changes and their impacts on FEW interactions, providing valuable insights for sustainable urban planning. These integrated methods, when combined with SD modeling, enable a more comprehensive approach to analyzing and optimizing urban FEW nexus dynamics, addressing sustainability challenges, improving resource management, and informing policy decisions across various urban contexts.

3.5 Post-Modeling Analytical Approaches in Urban FEW Nexus

Following the development of system dynamics (SD) models in urban water–energy–food (FEW) nexus research, a suite of post-modeling analytical approaches has been employed to enhance system understanding and inform policy design. As shown in Table 4.5, these approaches primarily include scenario analysis, sensitivity analysis, optimization, and decision-making support, each addressing distinct analytical objectives within the broader modeling framework.

Scenario and sensitivity analyses are the most frequently adopted methods, appearing in 27 and 7 studies, respectively. Scenario analysis is employed to explore the impacts of alternative future conditions or interventions by systematically varying key drivers such as technological advancement, population dynamics, resource availability and pricing, behavioral shifts, and economic growth. Some studies utilize established scenario frameworks, such as the Global Scenario Group (GSG) scenarios, which present contrasting development pathways ranging from business-as-usual to sustainability-focused transitions. Sensitivity analysis, on the other hand, is used to identify the most influential parameters within the model. By evaluating the responsiveness of system outputs to variations in input variables (e.g., demographic trends, behavioral patterns, or implementation levels of policies), sensitivity testing enhances model robustness and supports validation. Additionally, it helps prioritize critical intervention points and assess potential future demand across the FEW sectors.

Optimization and decision-making support methods have been applied in seven and two studies, respectively, with the shared aim of identifying desirable development pathways and supporting effective policy formulation. Optimization

Table 4.5 Integrated methods applied in urban FEW system dynamics modeling across different subjects

Subject	Methods	Amount
Community FEW management	Urban Metabolism Analysis	1
FEW uses of urban agriculture	Life Cycle Assessment	3
Multicity FEW management	Catastrophe progression	1
	Coupled Coordination Degree Analysis	1
Urban FEW management	AI	1
	Delphi Method—decision-making	1
	Metabolic Analysis	1
	Multi-criteria Decision-Making	1
	Optimization Tool	1
	Cost–Benefit–Risk Tradeoff Analysis	1
Urban FEW uses in building	Life Cycle Assessment	1
Urban land use	CLUE-S	1
Urban water management	Life Cycle Assessment	1

approaches are used to maximize synergies across FEW subsystems through scenario comparison and single- or multi-objective frameworks, allowing researchers to evaluate trade-offs and derive solutions that balance resource efficiency, environmental outcomes, and policy effectiveness. Decision-making support tools, such as TOPSIS (Technique for Order Preference by Similarity to Ideal Solution), are employed to rank policy alternatives based on multiple criteria including cost, environmental impact, and social equity. These tools help translate model outputs into actionable insights, aiding strategic planning under conditions of complexity and uncertainty.

4 Future Directions for Urban FEW SD Model Advancement

4.1 Comprehensive Assessment and Emerging Gaps

This systematic review and temporal analysis of urban FEW system dynamics modeling reveals a predominant focus on macro-level urban sustainability, especially in areas such as water resource management, carbon mitigation, and agriculture. These trends are reflected in the timeline visualization, where Cluster 0 ("energy–carbon nexus modeling") and keywords such as "water management" and "system dynamics model" demonstrate early and sustained prominence. However, applications at the microscale, such as buildings, households, and communities, remain underexplored. This is a significant omission, as household activities substantially contribute to urban energy use and greenhouse gas emissions (Hussien et al. 2017), and behavior-driven consumption patterns are central to demand-side management. Incorporating household-level feedback loops into system dynamics (SD) models could improve responsiveness to behavioral changes and enable more targeted policy interventions.

While comparative studies across multiple cities are increasing, most models continue to treat cities as isolated systems. The absence of terms such as "regional modeling" and "intercity cooperation" in the co-occurrence timeline and burst analysis indicates a persistent gap in addressing cross-boundary resource flows. Yet, urban FEW demands are heavily reliant on interregional infrastructure, such as water transfers, electricity grids, and food distribution networks, which extend beyond municipal limits. Integrating regional dynamics into SD frameworks would capture intercity interdependencies and support the modeling of shared governance, resource allocation, and conflict resolution mechanisms.

More and more education and scientific research, medical services, and various service industries (accommodation, food and beverage, entertainment, etc.) have concentrated in Tokyo, resulting in increased electricity needs. It is a typical example of a railway-oriented city, where 76% of total fuel consumption comes from transport. In addition, there is a large environmental burden in agriculture and manufacturing; because the local food supply cannot meet local demand, so food must be imported from other areas. As for water consumption, thermal power generation increased due to the shutdown of nuclear power stations after the Great East Japan Earthquake. To maintain a steady supply of electricity, the thermal power generation needed a lot of water for cooling, resulting in a doubling of water consumption.

Additionally, land-use system dynamics models often include FEW components in isolation, failing to represent their inter-sectoral feedback. For instance, urban expansion affects energy consumption, transportation demand, available farmland, and watershed runoff. Future work should embed a more integrated FEW perspective into land-use SD models to better identify trade-offs and synergies for adaptive planning under climate and demographic uncertainties.

4.2 Subsystem Integration in Urban FEW Modeling

System dynamics modeling offers strong potential for integrating diverse urban subsystems, but the extent and quality of this integration vary significantly. Water subsystems, particularly supply, demand, and wastewater treatment, are most frequently modeled, as evidenced by keyword prominence in Cluster 5 and Table 4.2 (e.g., "water resources," "water pollution"). However, advanced strategies such as decentralized reuse, stormwater harvesting, and desalination are rarely incorporated. Melbourne (decentralized stormwater reuse) offers transferable templates for improving water resilience in SD models (Du et al. 2022).

In the energy domain, while supply and demand dynamics are commonly represented, few models incorporate the interplay of policy incentives, technological diffusion, and user behavior. Similarly, food systems are often limited to modules on production and waste management, neglecting variables such as dietary transitions, vertical farming, and food logistics. These gaps overlook key leverage points for sustainability in dense urban settings where food–energy interactions are highly spatialized.

Another significant challenge in urban FEW system dynamics modeling is the oversimplification of cross-sector interdependencies, particularly the cascading effects that arise from sectoral disruptions. While some models attempt to represent these linkages, static or linear assumptions are used widely, failing to capture the dynamic and nonlinear nature of real-world feedback. In practice, disturbances in one sector, such as a power outage, can initiate cascading failures that compromise water treatment and distribution systems, which in turn disrupt agricultural irrigation and urban food supply chains. These cascading disruptions underscore the need for models that can simulate complex intersectoral dynamics and system-wide vulnerabilities. The recent emergence of modeling themes such as "scenario analysis" and "carbon footprint" in the literature (2023–2025) reflects growing scholarly attention to such complexities, particularly in relation to sustainability trade-offs and risk assessment. To address these challenges, future system dynamics models should integrate probabilistic methods and agent-based modeling to enable adaptive response simulations, uncertainty propagation, and resilience testing under compound shock scenarios.

4.3 Advancements and Challenges in Integrated Modeling Approaches

Recent methodological advances have introduced tools such as Life Cycle Assessment (LCA) and Urban Metabolism Analysis into FEW SD modeling. LCA enables detailed assessment of environmental trade-offs at the process level (e.g., energy sources, food supply chains), while urban metabolism supports macroscale tracking of material and energy flows. Despite their potential synergy, cluster and co-occurrence analyses show limited methodological integration.

Artificial intelligence (AI) offers a promising frontier for dynamic urban modeling. Machine learning and reinforcement learning techniques can enhance real-time forecasting, optimization, and automated scenario generation. While current use is limited, recent work by Khatun et al. (2022) highlights how AI–SD hybrids can improve decision-making under uncertainty. These models can adapt to new data, refine parameter estimates dynamically, and better simulate nonlinear feedback. Although AI-related terms are not prominent in the keyword analysis, their increasing relevance suggests they should be more fully integrated into future FEW modeling efforts.

Lastly, while optimization and cost–benefit analysis tools are commonly used in urban FEW planning, they often overlook system-wide interactions and equity considerations. Most models prioritize economic efficiency, ignoring co-benefits and externalities. A more forward-looking approach involves multi-objective optimization frameworks that balance environmental, social, and economic outcomes—particularly those aligned with justice-centered urban policy. This would significantly enhance the policy relevance of SD models for managing trade-offs in resource-constrained urban environments.

4.4 Toward an Ideal Urban FEW System Dynamics Model

Advancing system dynamics modeling for the urban FEW nexus requires the development of more integrated and adaptive frameworks that can represent both physical and institutional dimensions of urban systems. Core model components should reflect the essential interactions among water, energy, food, land use, and governance. Besides, the model should also allow for the inclusion of emerging technologies and context-specific dynamics to enhance decision support.

Water subsystems should ideally include not only municipal supply and treatment infrastructure but also decentralized reuse systems, stormwater harvesting, and the effects of aging infrastructure, particularly for cities facing water stress and climate variability. Energy subsystems benefit from expanding beyond static supply and demand balances to incorporate renewable integration, electrification scenarios, and responses to pricing mechanisms or policy interventions. Food systems can be enhanced by incorporating elements such as urban agriculture, dietary shifts, food supply chain disruptions, and waste management practices, which together shape resource flows and sustainability outcomes. Land-use components should account for the interplay between urban expansion, ecosystem services, and green infrastructure to enable a more holistic evaluation of development pathways and environmental trade-offs. While these physical subsystems form the foundation of effective urban FEW models, the integration of socioeconomic and institutional feedback is equally important. Capturing behavioral responses, policy evolution, stakeholder priorities, and governance structures allows models to more accurately simulate long-term trajectories and policy impacts.

An essential complement to model development is the application of post-modeling analytical approaches. It enables to convert system dynamics outputs into actionable policy insights. Optimization techniques, such as single- and multi-objective scenario analyses, are increasingly used to identify efficient development pathways, evaluate trade-offs, and maximize synergies across the FEW nexus. These methods support decision-making by balancing resource efficiency, environmental goals, and policy effectiveness. Additionally, decision-support tools such as multi-criteria analysis (e.g., TOPSIS) enable the ranking of policy alternatives based on criteria including cost, sustainability, and social equity. Integrating these analytical layers enhances the strategic value of system dynamics models, ensuring they not only simulate system behavior but also inform urban resource management.

To further enhance the responsiveness, realism, and decision-support capabilities of urban food–energy–water (FEW) system dynamics models, the integration of advanced digital technologies is increasingly valuable. Artificial intelligence (AI) techniques, such as machine learning and reinforcement learning, offer powerful tools for incorporating real-time or high-frequency data, improving forecasting precision, and automating scenario development under uncertainty. Recent applications of generative AI models—such as Generative Adversarial Networks (GANs) and transformers—demonstrate their ability to autonomously generate urban scenarios and spatial datasets, significantly reducing the time and cost associated with infrastructure simulation and planning (Xu et al. 2024).

Separately, digital twin platforms offer dynamic, virtual representations of urban infrastructure systems that enable high-fidelity, interactive simulations of evolving physical and operational conditions. These platforms are increasingly used to support real-time monitoring, predictive analytics, and the modeling of interdependent systems. For example, an energy system digital twin was developed to simulate the behavior of distributed renewable energy and electric vehicle integration using stochastic modeling techniques to account for system variability (Liu and Zhang 2024). In a separate case, Shehadeh et al. (2024) employed a digital twin framework in an urban refugee setting to assess water scarcity, energy demand, and the implementation of solar technologies, drawing on both real-time data and community insights to inform sustainability interventions (Shehadeh et al. 2024). These applications highlight the growing role of AI and digital twins in producing more adaptive, data-informed, and context-sensitive FEW system dynamics models for urban planning and resilience building.

5 Conclusions

This study examines the role of system dynamics (SD) modeling in advancing the understanding of the urban food–energy–water (FEW) nexus, emphasizing its potential to address the complexities of interdependent resource systems. Through a systematic review, we identified key research areas, assessed the integration of subsystems in FEW models, and analyzed the methodological approaches applied in SD-based studies. By synthesizing insights on subsystem interactions, evaluating the strengths and limitations of integrated modeling, and proposing interdisciplinary advancements, this study contributes to the refinement of FEW system modeling.

Despite its effectiveness in capturing FEW interactions, SD modeling still faces challenges in representing cross-sectoral feedback, integrating multi-scale dynamics, and adapting to evolving urban conditions. Many models focus on specific components—such as water supply, energy consumption, and food production—while overlooking behavioral, socioeconomic, and governance factors. Additionally, limited adoption of emerging technologies, including artificial intelligence (AI) and real-time data integration, constrains SD models' capacity for predictive, adaptive, and scenario-based decision-making. Overcoming these limitations is essential to ensuring SD modeling remains a robust tool for sustainable urban planning and resource management.

Enhancing SD modeling for the urban FEW nexus requires incorporating advanced analytical techniques that address key challenges—interdisciplinary integration, adaptive modeling, and policy implementation. AI-driven SD models, agent-based modeling for capturing behavioral dynamics, and optimization frameworks that account for economic, environmental, and social trade-offs can improve predictive accuracy and decision support. Engaging stakeholders, policymakers, and urban planners through participatory approaches will further strengthen the practical relevance of these models. Future research should focus on expanding multi-scale modeling frameworks, improving climate adaptation strategies, and fostering data-driven approaches to support real-time decision-making.

Given the complexity of urban FEW systems, a multi-method approach is necessary to navigate resource management challenges. Effective SD models must not only provide theoretical insights but also deliver actionable guidance for policymakers. While SD modeling has established a solid foundation for FEW research, its continued evolution must align with technological advancements, interdisciplinary methodologies, and policy priorities. As urban sustainability challenges intensify, refining SD modeling to be more predictive, adaptive, and policy-relevant will be crucial in ensuring long-term resource security, efficiency, and resilience.

Acknowledgments The content of this chapter comes from part of the M-NEX (m-nex.net) project, which was under a grant of the Sustainable Urbanization Global Initiative (SUGI)—Food–Water–Energy Nexus by Belmont Forum and JST (Japan Science and Technology Agency).

References

Allouche J, Middleton C, Gyawali D (2015) Technical veil, hidden politics: interrogating the power linkages behind the nexus. Water Altern 8(1):610–626. https://www.water-alternatives.org/index.php/volume8/v8issue1/275-a8-1-4

Baganz GF, Schrenk M, Körner O, Baganz D, Keesman KJ, Goddek S et al (2021) Causal relations of upscaled urban aquaponics and the food-water-energy nexus—a Berlin case study. Water 13(15):2029

Brouwer F, Hofkes M, Linderhof V (2020) System dynamics modeling for the food-water-energy nexus: application to Mexico City. Water 12(1):245. https://doi.org/10.3390/w12010245

Chan AH (2015) Connecting cities and their environments: harnessing the water-energy-food nexus for sustainable urban development. Change and Adaptation in Socio-Ecological System, 2(1).

Das DK, Chirisa I (2021) Exploring the water-nutrient-food nexus for an African City region: linking the Chivero Lake and Harare city region, Zimbabwe. Int Rev Spatial Plann Sustainable Dev 9(4):82

Du S, Liu G, Li H, Zhang W, Santagata R (2022) System dynamic analysis of urban household food-energy-water nexus in Melbourne (Australia). J Clean Prod 379:134675

Egieya JM, Parker Y, Hofmann VS, Daher B, Gorgens J, Goosen NJ (2024) Predictive simulation of the water-energy-food nexus for the City of Cape Town. Sci Total Environ 934:173289

Fagan JE, Reuter MA, Langford KJ (2007) Informing water policy using dynamic, process-based system modelling. MODSIM 2007 – International Congress on Modelling and Simulation, p 163
Fang C, Cui X, Li G, Bao C, Wang Z, Ma H, Sun S, Liu H, Luo K, Ren Y (2019) Modeling regional sustainable development scenarios using the urbanization and eco-environment coupler: case study of Beijing-Tianjin-Hebei urban agglomeration, China. Sci Total Environ 689:820
Foran T (2015) Node and regime: interdisciplinary analysis of water–energy–food nexus in the Mekong region. Water Altern 8(1):655–674. https://www.water-alternatives.org/index.php/alldoc/articles/vol8/v8issue1/278-a8-1-6
Francisco ÉC, Ignácio PSDA, Piolli AL, Dal Poz MES (2023) Food-energy-water (FEW) nexus: sustainable food production governance through system dynamics modelling. J Clean Prod 386:135825
Fu B, Wu M, Che Y, Yang K (2017) Effects of land-use changes on city-level net carbon emissions based on a coupled model. Carbon Manag 8(3):245
Hu M-C, Fan C, Huang T, Wang C-F, Chen Y-H (2019) Urban metabolic analysis of a food-water-energy system for sustainable resources management. Int J Environ Res Public Health 16(1):90
Huang A, Chang F-J (2021) Prospects for rooftop farming system dynamics: an action to stimulate water-energy-food nexus synergies toward green cities of tomorrow. Sustainability (Switzerland) 13(16):9042
Hussien WA, Memon FA, Savic DA (2017) An integrated model to evaluate water-energy-food nexus at a household scale. Environ Model Softw 93:366
Khatun MT, Hiekata K, Takahashi Y, Okada I (2022) Design and management of software development projects under rework uncertainty: a study using system dynamics. J Decis Syst 32(2):265–288. https://doi.org/10.1080/12460125.2021.2023257
Kumar P, Natarajan R, Ashok K (2021) Sustainable alternative futures for urban India: the resource, energy, and emissions implications of urban form scenarios. Environ Res Infrastruct Sustain 1(1):11004
Kurian M (2017) The water-energy-food nexus: trade-offs, thresholds and transdisciplinary approaches to sustainable development. Environ Sci Pol 68:97–106. https://doi.org/10.1016/j.envsci.2016.11.006
Laspidou C, Kokkinaki A, Papadopoulou M (2020) Urban resilience through integrated water-energy-food management: the nexus project in Thessaly. Sci Total Environ 700:134295. https://doi.org/10.1016/j.scitotenv.2019.134295
Li X, Zhang L, Hao Y, Zhang P, Xiong X, Shi Z (2022) System dynamics modeling of food-energy-water resource security in a megacity of China: insights from the case of Beijing. J Clean Prod 355:131773
Li W, Wang J, Zhao Y, Jiang S, Zhu Y, Qi T, Ling M, Baker JS, Xiong Y, He F, Li H, He G, Wang H (2024) Development path for China's mega-large urban agglomerations based on the water-energy-food nexus. J Clean Prod 472:143481
Liu K, Zhang S (2024) Smart cities stochastic secured energy management framework in digital twin: policy frameworks for promoting sustainable urban development in smart cities. Sustain Energy Technol Assess 65:103720. https://doi.org/10.1016/j.seta.2024.103720
Liu G, Du S, Gao Y, Xiong X, Lombardi G, Meng F, Chen Y, Chen C (2024a) A study on energy-water-food-carbon nexus in typical Chinese northern rural households. Energy Policy 188:114100
Liu H, Yan W, Kobayashi H (2024b) Modelling the system dynamics of household food, water, and energy nexus effects. Heliyon 10(4):e25886
Mirindi D, Sušnik J, Masia S, Jewitt G (2024) A system dynamics modelling assessment of water-energy-food resource demand futures at the city scale: Goma, Democratic Republic of Congo. World Dev Sustain 5:100159
Ouyang Y, Cai Y, Xie Y, Yue W, Guo H (2021) Multi-scale simulation and dynamic coordination evaluation of water-energy-food and economy for the Pearl River Delta city cluster in China. Ecol Indic 130:108155
Parsa A, Van De Wiel M, Schmutz U, Fried J, Black D, Roderick I (2023) Challenging the food waste hierarchy. J Environ Manag 344:118554

Parsa A, Van De Wiel M, Schmutz U, Taylor I, Fried J (2024) Balancing people, planet, and profit in urban food waste management. Sustainable Prod Consumption 45:203

Procter AC, Kaplan PÖ, Araujo R (2016) Net zero Fort Carson: integrating energy, water, and waste strategies to lower the environmental impact of a Military Base. J Ind Ecol 20(5):1134

Qi C, Zhang L, Wang C, Liu J (2022) System dynamics modeling for sustainable urban resource management: a review of applications to the FEW nexus. Kybernetes 49(2):460–504

Scott CA, Kurian M, Wescoat JL (2015) The water-energy-food nexus: enhancing adaptive capacity to complex global challenges. In: Governing the nexus: water, soil and waste resources considering global change. Springer, Cham, pp 1–19. https://doi.org/10.1007/978-3-319-05747-7_2

Shehadeh A, Alshboul O, Arar M (2024) Enhancing urban sustainability and resilience: employing digital twin technologies for integrated WEFE Nexus management to achieve SDGs. Sustainability 16(17):7398. https://doi.org/10.3390/su16177398

Shiu HY, Lee M, Lin ZE, Chiueh PT (2023) Dynamic life cycle assessment for water treatment implications. Sci Total Environ 860:160224

Smajgl A, Ward J, Pluschke L (2016) The water–food–energy nexus—realising a new paradigm. J Hydrol 533:533–540. https://doi.org/10.1016/j.jhydrol.2015.12.033

Soares Dal Poz ME, de Arruda Ignácio PS, Azevedo A, Francisco EC, Piolli AL, Gheorghiu da Silva G, Pereira Ribeiro T (2022) Food, energy and water nexus: an urban living laboratory development for sustainable systems transition. Sustainability (Switzerland) 14(12):7163

Song S, Hou Y, Lim RBH, Gaw LYF, Richards DR, Tan HTW (2022) Comparison of vegetable production, resource-use efficiency and environmental performance of high-technology and conventional farming systems for urban agriculture in the tropical city of Singapore. Sci Total Environ 807:150621

United Nations (2018) World urbanization prospects: the 2018 revision. Department of Economic and Social Affairs, Population Division. https://population.un.org/wup/

Valencia A, Hossain MU, Chang N-B (2022a) Building energy retrofit simulation for exploring decarbonization pathways in a community-scale food-energy-water-waste nexus. Sustain Cities Soc 87:104173

Valencia A, Zhang W, Chang N-B (2022b) Sustainability transitions of urban food-energy-water-waste infrastructure: a living laboratory approach for circular economy. Resour Conserv Recycl 177:105991

Wang B, Tian Y, Li X, Li C (2021) Analysis and prediction of sustainable utilization of water resources in Chengde city based on system dynamics model. Water (Switzerland) 13(24):3534

Wang C, Chen Y, Sun M, Wu J (2023) Potential of technological innovation to reduce the carbon footprint of urban facility agriculture: a food–energy–water–waste nexus perspective. J Environ Manag 339:117806

Wen C, Dong W, Zhang Q, He N, Li T (2022) A system dynamics model to simulate the water-energy-food nexus of resource-based regions: a case study in Daqing City, China. Sci Total Environ 806:150497

Xu H, Omitaomu F, Sabri S, Zlatanova S, Li X, Song Y (2024) Leveraging generative AI for urban digital twins: a scoping review on the autonomous generation of urban data, scenarios, designs, and 3D city models for smart city advancement. Urban Inf 3(1):29

Yu X, Shan L, Wu Y (2021) Land use optimization in a resource-exhausted city based on simulation of the f-e-w nexus. Land 10(10):1013

Zhang L, Zhang P, Chang Y, Xu M, Hao Y, Liang S, Liu G, Yang Z (2019) Food-energy-water (FEW) nexus for urban sustainability: a comprehensive review. Resour Conserv Recycl 142:215. https://consensus.app/papers/foodenergywater-few-nexus-for-urban-sustainability-a-zhang-zhang/3ae1fcb68c2b59e7bd76caabdab33790/?utm_source=chatgpt

Zhou Y, Chang F-J, Chang L-C, Herricks E (2024) Elevating urban sustainability: an intelligent framework for optimizing water-energy-food nexus synergies in metabolic landscapes. Appl Energy 360:122849

Chapter 5
Assessing the Impact of Disasters on Urban Energy Structure and Carbon Emissions

Xujie Hu and Wanglin Yan

Abstract This chapter analyzes the impact of disasters in Tokyo by evaluating the changes in energy structure and carbon emissions in 2011, 2015, and 2020 through deep analysis of direct and indirect consumption and emissions through input–output tables. Unexpected events can have profound impacts on urban energy supply and consumption systems. It is crucial to assess the impact of such disasters on urban energy use and carbon emissions. The most typical disasters during the past decades are the Great East Japan Earthquake and the COVID-19. Both has significantly influenced urban energy supply and consumption. This research, using Tokyo as an example, quantifies the disaster's impact on energy use and carbon emissions. The direct energy consumption and the nexus effects with carbon emissions are quantified into 19 categories. Additionally, disaster's impacts on urban systems are evaluated through comparative analysis. Finally, nexus strategies for energy conservation and emissions reduction are identified. The research will clarify the common and different impacts on energy structures and carbon emissions caused by the Great East Japan Earthquake and COVID-19. The findings will provide scientific evidence for governments making implemental policies for decarbonization and preparing for unprecedented events in future.

Keywords Disasters impact · Energy consumption · Carbon emissions · Input–output analysis · Tokyo

X. Hu (✉)
Graduate School of Media and Governance, Keio University, Fujisawa, Kanagawa, Japan
e-mail: huxujie@keio.jp

W. Yan
Faculty of Environment and Information Studies/Graduate School of Media and Governance, Keio University, Fujisawa, Kanagawa, Japan

W. Yan (ed.), *Urban Nexus Approach for Sustainable Cities*,
https://doi.org/10.1007/978-981-95-6156-8_5

1 Introduction

1.1 Research Background

Urban environments are increasingly susceptible to various disasters that can significantly disrupt energy structures and carbon emissions. Tokyo, as one of the largest and most densely populated cities in the world, is a critical case for studying the impact of both natural and human-induced disasters on urban energy systems. In the past decade, two unprecedented events have profoundly affected Tokyo's energy landscape: the Great East Japan Earthquake in 2011 and the global COVID-19 pandemic in 2020. Both events have caused widespread disruptions to energy supply, consumption, and carbon emissions, making it essential to assess their impacts from an academic and policy perspective.

The Great East Japan Earthquake, which struck on March 11, 2011, was one of the most severe natural disasters in Japan's history, causing a tsunami and the Fukushima nuclear disaster. The event led to significant changes in Japan's energy policies, particularly in the reduction of nuclear power and increased reliance on fossil fuels. This shift in energy structure had a cascading effect on carbon emissions, as Japan had to rely more on coal, oil, and natural gas for its energy needs. Tokyo, being Japan's largest urban center, was at the forefront of these changes, experiencing both direct and indirect impacts on its energy consumption and carbon output.

In contrast, the COVID-19 pandemic in 2020 presented a different kind of disaster—one that disrupted economic activity, human mobility, and industrial production. The pandemic led to a sharp reduction in energy demand across various sectors, especially in transportation and commercial activities, resulting in a temporary decline in carbon emissions. However, the long-term effects of the pandemic on urban energy structures are still unclear, as governments around the world have implemented stimulus packages and recovery plans that may either accelerate or hinder efforts toward decarbonization. Understanding how such an event influences energy structures and carbon emissions in a megacity like Tokyo is critical for future policy development.

1.2 Literature Review

The impact of disasters on urban energy systems and carbon emissions is a growing area of research that spans multiple disciplines, including environmental science, economics, and urban planning. This review synthesizes key findings from previous studies that address the disruptions caused by disasters to energy structures, the resulting changes in carbon emissions, and the implications for policy and urban resilience.

1.2.1 Input–Output Analysis (IOA)

Input–Output Analysis (IOA) has become a prominent tool in environmental studies to measure and understand the economic and environmental interdependencies across urban systems, particularly in evaluating energy consumption and carbon emissions. By mapping the flow of goods and services across sectors, IOA captures both the direct and indirect impacts of activities on resource use and emissions. This literature review covers the evolution of IOA in environmental studies, its methodologies and modifications, and its application in urban energy and carbon analysis. It highlights the advantages and limitations of IOA and identifies trends in its application for disaster impact analysis on urban systems.

Evolution and foundation of Input–Output Analysis in environmental studies initially developed by economist Wassily Leontief (1936). IOA's purpose was to examine the interdependencies of sectors within an economy, using a structured table that quantifies transactions between industries. IOA models have evolved to incorporate environmental considerations, recognizing the interconnected nature of economic activities and environmental outcomes (Leontief 1974). Leontief's work paved the way for the development of environmentally extended input–output analysis (EE-IOA), which integrates environmental metrics such as energy use and pollutant emissions, making it a critical tool in environmental impact assessment (Lenzen et al. 2003).

IOA gained traction in environmental research in the 1970s, with efforts to quantify indirect effects of economic activities on resource use. This shift was motivated by the realization that direct measurements of energy consumption and emissions, especially in urban environments, failed to account for interindustry relationships. EE-IOA allows analysts to estimate both the immediate environmental impact within a given sector and the ripple effects on other sectors, making it valuable for assessing urban energy flows and carbon footprints (Wiedmann et al. 2006).

1.2.2 Methodologies and Modifications of Input–Output Analysis

IOA models typically start with a basic input–output table, which is a matrix representing financial transactions between sectors. To account for environmental factors, these tables are extended with additional data on sectoral energy consumption, emissions, and other environmental indicators. The process generally involves calculating a Leontief inverse matrix, which reveals the total economic output needed to satisfy one unit of demand in each sector. The modification of IOA models to include environmental dimensions, such as carbon emissions and energy use, enables researchers to quantify indirect energy requirements and carbon emissions induced by urban sectors (Miller and Blair 2009).

Over time, researchers have developed multi-region input–output (MRIO) models, which capture interactions across regions, and hybrid IO models, which combine monetary and physical units. MRIO models are particularly useful for urban studies in globalized economies, where cities' consumption patterns are intertwined with production systems in distant regions (Tukker and Dietzenbacher 2013). For instance, when Tokyo consumes energy-intensive products imported from another region, MRIO can trace the carbon emissions associated with these imports, offering a comprehensive view of the city's environmental footprint (Duarte et al. 2016). Hybrid IO models are also increasingly used to combine environmental metrics with economic ones, allowing for more precise energy and emission tracking within the urban context.

1.2.3 Application of IOA in Urban Energy and Carbon Emissions Analysis

One of IOA's core strengths lies in its ability to highlight indirect emissions, which are often overlooked in conventional energy accounting. Urban areas generally consume vast amounts of resources and energy, much of which is imported. Thus, the indirect emissions—those associated with the production, transport, and disposal of goods and services—are substantial (Wiedmann and Minx 2008). IOA is particularly effective in capturing these indirect emissions, helping to inform sustainable urban planning strategies that reduce both direct and indirect environmental impacts (Su and Ang 2013).

In their study on Tokyo's carbon emissions, Takase et al. (2005) used IOA to evaluate how energy consumption across different urban sectors contributed to the city's overall carbon footprint. By incorporating indirect emissions, they found that sectors such as construction and services had a far greater impact on emissions than previously assumed. Similarly, Baiocchi and Minx (2010) applied IOA in urban areas of the UK and found that, while direct emissions in households and commercial sectors were substantial, indirect emissions from imported goods accounted for a large part of the cities' total emissions.

Furthermore, dynamic IO models, which incorporate time-based changes, are increasingly used to assess how events like natural disasters influence urban energy structures and emissions over time. For instance, Hu and Yan (2024) analyzed the impact of the 2011 Great East Japan Earthquake on carbon emissions using a dynamic IO framework. They found that the earthquake had a long-lasting impact on energy demand, as well as emissions, by shifting the regional energy mix toward fossil fuels in the short term, while later facilitating shifts toward renewable energy as part of disaster recovery strategies.

1.2.4 IOA in Assessing Disaster Impacts on Urban Energy and Emissions

One of the critical applications of IOA in urban environmental studies is disaster impact analysis. Disasters such as earthquakes and pandemics can significantly disrupt urban energy systems and shift consumption patterns. These disruptions create immediate and long-term changes in energy demand, often resulting in indirect environmental impacts (Koks et al. 2016). For instance, the Great East Japan

Earthquake caused widespread power outages, reducing energy demand in affected areas while simultaneously increasing energy consumption in surrounding regions due to migration and supply chain adjustments (Hu and Yan 2024). IOA models, particularly in multi-regional frameworks, allow researchers to capture these indirect effects.

Similarly, during the COVID-19 pandemic, researchers used IOA to examine shifts in urban energy use and carbon emissions, driven by lockdowns and changes in consumer behavior. Studies found that while direct emissions from transportation dropped, there was a surge in emissions from residential sectors due to increased electricity demand for remote work and online services (Liu et al. 2020). IOA has thus proven essential for understanding the broader, multi-sectoral impact of such disruptions on urban energy consumption and emissions.

1.2.5 Advantages and Limitations of IOA in Urban Environmental Research

IOA provides significant advantages in urban environmental research, primarily by enabling a detailed view of both direct and indirect impacts across sectors. This ability is essential for understanding cities' comprehensive carbon footprints, as urban areas often outsource substantial portions of their energy-intensive activities. IOA's focus on sectoral interdependencies also makes it valuable for studying the effects of disasters, where disruptions in one sector can have cascading effects on others (Minx et al. 2009).

However, IOA has notable limitations, particularly in its reliance on static assumptions about production functions and sectoral relationships, which can limit its accuracy in dynamic urban environments. While dynamic IO models partially address these limitations, they are often complex and require extensive, high-quality time-series data that may not always be available (Hu and Yan 2024). Additionally, standard IOA lacks the spatial granularity necessary for capturing local environmental impacts accurately. In response, researchers have begun to integrate IOA with other methodologies, such as life cycle assessment (LCA), to address these limitations by adding spatial and temporal specificity (Suh 2004).

1.2.6 Trends and Future Directions in IOA for Urban Disaster Analysis

Recent advancements in IOA are moving toward multi-scale models that combine national, regional, and urban IO data to better capture the localized effects of disasters on urban systems. Additionally, with growing attention to climate resilience, there is an increasing focus on adapting IOA models to account for sectoral vulnerabilities, infrastructure resilience, and adaptive capacity in the face of disasters (Koks and Thissen 2016). The integration of IOA with resilience-based frameworks, such as the urban resilience assessment, is anticipated to further improve its applicability for evaluating how disasters affect urban energy systems.

There is also a push toward utilizing big data and real-time energy tracking to enhance IOA's accuracy and dynamic capabilities. By incorporating real-time data, IOA models can better capture immediate and evolving impacts of disruptions, providing valuable insights for policymakers and urban planners (Miller and Blair 2009).

1.2.7 Summary

Input–Output Analysis remains an invaluable tool in environmental research, particularly in understanding urban energy and carbon emissions. Its ability to capture direct and indirect environmental impacts across sectors has made it indispensable for assessing the complex interdependencies within urban systems. In the context of disasters, IOA helps illuminate how immediate disruptions and longer-term recovery efforts affect energy use and emissions, providing insights crucial for planning sustainable, resilient cities. However, IOA must continue evolving to address its limitations, particularly through dynamic modeling, integration with other assessment tools, and incorporation of real-time data. By embracing these advancements, IOA will remain a cornerstone in assessing urban resilience to both natural and human-induced disasters.

1.3 Research Objectives

The primary objective of this research is to analyze the impact of disasters on Tokyo's energy structure and carbon emissions by evaluating changes in both direct and indirect energy consumption and emissions across different economic sectors. Using input–output tables, the study will quantify the disaster-induced shifts in energy use and carbon output for the years 2011, 2015, and 2020. Specifically, the research aims to: (1) Quantify direct energy consumption and carbon emissions in Tokyo, categorized into 11 sectors (e.g., residential, transportation, industrial); (2) Calculate indirect energy consumption and emissions across different categories, considering energy conversion losses; (3) Compare the impacts of the Great East Japan Earthquake and COVID-19 on Tokyo's energy structure and carbon emissions.

Identify commonalities and differences in the effects of natural and human-induced disasters on urban energy systems. Propose strategies for energy conservation and emissions reduction in the aftermath of such disasters. By focusing on Tokyo, this research will provide valuable insights into how a megacity can respond to external shocks and adapt its energy systems to ensure sustainability and resilience.

1.4 Significance of the Research

The significance of this research lies in its potential to inform urban energy policy in the face of disasters. As cities around the world grapple with the dual challenges of climate change and increasing disaster risk, understanding how these events impact energy consumption and carbon emissions is crucial for developing effective mitigation and adaptation strategies. This study will offer scientific evidence that can help policymakers design decarbonization strategies tailored to urban environments, taking into account the unique vulnerabilities and opportunities that arise during disasters.

Moreover, by comparing the effects of a natural disaster (the Great East Japan Earthquake) with those of a global pandemic (COVID-19), the research will shed light on the different mechanisms through which disasters influence energy systems. This comparative analysis will enhance our understanding of how urban energy systems can be made more resilient to both.

2 Methods

2.1 Research Framework

Figure 5.1 provides a comprehensive framework for assessing the impact of disasters on energy use and carbon emissions in Tokyo. This framework is designed to systematically evaluate how major disruptive events, such as earthquakes and pandemics, affect urban energy systems, leading to shifts in consumption patterns, supply dynamics, and emissions trends. By integrating multiple analytical approaches, the framework aims to offer a detailed understanding of the relationship between disaster events and energy sustainability, ultimately guiding the development of effective mitigation and adaptation strategies.

The first step in this framework involves quantifying direct energy consumption and carbon emissions using input–output (IO) tables. The energy consumption and emissions are categorized into 19 distinct sectors, ranging from manufacturing and transportation to commercial activities and public services. This classification enables a sector-specific evaluation, allowing for the identification of industries most vulnerable to disruptions and those contributing the most to carbon emissions. The use of IO tables facilitates a detailed breakdown of energy inputs and outputs, offering insights into how changes in economic activity translate into variations in energy demand and emissions.

The next component of the framework involves a comparative analysis of the impacts of two significant disasters: the Great East Japan Earthquake of 2011 and the COVID-19 pandemic of 2020. By comparing these two distinct crises, the framework highlights the different mechanisms through which disasters influence urban energy structures, revealing both short-term shocks and long-term structural

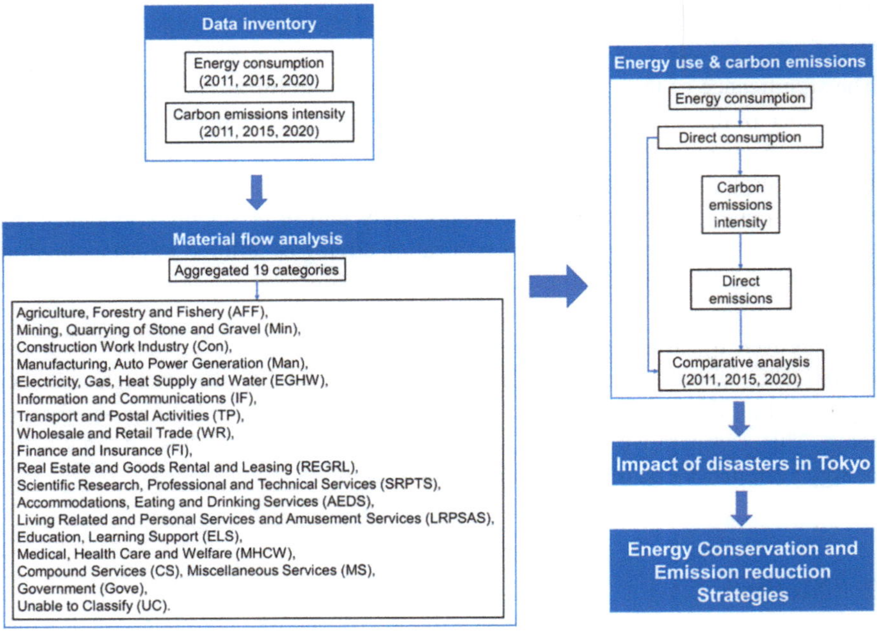

Fig. 5.1 Research framework

changes. This comparative approach is essential for understanding the varying nature of disruptions and their implications for future resilience planning.

Finally, the framework identifies potential strategies for energy conservation and carbon emissions reduction in response to these disasters. Overall, this framework serves as a valuable tool for policymakers, researchers, and urban planners, seeking to understand the intricate links between disaster events, energy consumption, and environmental sustainability. By integrating quantitative assessments with comparative analysis, it provides a robust foundation for developing long-term strategies that enhance Tokyo's energy resilience and contribute to broader global efforts in sustainable urban development.

2.2 *Direct Consumption and Emissions*

Energy consumption can be calculated directly for each category. The types of energy in Energy Consumption Statistics by Prefecture from the Agency for Natural Resources and Energy mainly include coal, coal products, oil, oil products, natural gas, town gas, electricity, and heat. In this chapter, carbon emissions are calculated based on these eight types of energy resources.

$$e_i = \sum E_{i,j} \times \mathrm{CAL}_j \tag{5.1}$$

$$c_i = \sum e_{i,j} \times \text{Emissions}_j \tag{5.2}$$

where e_i and c_i are direct energy consumption (TJ) and carbon emissions (tC) of each category, CALj is the calorific value of different energy (J), and Emissionsj indicates the carbon intensity (tC per TJ).

3 Data Processing

To assess resource consumption and CO_2 emissions, we used the Tokyo Metropolitan Government Statistics Division's economic input–output table (IO table) with different structures (38 sectors and 191-sector structures) for 2011 and 2015. The IO tables include intermediate demand, intermediate supply, final demand, imports, gross value-added, and total production value (Fig. 5.1). Intermediate sectors include many industries, such as agriculture, mining, and manufacturing. Final demand includes households, government, and export. Energy-use data, including four types of fuel, were gathered from the Tokyo Metropolitan Government's Bureau of Environment. Information on water use of each sector was acquired and analyzed from the Tokyo Metropolitan Government's Waterworks Bureau.

Tokyo's monetary input–output Tables (38-sectors) were aggregated to a 19-sector table for assessing urban energy consumption and carbon emissions, as follows:

Agriculture, Forestry, and Fishery (AFF), Mining, Quarrying of Stone and Gravel (Min), Construction Work Industry (Con), Manufacturing, Auto Power Generation (Man), Electricity, Gas, Heat Supply, and Water (EGHW), Information and Communications (IF), Transport and Postal Activities (TP), Wholesale and Retail Trade (WR), Finance and Insurance (FI), Real Estate and Goods Rental and Leasing (REGRL), Scientific Research, Professional and Technical Services (SRPTS), Accommodations, Eating, and Drinking Services (AEDS), Living Related and Personal Services and Amusement Services (LRPSAS), Education, Learning Support (ELS), Medical, Health Care, and Welfare (MHCW), Compound Services (CS), Miscellaneous Services (MS), Government (Gove), Unable to Classify (UC) (Fig. 5.2).

4 Results

4.1 Direct Energy Consumption by Category in Tokyo

Figures 5.3, 5.4, and 5.5 illustrate the changes in direct energy consumption across various categories from 2011 to 2020. These figures provide insights into the shifting energy consumption patterns within Tokyo over the years, revealing how different sectors have responded to disaster challenges.

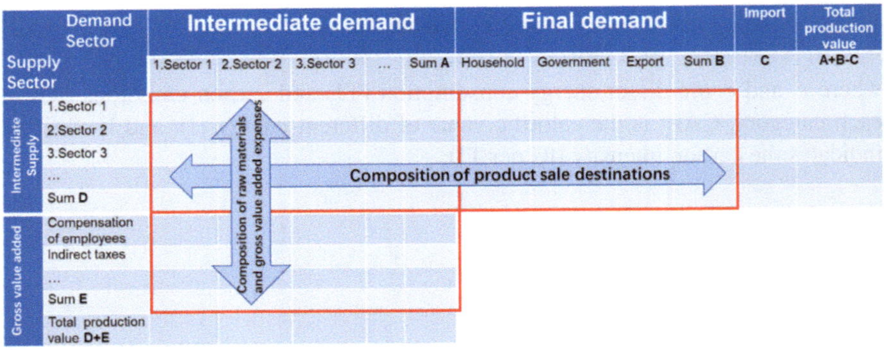

Fig. 5.2 Tokyo economic input–output table

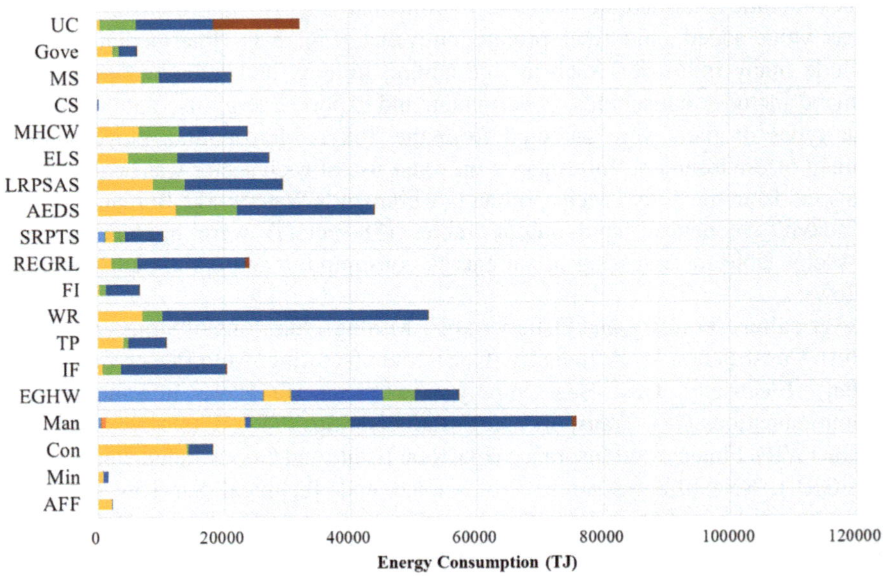

Fig. 5.3 Energy consumption in Tokyo, 2011

4.1.1 Key Sectors and Trends

In 2011, the largest energy consumption was observed in the Manufacturing (Man) sector. This reflects the significant industrial activity during this period, as manufacturing processes are typically energy-intensive, requiring substantial inputs to sustain operations. However, over the decade, energy consumption in this sector experienced a notable decline. This reduction can be attributed to various factors, including advancements in energy efficiency, structural shifts in the economy toward less energy-intensive industries, and policies aimed at reducing industrial energy consumption and carbon emissions.

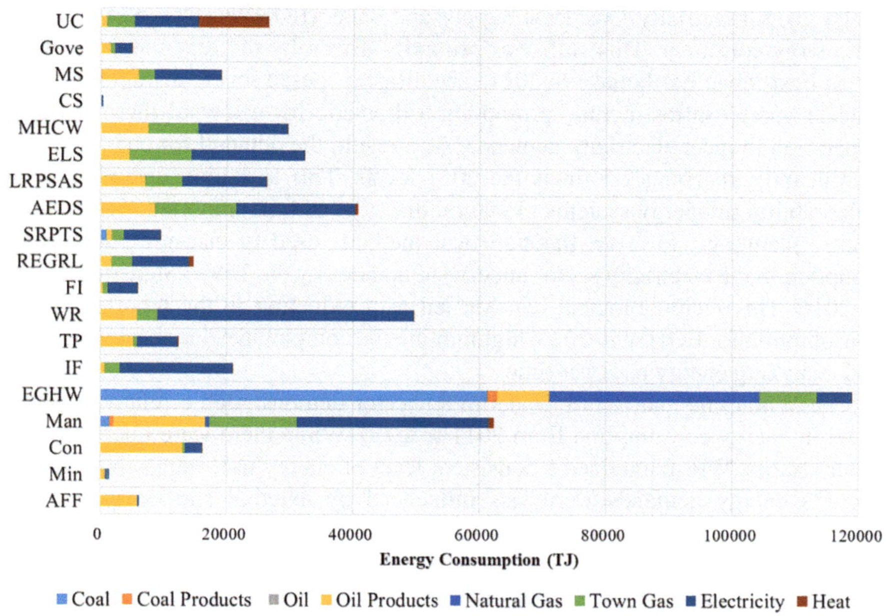

Fig. 5.4 Energy consumption in Tokyo, 2015

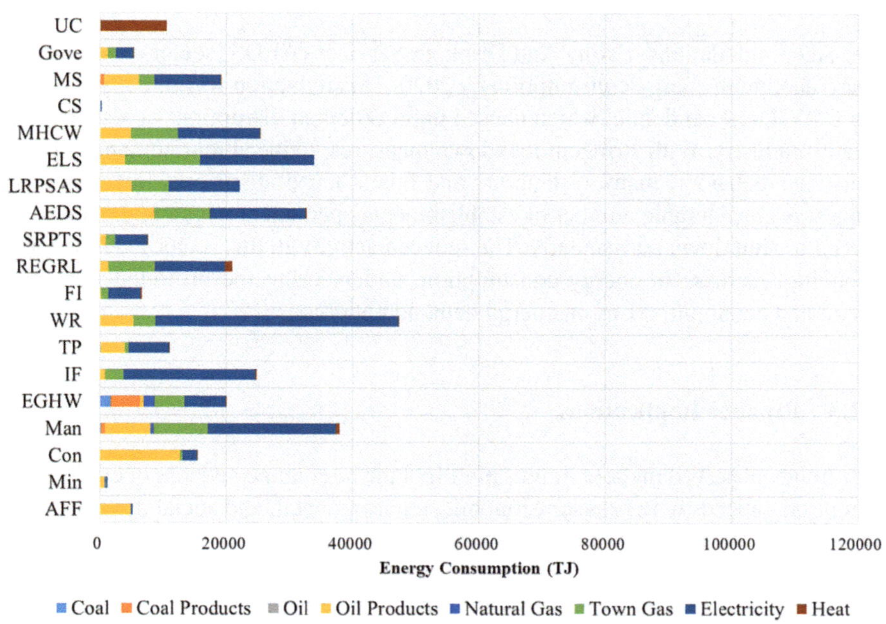

Fig. 5.5 Energy consumption in Tokyo, 2020

By 2015, Electricity, Gas, Heat Supply, and Water (EGHW) emerged as the largest energy consumer. This shift was primarily driven by the repercussions of the Great East Japan Earthquake in 2011. The disaster caused severe disruptions to the nuclear energy infrastructure, prompting a dramatic pivot toward thermal power generation to meet electricity demands. As a result, the demand for coal increased significantly, reaching 2.5 times the 2011 levels. This transition underscores the vulnerability of energy systems to natural disasters and the importance of resilient energy planning. However, the statistical methods used to measure energy consumption in the electricity sector underwent updates by the Tokyo Statistics Bureau in 2016. This methodological revision led to a reduction in the reported energy consumption for EGHW in 2020, highlighting the complexities involved in tracking and analyzing energy use over time.

The Wholesale and Retail Trade (WR) sector demonstrated a relatively stable trend in energy consumption from 2011 to 2020. While there were fluctuations in other sectors, WR maintained a consistent level of energy use, suggesting that this sector's energy demands were less influenced by external shocks or systemic changes during the observed period. By 2020, WR emerged as the largest energy consumer. This shift likely reflects broader economic and social trends, such as the growth of e-commerce and logistical operations, which increased the energy demands associated with warehousing and transportation.

4.1.2 Impact of the Pandemic on Energy Consumption

The Accommodations, Eating, and Drinking Services (AEDS) sector experienced a sharp decline in energy consumption by 2020. This reduction was directly linked to the COVID-19 pandemic, which caused unprecedented disruptions across the hospitality industry. With stringent lockdown measures, reduced tourism, and shifts in consumer behavior, many restaurants and hotels faced significant financial challenges. A considerable number of establishments operated at a loss, and many were forced to shut down permanently. The reduced activity in this sector led to a corresponding decrease in energy consumption, underscoring the profound impact of social and economic crises on energy demand patterns.

4.1.3 Broader Implications

The trends observed in these figures highlight the interconnectedness of energy consumption patterns with broader economic, environmental, and social dynamics. For instance, the decline in manufacturing energy consumption aligns with global trends toward deindustrialization and increased energy efficiency. However, the rise and subsequent fall of EGHW's energy consumption illustrate the direct and indirect impacts of external shocks such as natural disasters and methodological changes in statistical reporting.

Similarly, the relatively stable energy use in the WR sector contrasts sharply with the volatility seen in AEDS, reflecting differing levels of resilience and adaptability among sectors. The pandemic's impact on AEDS underscores the vulnerability of sectors heavily reliant on physical presence and direct customer interaction, while the stability in WR suggests a more robust adaptation to changing circumstances, potentially aided by technological advancements and shifts toward online commerce.

4.2 Direct Carbon Emissions by Category in Tokyo

Direct carbon emissions from various categories in Tokyo can be quantified for six energy sources: coal, coal products, oil, oil products, natural gas, and town gas. Analyzing data from 2011 to 2020 reveals dynamic shifts in carbon emission patterns across sectors and energy sources, reflecting economic, environmental, and social influences as indicated in Figs. 5.6, 5.7, and 5.8.

4.2.1 Key Sectors and Trends

In 2011, Tokyo's direct carbon emissions were dominated by petroleum products, which accounted for 48.65% of total emissions. This was followed by urban gas at 24.58% and coal at 17.98%. The heavy reliance on petroleum products highlighted

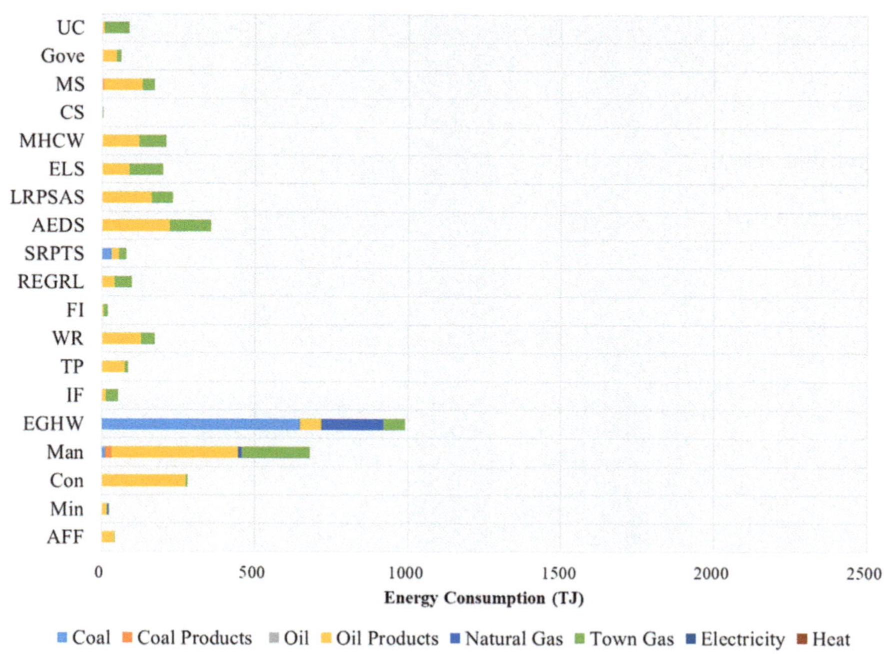

Fig. 5.6 Carbon emissions in Tokyo, 2011

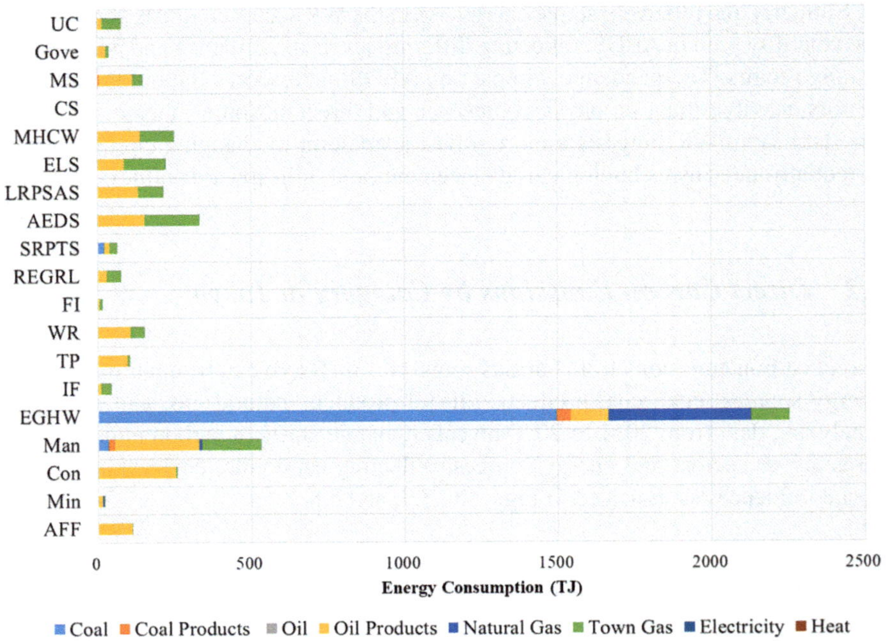

Fig. 5.7 Carbon emissions in Tokyo, 2015

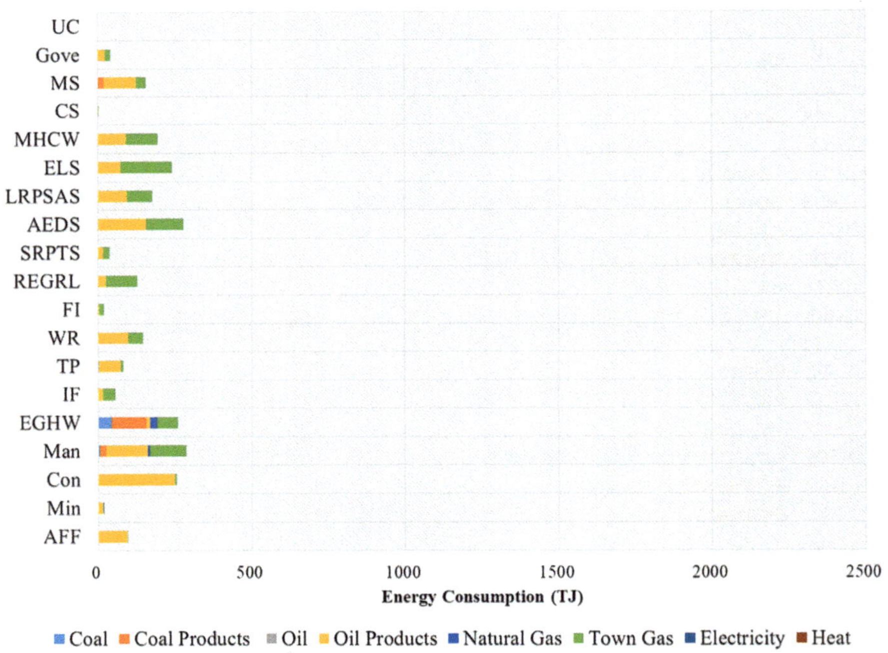

Fig. 5.8 Carbon emissions in Tokyo, 2020

the significant energy demands of sectors such as Manufacturing (Man) and Transport and Postal Activities (TP). Manufacturing's emissions mainly originated from oil products and urban gas, while TP's reliance on petroleum products was particularly pronounced due to transportation fuel needs.

By 2015, the energy landscape in Tokyo had shifted. Carbon emissions from coal rose sharply from 17.98% in 2011 to 31.48%, making it a more significant contributor to Tokyo's emissions profile. Despite this increase, petroleum products remained the primary source of emissions. The spike in coal usage was largely driven by the 2011 Great East Japan Earthquake, which disrupted nuclear energy production and led to a heavy reliance on thermal power generation to meet electricity demands. This transition caused coal-based emissions in the Electricity, Gas, Heat Supply, and Water (EGHW) sector to rise significantly, with coal emissions in 2015 being 2.5 times those in 2011.

The EGHW sector emerged as the largest direct carbon emitter from 2011 to 2015, primarily due to its dependence on coal and natural gas. More than 80% of its emissions in 2015 stemmed from these two energy sources. Manufacturing (Man) remained the second-largest emitter, with its emissions predominantly coming from oil products and urban gas.

By 2020, a marked reduction in Tokyo's overall carbon emissions was observed, attributable to the COVID-19 pandemic. The pandemic's impact was particularly evident in sectors such as Manufacturing (Man), Scientific Research, Professional and Technical Services (SRPTS), and Accommodations, Eating, and Drinking Services (AEDS). These sectors experienced significant declines in emissions due to decreased activity levels. Manufacturing's reduction was the most pronounced, reflecting widespread industrial slowdowns. Similarly, the SRPTS and AEDS sectors saw substantial drops, driven by reduced research activities, limited professional services, and the shutdown of numerous restaurants and hotels.

In contrast, certain sectors, including Information and Communications (IF), Finance and Insurance (FI), and Real Estate and Goods Rental and Leasing (REGRL), exhibited increased emissions during the same period. This growth can be linked to heightened digital activity, financial operations, and real estate demands amid the pandemic, underscoring the divergent impacts of the crisis on different economic activities.

4.2.2 Sectoral Analysis of Emissions Trends

The Man's emissions steadily declined from 2011 to 2020. This trend reflects a combination of improved energy efficiency, structural economic shifts toward less energy-intensive industries, and pandemic-induced disruptions in 2020. Despite being the second-largest emitter in 2015, its contribution to Tokyo's total emissions has diminished over the decade.

The EGHW sector's emissions trajectory highlights its vulnerability to external shocks. The 2011 earthquake triggered a substantial shift to thermal power generation, leading to a surge in coal-based emissions. However, by 2020, emissions from this sector had decreased, partially due to the adoption of updated statistical methods by the Tokyo Statistics Bureau in 2016, which altered how electricity sector emissions were calculated.

The TP sector's emissions exhibited a gradual decline, mirroring trends in fuel efficiency improvements and shifts toward alternative transportation modes. The pandemic further accelerated this reduction in 2020 as mobility restrictions curtailed transport activities.

The AEDS sector faced unprecedented challenges during the pandemic, with many establishments operating at a loss or closing permanently. This led to a significant drop in energy consumption and associated carbon emissions, illustrating the sector's sensitivity to socioeconomic disruptions.

The Information and Communications (IF), Finance and Insurance (FI), and Real Estate and Goods Rental and Leasing (REGRL) sectors experienced growth in emissions during the pandemic, driven by increased demand for digital infrastructure, financial services, and real estate activities. The rise in emissions from IF reflects the growing reliance on data centers and communication networks during periods of remote work and online interaction.

4.2.3 Broader Implications

The observed trends underscore the complex interplay between natural disasters, economic activities, and societal changes in shaping carbon emission patterns. The sharp increase in coal emissions following the 2011 earthquake highlights the need for resilient and sustainable energy systems capable of adapting to crises. Simultaneously, the pandemic-induced shifts in sectoral emissions emphasize the importance of flexibility and adaptability in urban energy management.

5 Discussion

5.1 Direct Energy Consumption Patterns in Tokyo (2011–2020)

The analysis of direct energy consumption in Tokyo from 2011 to 2020 highlights critical shifts in sectoral energy use, driven by external events such as the Great East Japan Earthquake and the COVID-19 pandemic, as well as broader economic transitions. In 2011, manufacturing dominated energy consumption, reflecting Tokyo's substantial industrial base. However, a significant decline in manufacturing energy consumption over the decade underscores the impact of energy efficiency improvements, structural economic changes, and environmental policies targeting industrial emissions. This trend mirrors global patterns of deindustrialization and suggests that Tokyo's economy is increasingly transitioning toward less energy-intensive industries.

The emergence of the Electricity, Gas, Heat Supply, and Water (EGHW) sector as the largest energy consumer by 2015 is notable. This shift was primarily due to the earthquake-induced disruption of nuclear power, necessitating a reliance on thermal power generation, particularly coal. While the rapid adaptation to coal-based power reflects the sector's critical role in maintaining energy security, it also underscores the vulnerability of Tokyo's energy system to natural disasters. The subsequent decline in reported energy consumption for EGHW in 2020, partly attributed to methodological revisions, illustrates the challenges of accurately tracking energy use and the need for robust data practices in energy monitoring.

The relative stability of energy consumption in the Wholesale and Retail Trade (WR) sector contrasts sharply with the fluctuations seen in other industries. This stability suggests inherent resilience and adaptability, possibly supported by the sector's digital transformation and growing logistical needs tied to e-commerce. By 2020, WR had surpassed other sectors in energy consumption, emphasizing the rising importance of distribution and warehousing in Tokyo's energy landscape.

The sharp decline in energy use within the Accommodations, Eating, and Drinking Services (AEDS) sector by 2020 reflects the devastating impact of the pandemic on sectors reliant on physical presence and consumer interaction. Lockdowns, reduced tourism, and shifting consumer behavior drastically reduced energy demand, highlighting the socioeconomic vulnerabilities of such industries. Overall, the observed energy consumption trends reveal a complex interplay between systemic resilience, economic activity, and external disruptions, shaping Tokyo's energy profile over the decade.

5.2 *Direct Carbon Emission Patterns in Tokyo (2011–2020)*

Direct carbon emissions in Tokyo from 2011 to 2020 exhibit distinct shifts across energy sources and sectors, reflecting economic, environmental, and external influences. In 2011, petroleum products were the dominant emission source, underscoring the energy-intensive nature of sectors such as Manufacturing (Man) and Transport and Postal Activities (TP). Manufacturing's reliance on oil products and urban gas, coupled with TP's heavy dependence on transportation fuels, highlights these sectors' significant contributions to Tokyo's emissions.

The sharp increase in coal emissions by 2015, driven by the pivot to thermal power generation following the 2011 earthquake, marks a critical transition in Tokyo's carbon profile. The EGHW sector's increasing reliance on coal illustrates the systemic challenges posed by nuclear energy disruptions and the pressing need for resilient energy planning. However, the sector's reduced emissions by 2020, influenced by both statistical updates and evolving energy sources, suggest an improving trajectory toward sustainability.

The pandemic-induced reductions in emissions across Manufacturing, Scientific Research, Professional and Technical Services (SRPTS), and AEDS sectors by 2020 further highlight the interdependence of economic activity and carbon output. Manufacturing's pronounced decline reflects broader industrial slowdowns, while the AEDS sector's challenges underscore the disproportionate impact of socioeconomic disruptions on certain industries.

Conversely, increased emissions in Information and Communications (IF), Finance and Insurance (FI), and Real Estate and Goods Rental and Leasing (REGRL) sectors during the pandemic underscore the rising significance of digital infrastructure and financial operations. This divergence highlights how shifts toward remote work and digital activity can reshape carbon emission profiles, presenting new challenges for sustainable urban energy management.

The trends in Tokyo's emissions emphasize the dynamic interplay between external crises, sectoral resilience, and adaptive capacity. They underscore the importance of a diversified and flexible energy system capable of mitigating carbon emissions while adapting to economic and environmental shifts.

6 Conclusions

The examination of Tokyo's direct energy consumption and nexus effect with carbon emissions from 2011 to 2020 provides valuable insights into the city's evolving energy dynamics. Key trends indicate a transition from industrial dominance to service-oriented energy consumption, shaped by both systemic shifts and external disruptions such as the 2011 Great East Japan Earthquake and the COVID-19 pandemic. The manufacturing sector's steady decline in energy consumption and emissions reflects broader economic transformations and policy-driven improvements in energy efficiency. This trend signifies Tokyo's alignment with global efforts to transition toward less energy-intensive economic structures.

The EGHW sector's trajectory highlights the vulnerability of energy systems to external shocks. The sharp rise in coal-based energy consumption post-earthquake and the subsequent decline, driven by statistical updates and changes in energy sourcing, underscore the need for resilient and sustainable energy planning. Similarly, the stability in the WR sector and the pandemic-induced reductions in the AEDS sector illuminate the varied responses of different industries to systemic challenges. These findings emphasize the importance of adaptability and innovation in ensuring energy resilience.

The divergent impacts of the pandemic on carbon emissions across sectors underscore the complexities of urban energy management. While certain sectors experienced significant reductions due to decreased activity, others saw growth linked to digital and financial operations. These patterns highlight the critical role of digitalization and the potential for technology-driven transformations in achieving sustainable urban energy systems.

Overall, Tokyo's energy and carbon emission trends underscore the interconnectedness of economic activities, external disruptions, and environmental objectives.

The findings call for enhanced nexus policies that prioritize energy efficiency, resilience, and sustainability while addressing single challenges of diverse sectors. By fostering innovation, embracing digitalization, and reinforcing systemic resilience, Tokyo can continue to progress toward a sustainable and adaptive urban energy framework.

Acknowledgments This chapter presents a key concept from research supported by the Taikichiro Mori Memorial Research Grants and Keio University's Doctoral Student Grant-in-Aid Program through the Ushioda Memorial Fund. We express our sincere gratitude for this support.

References

Baiocchi G, Minx JC (2010) Understanding changes in the UK'S CO2 emissions: a global perspective. Environ Sci Technol 44:1177

Duarte J, Pinto-Gouveia J, Cruz B (2016) Relationships between nurses' empathy, self-compassion and dimensions of professional quality of life: a cross-sectional study. Int J Nurs Stud 60:1–11

Hu X, Yan W (2024) Quantifying the impact of earthquakes on urban energy consumption and carbon emissions in Tokyo from a nexus perspective. Environ Res Commun 6(7):075020

Koks EE, Thissen M (2016) A multiregional impact assessment model for disaster analysis. Econ Syst Res 28(4):429–449

Koks EE, Carrera L, Jonkeren O, Aerts JC, Husby TG, Thissen M et al (2016) Regional disaster impact analysis: comparing input–output and computable general equilibrium models. Nat Hazards Earth Syst Sci 16(8):1911–1924

Lenzen M, Murray SA, Korte B, Dey CJ (2003) Environmental impact assessment including indirect effects—a case study using input–output analysis. Environ Impact Assess Rev 23(3):263–282

Leontief WW (1936) Quantitative input and output relations in the economic systems of the United States. Rev Econ Stat 18(3):105–125

Leontief W (1974) Sructure of the world economy: outline of a simple input-output formulation. Am Econ Rev 64(6):823–834

Liu Z, Ciais P, Deng Z, Lei R, Davis SJ, Feng S et al (2020) Near-real-time monitoring of global CO2 emissions reveals the effects of the COVID-19 pandemic. Nat Commun 11(1):5172

Miller RE, Blair PD (2009) Input-output analysis: foundations and extensions. Cambridge university press

Minx JC, Wiedmann T, Wood R, Peters GP, Lenzen M, Owen A et al (2009) Input–output analysis and carbon footprinting: an overview of applications. Econ Syst Res 21(3):187–216

Su B, Ang BW (2013) Input–output analysis of CO2 emissions embodied in trade: competitive versus non-competitive imports. Energy Policy 56:83–87

Suh S (2004) Functions, commodities and environmental impacts in an ecological–economic model. Ecol Econ 48(4):451–467

Takase K, Kondo Y, Washizu A (2005) An analysis of sustainable consumption by the waste input-output model. J Ind Ecol 9(1–2):201–219

Tokyo Metropolitan Government, Bureau of General Affairs, Statistics Division. (2011). Tokyo input–output tables 2011. Tokyo Metropolitan Government. https://www.toukei.metro.tokyo.lg.jp/sanren/2011/sr11t1.htm

Tukker A, Dietzenbacher E (2013) Global multiregional input–output frameworks: an introduction and outlook. Econ Syst Res 25(1):1–19

Wiedmann T, Minx J (2008) A definition of 'carbon footprint'. Ecol Econ Res Trends 1(2008):1–11

Wiedmann T, Minx J, Barrett J, Wackernagel M (2006) Allocating ecological footprints to final consumption categories with input–output analysis. Ecol Econ 56(1):28–48

Part III
Nexus Analyses for Urban Food-Energy-Water Complexity

Chapter 6
Urban Agricultural Activities in Residential Neighborhood with a Mixture of Agricultural Land in Tokyo Suburbs

Keidai Kishimoto and Wanglin Yan

Abstract Involving city dwellers in urban agricultural (UA) activities, such as farming at allotment gardens and shopping at farmers' markets, provides valuable ecosystem services. There have been reported mismatches between the supply and demand of these services, highlighting the need for understanding where people access them in urban context. This chapter aims to enhance our understanding of city dwellers' potential access to ecosystem services in UA activities, focusing on Tokyo's unique landscape of residential neighborhood with a mixture of agricultural land. The chapter covers the following topics: (1) tracing the two historical phases of urban growth and urban maturity that has shaped Tokyo's unique landscape, and (2) mapping the access of city dwellers to UA activities on the neighborhood scale, using case studies of allotment gardens and communal and individual farmers' markets. Based on these analyses, we propose a nexus-based management that aligns supply and demand, integrating the priorities of both the agriculture and housing sectors to address urban sustainability.

Keywords Urban agriculture · Ecosystem services · Access · Residential neighborhood with a mixture of agricultural land

---

K. Kishimoto (✉)
School of Human Science and Environment, University of Hyogo, Himeji City, Hyogo, Japan
e-mail: keidai@keio.jp

W. Yan
Faculty of Environment and Information Studies/Graduate School of Media and Governance, Keio University, Fujisawa, Kanagawa, Japan

W. Yan (ed.), *Urban Nexus Approach for Sustainable Cities*,
https://doi.org/10.1007/978-981-95-6156-8_6

1 Introduction

In the last decade, urban agriculture (UA) has attracted the attention of society, policymakers, and scholars due to its benefits to the well-being of city dwellers and urban sustainability (International Resource Panel 2021). Such benefits include not only the production of fresh vegetables and fruits but also cross-cutting ecosystem services, e.g., water runoff mitigation, biodiversity conservation, and landscape maintenance (Lovell 2010; Peng et al. 2015; Zasada 2011). Furthermore, involving city dwellers in UA activities—such as farming at allotment gardens and shopping at farmers' markets—fosters communications, recreation, exercise, and learning farming and cooking techniques (Guitart et al. 2012; Ruggeri et al. 2016; Cabral et al. 2017; Teig et al. 2009; Veen et al. 2016; Sakurai 2006). Notably, during the COVID-19 pandemic, the demand for UA activities highlighted the ecosystem services related to recreation, relaxation, and exercise (Music et al. 2021; Iida et al. 2023; Kamiyama et al. 2023).

Recent research on ecosystem services has emphasized the importance of matching supply and demand (Baró et al. 2015). Ecosystem services are conceptualized as processes from supply by natural ecosystems to demand by social ecosystems (Zhang et al. 2022; Spangenberg et al. 2014; Potschin-Young et al. 2018). On the supply side, various types of UA, such as rooftop gardens, community gardens, and allotment gardens (Taylor and Taylor Lovell 2012; Pulighe and Lupia 2016; Hsiao 2021), exhibit varying capacities to supply ecosystem services due to factors such as soil quality and vegetation types. On the demand side, city dwellers' demand for these services vary depending on demographic factors such as age and income levels. People can fully access ecosystem services only when the supply of ecosystem services in a particular area corresponds to the demand of the local population for ecosystem services. However, mismatches are reported in urban areas (Baró et al. 2015) due to fewer opportunities for involving city dwellers in UA activities. This highlights the need for understanding where people access ecosystem services in urban context.

Existing studies have explored residents' access to UA activities, often focusing on proximity from urban areas (Zasada 2011). For example, city proximity to farmers' market and tourist farms was acknowledged as an important driver in the Ruhr metropolitan area in Germany (Pölling 2016). In contrast, a variety of UA activities are implemented away from cities in Tuscany in Italy (Lange et al. 2013). People in inner London (UK) has significantly less opportunities for joining allotment gardens—they must wait for a several years in waiting list—compared to those in outer areas (Fletcher and Collins 2020). Furthermore, despite the growing interest in UA of city dwellers, UA has constantly declined due to conversion into built-up areas in recent years (Dobson et al. 2020; Fletcher and Collins 2020; Spilková and Vágner 2016). Most of these research have mainly been carried out in Europa and North America, where a clear distinction exists between UA and inner-city due to strict zoning system that controls built-up areas and agricultural land (Piorr et al. 2011).

Fig. 6.1 Typical landscape in Tokyo

In contrast, the megacities of East and Southeast Asia such as Tokyo, Bangkok, and Manila present different spatial dynamics. In these cities, the existing studies suggest that city dwellers may be relatively accessible to UA activities (Yokohari et al. 2000; Hara et al. 2005; Murakami et al. 2005). These megacities have experienced complex zoning system, its change, and population growth (Yagi and Garrod 2018; Yokohari et al. 2000; Li et al. 2017). This has resulted in a unique sprawled landscape where agricultural and residential land are positioned side by side as shown in Fig. 6.1 (Yokohari et al. 2000; Hara et al. 2005; Murakami et al. 2005), namely, residential neighborhoods with a mixture of agricultural land. Currently, such land use arrangements are prevalent across metropolitan area, with the exception of urban centers and subcenters (Sioen et al. 2018). Consequently, city dwellers in East and Southeast Asian megacities seem to be often located closer to UA than those in European cities. This proximity allows for greater opportunities to access ecosystem services supplied by UA activities. For instance, involving UA activities in walkable neighborhood in Tokyo positively influence subjective well-being, physical activity, and food security under the COVID-19 pandemic (Iida et al. 2023). Suburban areas in Tokyo with a high ratio of households with children provide more opportunities to access UA activities, such as hands-on training programs led by instructors (Kishimoto and Yan 2023). Exploring the potential of the access between supply of and demand for ecosystem services of UA activities in these unique landscapes will provide valuable information to local governments, farmers, and civil society organizations on policymaking.

Therefore, this chapter aims to enhance our understanding of the potential access of city dwellers to ecosystem services in UA activities, focusing on Tokyo's unique landscape of residential neighborhood with a mixture of agricultural land. To achieve this, the chapter covers the following topics: (1) tracing the two historical phases of urban growth and urban maturity that has shaped Tokyo's unique landscape, and (2) mapping the access of city dwellers to UA activities on the neighborhood scale, using case studies of allotment gardens and farmers' markets. Based on these analyses, we propose a nexus-based management that aligns supply and demand, integrating the priorities of both the agriculture and housing sectors to address urban sustainability.

2 Tracing Historical Phases Behind Tokyo's Unique Landscape

2.1 Urban Growth and Remaining Agricultural Land

The present urban structure of Tokyo has developed since the city was designated as the principal capital of Japan in 1603. From the seventeenth to the early nineteenth century, urbanization flourished around the present-day imperial palace, resulting in the largest city in the world by population, with approximately one million residents at that time. To support this large population, agriculture persisted in the fringe areas of the city center in Tokyo. From the late nineteenth century to the early twentieth century, Tokyo experienced rapid urbanization, particularly in the current suburban areas. This transformation was driven by the development of urban infrastructure, including railway networks such as the circular Yamanote Line and radial lines from subcenters to suburban areas (Liu et al. 2022; Okata and Murayama 2010). Especially after the Great Kanto Earthquake, suburbanization advanced further, influenced by modern urban planning, such as greenbelts and garden cities inspired by Howard.

After the World War II, rapid economic growth and massive migration into cities from the 1950s to the 1960s brought pollution problems and uncontrolled settlements. In this period, the government enacted the City Planning Act in 1968 to control land use and settlements through a zoning system. Zones designated as the Urbanization Promotion Area were encouraged to develop housing, apartments, and new towns by conversion from agricultural land to these land uses. Zones designated as the Urbanization Control Area were restricted from urban development. On the other hand, agricultural land still remains in the Urbanization Promotion Area in the suburban areas, because the zone was designated very extensively. Some farmers continue to farm their own agricultural land for business with tax incentives under the Act of Productive Green Space enacted in 1974 (Nagashima 1976).

The bubble economy and rising land prices led to suburbanization in the 1980s and early 1990s. During this period, significant urbanization pressure led to a shift in perception, viewing UA as unnecessary. Agricultural land not designated as Productive Green Space in the Urbanization Promotion Area was mainly converted into developed spaces. Most of the remaining agricultural land was preserved as reserved land for housing development through Productive Green Space program. Furthermore, a period of low economic growth from the late 1990s to the 2000s enabled city centers to turn into re-urbanization (Kanda et al. 2020; Ushijima 2012). Large area of agricultural land was converted into parking lots, parks, welfare facilities, and convenience stores in the 1990s and 2000s (Satake and Saio 2018; Takatori 2000). As a result, agricultural land was gradually subdivided, creating a mixed landscape of residential and agricultural lands (Fig. 6.2).

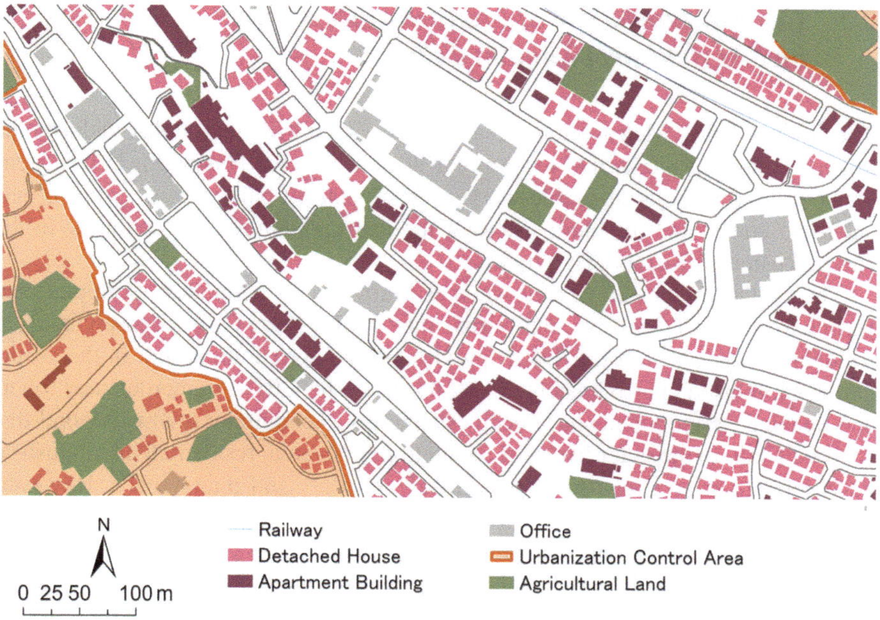

Fig. 6.2 Land use arrangement in residential neighborhood with a mixture of agricultural land. (Data source: Authors created using Data for City Planning in Kawasaki City and Digital National Land Information from Ministry of Land, Infrastructure, Transport and Tourism)

2.2 *Urban Maturity and Policy Shift Toward Mixed Use*

Over the past two decades, the population in Japan has shifted from growth to decline. This trend has resulted in a decrease in demand for reserved land for development. Additionally, city dwellers have adopted more diversified lifestyles, focusing on quality of life and well-being. On the other hand, agricultural land in Tokyo has decreased due to economic challenges, aging farmers, and a lack of successors. Consequently, the value of ecosystem services has garnered more attention from society, policymakers, and scholars than ever before.

As expectations for ecosystem services increased, the land use policies also underwent a paradigm shift. The City Planning Act enacted in 1968 recognized that agricultural land within Urbanization Promotion Area should be developed as built-up areas. However, the Act for Promotion of Urban Agriculture enacted in 2015 recognized that UA should be integrated within cities (Ministry of Agriculture, Forestry and Fisheries and Ministry of Land, Infrastructure, Transport and Tourism 2016). Moreover, in 2018, the City Planning Act was revised to establish the Countryside Residential District as a residential zone where the farming and residential environment coexist in harmony. In addition, in 2022, the Tokyo Metropolitan Government established guidelines for the strategic management of residential neighborhoods with a mixture of green space and agricultural land.

2.3 *Current Landscape of Mixture of Residential and Agricultural Land in Tokyo*

The Greater Tokyo Metropolitan Area encompasses a wide region that includes the Tokyo 23 special wards and extends 50–80 km outward (Fig. 6.3). Urban areas locate around stations to enable workers from densely built-up residential zones, suburban zones, and the outer fringe of the metropolitan area to commute to the central business district (CBD) by train.

The CBDs are located in areas such as Marunouchi, Shimbashi, Shinjuku, and Ikebukuro within Tokyo's special wards. Surrounding these CBDs, densely built-up residential zones have formed, leaving virtually no agricultural land. In the suburban zone, located approximately 10–40 km from the CBDs, numerous new towns and housing developments have been established as commuter towns. Beyond this suburban belt, in the outer fringe of the metropolitan area, residential and industrial developments have extended along railway corridors and major roads. However, these developed zones are often interspersed with or surrounded by agricultural land, reflecting a more dispersed pattern of land use.

In summary, urban land use tends to gradually decrease in intensity with increasing distance from the CBD, while agricultural land and other nonurban uses show a corresponding increase (Fig. 6.4). This spatial transition is observable not only at

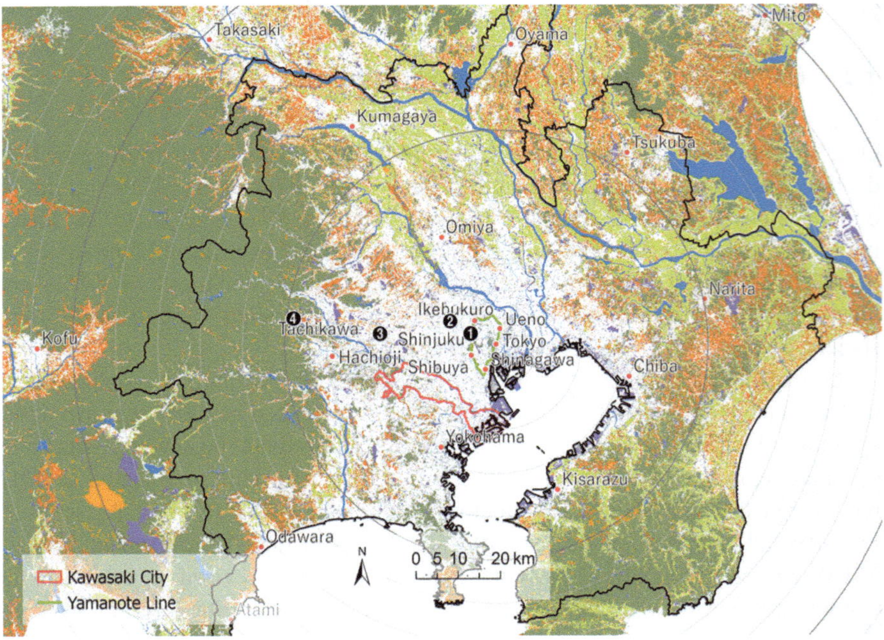

Fig. 6.3 Tokyo Metropolitan Area covers areas within 80 km of the CBD. ①–④ show the location of each map in Fig. 6.4. Kawasaki City is covered in Sect. 3. (Data source: Authors created using Digital National Land Information from Ministry of Land, Infrastructure, Transport and Tourism)

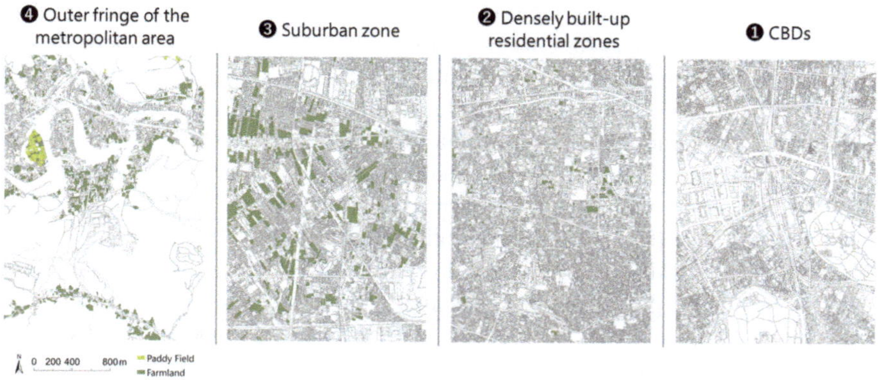

Fig. 6.4 Urban-rural gradient and residential neighborhood with a mixture of agricultural land. (Data source: Authors created using the Fundamental Geospatial Data from the Geospatial Information Authority of Japan and Land Percel Data from Ministry of Agriculture, Forestry and Fisheries)

the macroscale from the city center to the urban fringe but also at the microscale—from railway stations into their surrounding hinterlands. Consequently, land allocated for agricultural use has become increasingly concentrated in peripheral areas of the Tokyo metropolitan region, particularly in locations that are distant from major railway stations.

3 Mapping the Access of City Dwellers to UA Activities

3.1 UA Activities in Tokyo

Farming at allotment gardens and shopping at farmers' markets are popular form of UA activities in Tokyo.

Allotment gardens provide individual plots to tenant families for a limited period under a contract. For families with children, these gardens offer an opportunity to learn about agriculture and food through the experiences of farming and cooking the harvested produce (Pothukuchi 2004). Additionally, sharing information and collaborating with others can serve as a valuable stepping stone to social participation (Teig et al. 2009; Veen et al. 2016). Currently, in Japan, municipalities and Japan Agriculture (JA) develop gardens to meet the demands of citizens, offer leisure activities, promote agricultural knowledge, and make use of idle farmland (Oba and Obase 2001). Private companies also establish gardens in response to requests from local governments and demands of local residents (Kamo and Yamada 2020). Producers establish gardens to secure management and lease fees, stimulate the local economy (Tamura 2009), or address challenges in managing agricultural land (Miyake and Matsumoto 1997).

Farmers' markets provide opportunities for local producers to ship their produce or processed products and for consumers to purchase them. In Japan, there are two main types of farmers' markets: communal farmers' markets operated by JA, municipalities, or Michi-no-eki (roadside stations) where multiple producers ship their products, and individual farmers' markets where each producer sell their products. These farmers' market help distribute products that may not meet the strict standards or large volume required by market (Fujita et al. 2000; Kimura 2010; Oishi 2019). These farmers' markets motivate farmers by enhancing their income (Nakajima et al. 2011; Nishino 2016), diversifying sales channels, and facilitating interactions among producers and consumers (Iida et al. 2004). Consumers purchase local products, processed products, and rare varieties that are not marketed, and gain knowledge and interest in local agricultural products, growing processes, and cooking methods.

3.2 Access of City Dwellers to UA Activities in Kawasaki City

Kawasaki City locates in the suburbs of Tokyo, about 10 km from CBDs of Shibuya and Shinagawa, and a 20-min away by train. Numerous detached houses and apartment housings were developed for commuters, resulting in population density exceeding 10,000 per km^2. While the eastern part of the city is industrialized, the central and western parts are hilly, with a mixture of residential lands and agricultural lands. In Kawasaki City, 31 allotment gardens are supplied, offering a total of 2180 plots with a total area of 45,429 m^2, 13 communal farmers' markets, and 167 individual farmers' markets provide fresh products (Fig. 6.5).

Figure 6.6 shows the distribution of allotment gardens and the distance from them. The map shows that there are no allotment gardens in the eastern part of the city. The central and western parts of the city were covered with allotment gardens, but many of them were out of walking or biking distance from the city. Approximately 21% of the detached houses and 18% of apartment buildings are within an 800-m radius, which is a 15-min walk. Approximately 86% of the detached houses and 85% of apartment buildings are within a 4000-m radius, which is a 15-min bicycle. Most city dwellers can access allotment gardens by walk or bicycle.

Figure 6.7 shows the distribution of communal farmers' market and the distance from them. They are located from the center to the western part of the city, especially along railroad lines. In the western part of the city, in particular, there is a high density of communal farmers' markets, and there are contiguous areas that can be accessed on foot or by bicycle. Approximately 12% of the detached houses and 14% of apartment buildings are within an 800-m radius, which is a 15-min walk. Approximately 75% of the detached houses and 73% of apartment buildings are within a 4000-m radius, which is a 15-min bicycle. On foot, apartment buildings seem to be more accessible to communal farmers' market than detached houses.

Figure 6.8 shows the distribution of individual farmers' market and the distance from them. Unlike communal farmers' markets, these shops were densely located in

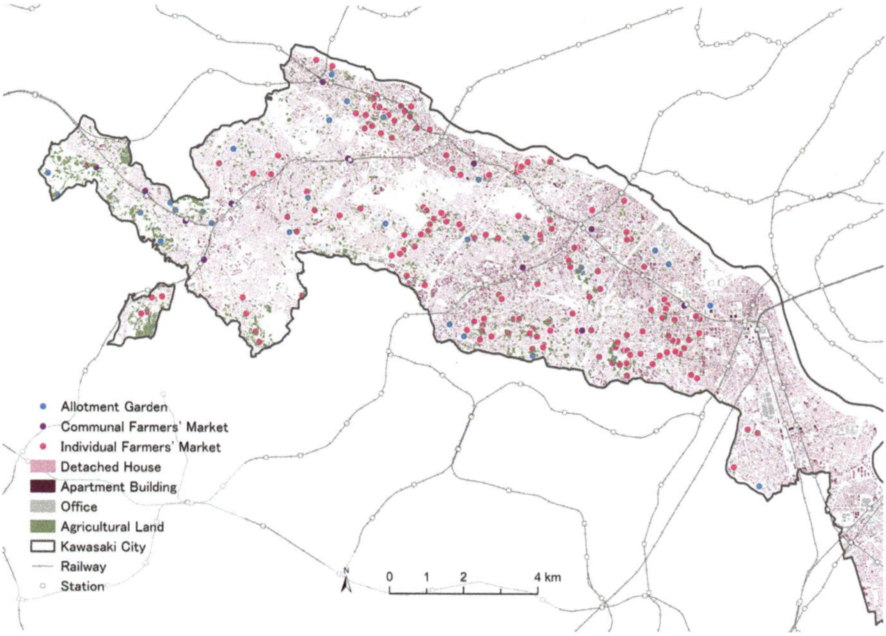

Fig. 6.5 Distribution of UA activities in Kawasaki City. (Data source: Authors created using Data for City Planning in Kawasaki City, Digital National Land Information from the Ministry of Land, Infrastructure, Transport and Tourism, list of allotment gardens from the Ministry of Agriculture, Forestry and Fisheries, and city bulletins)

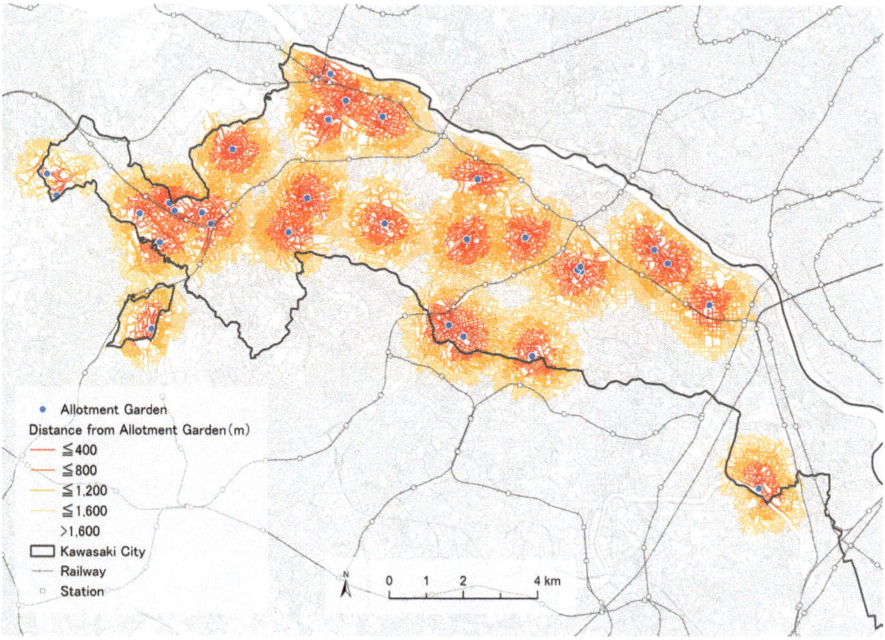

Fig. 6.6 Access to allotment gardens in Kawasaki City. (Data source: Authors created using list of allotment gardens from the Ministry of Agriculture, Forestry and Fisheries, Digital National Land Information from the Ministry of Land, Infrastructure, Transport and Tourism, and Digital Map from the Geospatial Information Authority of Japan)

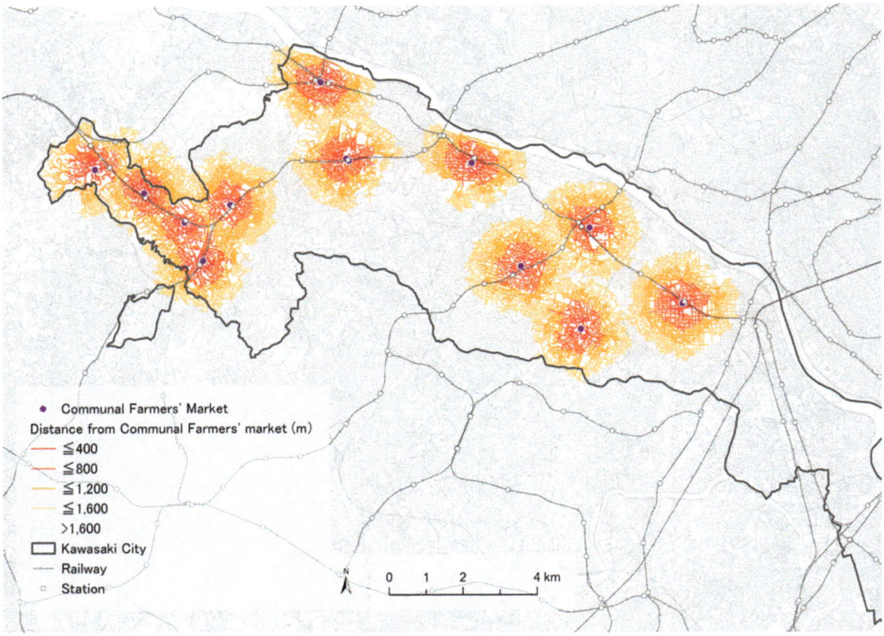

Fig. 6.7 Access to communal farmers' market. (Data source: Authors created using list of farmers' market from Kawasaki City, Digital National Land Information from the Ministry of Land, Infrastructure, Transport and Tourism, and Digital Map from the Geospatial Information Authority of Japan)

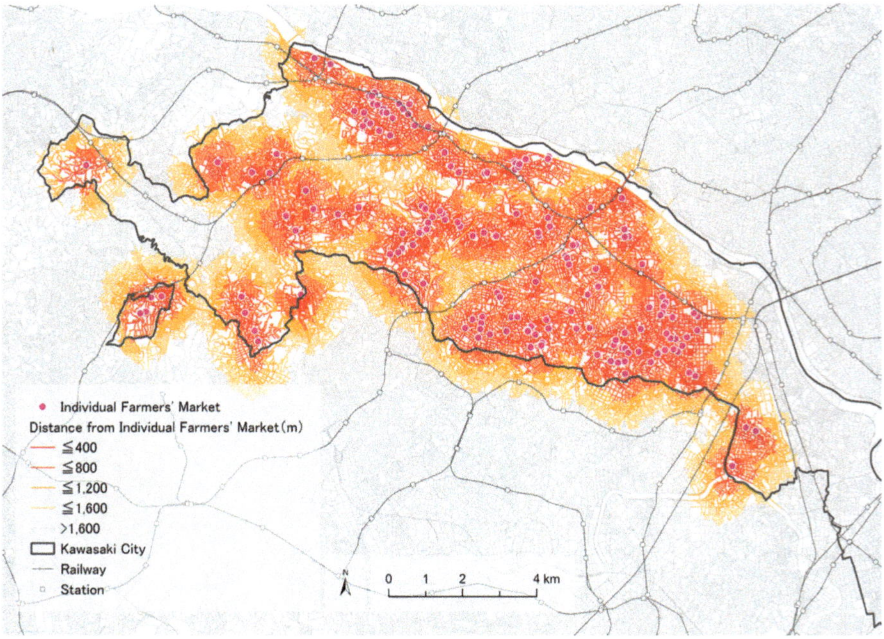

Fig. 6.8 Access to individual farmers' market. (Data source: Authors created using list of farmers' market from Kawasaki City, Digital National Land Information from the Ministry of Land, Infrastructure, Transport and Tourism, and Digital Map from the Geospatial Information Authority of Japan)

the hinterland of the railroad in the central part of the city, and were located in contiguous areas that could be reached on foot or by bicycle. In contrast, they were located in a fragmented manner in the western part of the city. Approximately 52% of the detached houses and 50% of apartment buildings are within an 800-m radius, which is a 15-min walk. Approximately 85% of the detached houses and 83% of apartment buildings are within a 4000-m radius, which is a 15-min bicycle. Half of city dwellers can access individual farmers' market by walking, and most of city dwellers can access by bicycle.

4 Conclusion

This chapter showed the process shaping a unique landscape in Tokyo, a residential neighborhood with a mixture of agriculture land, and the potential for easy access from city dwellers to UA activities under such landscape, taking case studies of allotment gardens, communal farmers' market, and individual farmers' market in Kawasaki City, 10 km away from the CBD.

Based on these analyses, we propose a nexus-based management that aligns supply and demand, sharing the situation of the community with various stakeholders such as citizens, producers, local government, JA, and private companies under the concept of area management, and designing the future landscape with a mixture of agricultural lands and residential lands. It is necessary to consider the lifestyles of city dwellers and the agricultural efforts of producers, local governments, and JA in an integrated manner. For example, the Tokyo Metropolitan Government has developed the agricultural scenic area development system, which aims to preserve agricultural landscapes by designating areas where a mixed landscape is relatively concentrated. In Setagaya City, Nerima City, and Suginami City, UA activities such as tours of forests and workshops have been conducted based on the landscape and citizen participation (Otsuka and Hashimoto 2019).

Acknowledgments This chapter is supported by the JSPS KAKENHI (grant number 22KJ2704, 25 K17753). Maps are all created by the author.

References

Baró F, Haase D, Gómez-Baggethun E, Frantzeskaki N (2015) Mismatches between ecosystem services supply and demand in urban areas: a quantitative assessment in five European cities. Ecol Indic 55:146–158

Cabral I, Keim J, Engelmann R, Kraemer R, Siebert J, Bonn A (2017) Ecosystem services of allotment and community gardens: a Leipzig, Germany case study. Urban For Urban Green 23:44–53

Dobson MC, Edmondson JL, Warren PH (2020) Urban food cultivation in the United Kingdom: quantifying loss of allotment land and identifying potential for restoration. Landsc Urban Plan 199(July):103803

Fletcher EI, Collins CM (2020) Urban agriculture: declining opportunity and increasing demand—how observations from London, U.K., can inform effective response, strategy and policy on a wide scale. Urban For Urban Green 55(November):126823

Fujita T, Naitoh S, Tomita K (2000) The structure of existence for agricultural direct sales place in urban suburbs. Agric Econ Stud 22:1–21

Guitart D, Pickering C, Byrne J (2012) Past results and future directions in urban community gardens research. Urban For Urban Green 11(January):364–373

Hara Y, Takeuchi K, Okubo S (2005) Urbanization linked with past agricultural Landuse patterns in the urban fringe of a deltaic Asian Mega-City: a case study in Bangkok. Landsc Urban Plan 73:16–28

Hsiao H (2021) Characteristics of urban gardens and their accessibility to locals and non-locals in Taipei City, Taiwan. Landsc Ecol Eng 17(January):41–53

Iida Y, Takahashi T, Hayashi N (2004) Effects on regional activations and farming incentive by farmers' market. J Rural Plan Assoc 23:211–216

Iida A, Yamazaki T, Hino K, Yokohari M (2023) Urban agriculture in walkable neighborhoods bore fruit for health and food system resilience during the COVID-19 pandemic. NPJ Urban Sustain 3(February):4. https://doi.org/10.1038/s42949-023-00083-3

International Resource Panel (2021) Urban agriculture's potential to advance multiple sustainability goals: an international resource panel think piece. United Nations Environment Programme, Nairobi. https://wedocs.unep.org/bitstream/handle/20.500.11822/38398/urban_agriculture.pdf

Kamiyama C, Hori K, Matsui T, Pretty J, Saito O (2023) Longitudinal analysis of home food production and food sharing behavior in Japan: multiple benefits of local food systems and the recent impact of the COVID-19 pandemic. Sustain Sci 18(5):2277–2291

Kamo R, Yamada T (2020) The motivation and management characteristics of entering Enterprise's allotment garden business in urban areas. Jpn J Farm Manag 58(3):57–62

Kanda H, Isoda Y, Nakaya T (2020) The transition of urban structure in Japan in the phase of decreasing population. Q J Geogr 72:91–106. (in Japanese with English abstract)

Kimura A (2010) Revitalization of local farming: a case study of locating farmers' markets in a suburban areas of a large City. Agric Market J Japan 19(1):24–30

Kishimoto K, Yan W (2023) Geospatial characteristics of allotment garden provision in Tokyo. Renew Agric Food Syst 38:e46

Lange A, Piorr A, Siebert R, Zasada I (2013) Spatial differentiation of farm diversification: how rural attractiveness and vicinity to cities determine farm households' response to the CAP. Land Use Policy 31:136–144

Li W, Wang D, Li H, Liu S (2017) Urbanization-induced site condition changes of peri-urban cultivated land in the black soil region of Northeast China. Ecol Indic 80(September):215–223

Liu Y, Nath N, Murayama A, Manabe R (2022) Transit-oriented development with urban sprawl? Four phases of urban growth and policy intervention in Tokyo. Land Use Policy 112(January):105854

Lovell ST (2010) Multifunctional urban agriculture for sustainable land use planning in the United States. Sustain Sci Pract Policy 2:2499–2522

Ministry of Agriculture, Forestry and Fisheries, and Ministry of Land, Infrastructure, Transport and Tourism (2016) Plan for promotion of urban agriculture. Ministry of Agriculture, Forestry and Fisheries

Miyake Y, Matsumoto Y (1997) Locational character of allotment gardens and landowners' attitudes: a case of Ichinomiya City in Aichi Pref. J Rural Plan Assoc 16(1):49–57

Murakami A, Zain AM, Takeuchi K, Tsunekawa A, Yokota S (2005) Trends in urbanization and patterns of land use in the Asian mega cities Jakarta, Bangkok, and metro Manila. Landsc Urban Plan 70:251–259

Music J, Finch E, Gone P, Toze S, Charlebois S, Mullins L (2021) Pandemic victory gardens: potential for local land use policies. Land Use Policy 109(October):105600

Nagashima H (1976) Urban development and its influence on rural areas around cities. Geogr Rev Jpn 49(5):314–326

Nakajima S, Murakami T, Sato K (2011) Evaluating the economic impact of farmers' markets using geographical information systems: an application of difference-in-differences estimation and spatial econometrics. Agric Inf Res 20(3):131–138

Nishino T (2016) Sustainable development of local farming through farm produce shops: case study of the Yoshii District of Takasaki, Gunma prefecture. E-J GEO 11(2):448–459

Oba R, Obase R (2001) A study on the allotment gardens in Tokyo area and the financial situation of landowners. J City Plan Inst Jpn 36:283–288

Oishi T (2019) Sustainability of upland agriculture in Okayama prefecture: a case study on a farmers' market in Kawakami area, Maniwa City. J Geogr 128(2):323–335

Okata J, Murayama A (2010) Tokyo's urban growth, urban form and sustainability. In: Sorensen A, Okata J (eds) Megacities: urban form, governance, and sustainability. Springer, Tokyo, pp 15–41

Otsuka M, Hashimoto S (2019) Evaluation of plan conformance under agricultural scenic area development system in Tokyo. Landsc Res Japan Online 13:8–15. https://doi.org/10.5632/jilaonline.13.8

Peng J, Liu Z, Liu Y, Hu X, Wang A (2015) Multifunctionality assessment of urban agriculture in Beijing City, China. Sci Total Environ 537(December):343–351

Piorr A, Ravetz J, Tosics I (2011) Peri-urbanisation in Europe: towards European policies to sustain urban-rural futures. University of Copenhagen/Academic Books Life Sciences, Copenhagen

Pölling B (2016) Comparison of farm structures, success factors, obstacles, clients' expectations and policy wishes of urban farming's main business models in North Rhine-Westphalia, Germany. Sustain Sci Pract Policy 8:446

Pothukuchi K (2004) Hortaliza: a youth 'nutrition garden' in southwest Detroit. Child Youth Environ 14(2):124–155

Potschin-Young M, Haines-Young R, Görg C, Heink U, Jax K, Schleyer C (2018) Understanding the role of conceptual frameworks: reading the ecosystem service cascade. Ecosyst Serv 29:428–440

Pulighe G, Lupia F (2016) Mapping spatial patterns of urban agriculture in Rome (Italy) using Google earth and web-mapping services. Land Use Policy 59:49–58

Ruggeri G, Mazzocchi C, Corsi S (2016) Urban gardeners' motivations in a metropolitan city: the case of Milan. Sustain Sci Pract Policy 8(October):1099

Sakurai S (2006) Development of a 'locally produced-locally consumed' movement based on the farmers' market and its impact on rural development. Agric Market J Japan 15(2):21–29

Satake H, Saio N (2018) Study on the conservation of farmland in urban areas based on transformation of productive green spaces and diversification of agricultural management: a case study of Nerima Ward, Tokyo. J City Plan Inst Jpn 53(3):522–528. (in Japanese with English abstract)

Sioen G, Terada T, Sekiyama M, Yokohari M (2018) Resilience with mixed agricultural and urban land uses in Tokyo, Japan. Sustainability 10(February):435

Spangenberg JH, von Haaren C, Settele J (2014) The ecosystem service cascade: further developing the metaphor. Integrating societal processes to accommodate social processes and planning, and the case of bioenergy. Ecol Econ 104:22–32

Spilková J, Vágner J (2016) The loss of land devoted to allotment gardening: the context of the contrasting pressures of urban planning, public and private interests in Prague, Czechia. Land Use Policy 52(March):232–239

Takatori Y (2000) The multi-functional system of urban agriculture in the suburbs of Tokyo: a case study in the Nishi-Ooizumi area of Nerima Ward, Tokyo. J Geogr 109(3):401–417. (in Japanese with English abstract)

Tamura T (2009) Study of farmer's intentions to setup an allotment garden utilizing the abandoned farmland: clarification of farmland lending and the requirements utilized for allotment garden. Pap Environ Inf Sci 23:385–390

Taylor JR, Taylor Lovell S (2012) Mapping public and private spaces of urban agriculture in Chicago through the analysis of high-resolution aerial images in Google earth. Landsc Urban Plan 108(October):57–70

Teig E, Amulya J, Bardwell L, Buchenau M, Marshall JA, Litt JS (2009) Collective efficacy in Denver, Colorado: strengthening neighborhoods and health through community gardens. Health Place 15(December):1115–1122

Ushijima C (2012) The urban life cycle in Tokyo 60 km zone and urban expansion and contraction. J Fac Lett Komazawa Univ 70:117–135. (in Japanese)

Veen EJ, Bock BB, Van den Berg W, Visser AJ, Wiskerke JSC (2016) Community gardening and social cohesion: different designs, different motivations. Local Environ 21(10):1271–1287

Yagi H, Garrod G (2018) The future of agriculture in the shrinking suburbs: the impact of real estate income and housing costs. Land Use Policy 76(July):812–822

Yokohari M, Takeuchi K, Watanabe T, Yokota S (2000) Beyond greenbelts and zoning: a new planning concept for the environment of Asian mega-cities. Landsc Urban Plan 47(April):159–171

Zasada I (2011) Multifunctional peri-urban agriculture: a review of societal demands and the provision of goods and services by farming. Land Use Policy 28(October):639–648

Zhang C, Li J, Zhou Z (2022) Ecosystem service cascade: concept, review, application and prospect. Ecol Indic 137(April):108766

Chapter 7
Decarbonization and Nexus Possibilities of Food Logistics Systems

Ayase Yonamine, Ryoko Okumura, and Wanglin Yan

Abstract This study quantitatively evaluates CO_2 emissions from the transportation and delivery stages of urban food retail operations, aiming to identify store-level logistical characteristics and structural challenges that influence emissions, and to explore potential strategies for building a more sustainable food system. Although transportation and delivery account for less than 1% of total emissions in the food supply chain, significant variation exists across retail stores due to differences in logistics structures and delivery efficiency. Using the fuel efficiency method outlined in government guidelines, we calculated annual emissions for 88 stores operated by a railway-affiliated supermarket chain in the Tokyo–Yokohama metropolitan area. A case study comparing store A and S revealed how sales volume and mixed-load delivery influence emission levels. Benchmarking against Seven & I Holdings highlighted inefficiencies in delivery density and emissions per unit of sales floor area. Further analysis showed that truck size restrictions at station-adjacent stores and the centralized location of the distribution center contribute to increased delivery frequency and longer transport distances. Based on these findings, we propose several improvement strategies, including route redesign, adoption of electric and hybrid vehicles, dynamic route optimization using AI and IoT, and the development of decentralized logistics hubs. These measures, combined with enhanced traceability and integrated resource management under the food–energy–water (FEW) nexus, offer a pathway toward more sustainable and low-carbon urban food systems.

A. Yonamine
Nikken Sekkei Research Institute, Fujisawa, Kanagawa, Japan
e-mail: ayase.ynmn@keio.jp

R. Okumura
Tokyu Research Institute, Shibuya, Tokyo, Japan
e-mail: ryoko-o@triinc.co.jp

W. Yan (✉)
Faculty of Environment and Information Studies/Graduate School of Media and Governance, Keio University, Fujisawa, Kanagawa, Japan
e-mail: w_yan@keio.jp

W. Yan (ed.), *Urban Nexus Approach for Sustainable Cities*,
https://doi.org/10.1007/978-981-95-6156-8_7

Keywords Retail industry · Supermarket · Logistics · Decarbonization · Traceability

1 Introduction

1.1 Food–Energy–Water Nexus in the Food Supply Chain

This chapter aims to clarify the structure of food transportation and delivery systems within the broader food system, and to examine the potential for decarbonization through the food–energy–water (FEW) nexus.

With the advancement of urbanization, the structure of food supply and demand has become increasingly complex. In the retail sector, particularly among supermarkets, the need to meet diverse consumer demands has led to widespread sourcing of food products, making the efficiency and environmental impact of food logistics critical issues. Estimates suggest that the food sector accounts for 19–29% of total greenhouse gas emissions (Campbell et al. 2017; Vermeulen et al. 2012). When focusing specifically on CO_2, this sector is responsible for approximately 10% of global emissions and about 7% of Japan's emissions (Oda and Uchiyama 2008; Ballot and Fontane 2010).

In response to these challenges, the concept of the FEW Nexus has gained attention in recent years (Yan and Nakayama 2021). Integrated management of food, energy, and water is expected to contribute to the development of sustainable urban systems. In the context of food logistics, this requires a combination of technological innovation and policy-driven approaches to reduce CO_2 emissions (Marrucci et al. 2020).

Previous studies have proposed several effective strategies for decarbonizing food logistics, including the use of life cycle assessment (LCA), the adoption of renewable energy, route optimization, and the deployment of electric and fuel-cell trucks. Meanwhile, regionally embedded distribution models such as Alternative Food Networks (AFNs) have shown lower environmental impact in upstream processes, but emissions from consumer travel in downstream processes remain a concern (Falasco et al. 2025; Rizet et al. 2010). These findings suggest that decarbonization in food logistics requires not only technological solutions but also a comprehensive approach that includes changes in consumer behavior and distribution structures.

In Japan, the food logistics system is characterized by geographical separation between production and consumption areas, and by the involvement of multiple stakeholders such as wholesalers, logistics centers, and retailers. This structure has led to longer transportation distances and increased intermediate inventory, complicating the evaluation of CO_2 emissions. The Ministry of Agriculture, Forestry and

Fisheries has also pointed out that the complexity of the food supply chain makes it difficult to quantitatively assess greenhouse gas (GHG) emissions, particularly due to variations in production processes and cultivation environments among producers. Moreover, carbon footprint assessments often rely on estimated values due to the difficulty of obtaining primary data, highlighting the need for enhanced traceability and digitalization for more accurate evaluations.

Improving traceability is not only essential for ensuring food quality and safety but also serves as a key tool for visualizing and reducing environmental impact. By enabling the tracking of food products from production through distribution, consumption, and disposal, traceability enhances transparency across the supply chain and facilitates the identification of emission sources (Mladen Krstić et al. 2023). While traceability systems are often adopted by small and medium-sized enterprises primarily for regulatory compliance, they have also been shown to contribute to quality control, operational efficiency, and consumer trust. Proposals for decentralized, lightweight traceability systems suggest the potential to resolve information asymmetries and improve the overall efficiency and sustainability of the supply chain (MAFF. 2021).

From the perspective of the FEW Nexus, efforts to decarbonize the food sector are advancing through the integrated and efficient use of food, energy, and water resources. For example, technologies are being developed to simultaneously optimize water use and improve energy efficiency in agricultural production (MAFF. 2021). However, resource efficiency across the entire supply chain remains insufficiently understood, and a combined approach involving comprehensive policy and technological innovation is needed. In particular, food transportation and delivery systems are highly complex due to seasonal and ingredient-based variations in sourcing locations, making it difficult to grasp their actual conditions.

Against this backdrop, the Ministry of Agriculture, Forestry and Fisheries formulated the "Green Food System Strategy" in May 2021 as a new policy initiative aimed at enhancing productivity and sustainability in the food, agriculture, and fisheries sectors through innovation. The strategy sets a goal of achieving zero CO_2 emissions in agriculture and related industries by 2050. To build a sustainable food system, it is essential to promote decarbonization across the entire food supply chain, visualize these efforts, and encourage financial flows toward climate action and sustainable consumption behavior.

The objective of this study is to quantitatively evaluate CO_2 emissions associated with the transportation and delivery stages in the food retail sector, identify structural challenges in store-level logistics systems, and propose potential measures for improvement. Through this, the study aims to contribute to the development of a sustainable urban food system.

1.2 Overview of CO_2 Emissions in the Food Supply Chain and the Role of the Transportation Sector

To comprehensively assess CO_2 emissions across the food supply chain, the classification framework of Scope 1, 2, and 3 greenhouse gas (GHG) emissions is widely recognized as effective. This international standard captures emissions throughout the life cycle of a product—from production to disposal. Scope 1 refers to direct emissions from sources owned or controlled by the company, such as fuel combustion and manufacturing processes. Scope 2 covers indirect emissions from the consumption of purchased electricity, heat, or steam. Scope 3 includes all other indirect emissions that occur in the upstream and downstream activities of the company's supply chain. Scope 3 is further divided into 15 categories, enabling companies to identify and prioritize key emission sources.

Understanding emissions through this framework allows for a holistic view of the supply chain and supports the development of effective reduction strategies (Fig. 7.1). In practice, many companies have begun setting science-based targets (SBTs) for achieving carbon neutrality and are actively exploring evaluation methods and countermeasures. For example, global food corporations such as Nestlé and Unilever are working to reduce Scope 3 emissions by analyzing the entire life cycle—from raw material procurement and transportation to consumer use and disposal—and visualizing their carbon footprints.

Traditionally, CO_2 emissions in the food supply chain have been assessed based on spatial units such as farmland, warehouses, and supermarkets. However, comprehensive evaluations across the entire supply chain have been limited. In Japan, the food distribution system is particularly complex and segmented due to the involvement of numerous intermediaries, making it difficult to trace distribution routes and optimize logistics.

Japan's total GHG emissions amount to approximately 1.17 billion tons of CO_2, with CO_2 emissions accounting for about 988 million tons. Among industry sectors, wholesale and retail account for 41.2 million tons of CO_2, making it the largest contributor within the "business and other" category. According to the 2021 GHG Emissions Calculation, Reporting, and Disclosure System, the food retail sector alone emitted 47,971 tons of CO_2 (out of a total of 26,243,143 tons).

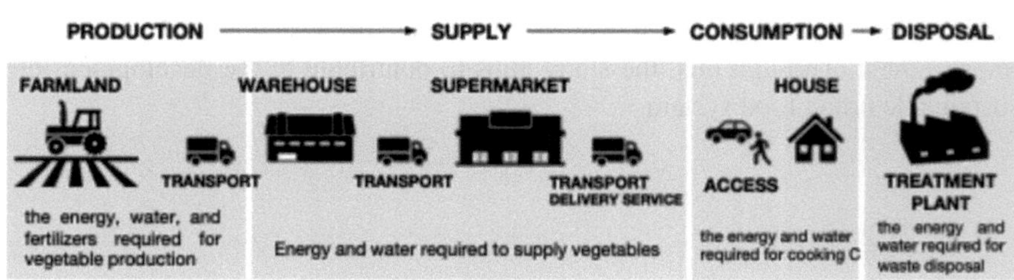

Fig. 7.1 CO_2 emission sources by segment in the food supply chain

Table 7.1 CO_2 emission sources by segment in the food supply chain

Scope 3 category		Retail industry (supermarket, convenience store)				
		Ito-Yokado	Yorkmart	AEON	FamilyMart	Lawson
1	Purchased goods and services (%)	66.5	94.7	–	75.3	75.3
2	Emissions from capital goods (%)	4.0	2.3	–	1.6	1.3
3	Emissions from fuel and energy (%)	2.4%	1.1	–	2.5	2.6
4	Upstream transportation and distribution (%)	1.1	0.6	–	1.5	2.4
5	Waste generated in operation (%)	1.0	0.5	–	2.4	0.4
6	Business commuting (%)	0.0	0.0	–	0.0	0.0
7	Employee commuting (%)	0.3	0.4	–	0.0	0.0
8	Upstream transportation (%)	–	–	–	–	–
9	*Downstream leased assets (%)*	1.0	0.1	–	–	–
10	Downstream transportation and distribution	–	–	–	–	–
11	Processing of sold products (%)	1.4	0.1	–	0.0	1.0
12	End of life treatment of sold products (%)	5.0	0.1	–	0.1	0.7
13	Downstream leased assets (%)	3.9	0.1	–	–	–
14	Franchises	–	–	–	–	–
15	Investments	–	–	–	–	–
Others		–	–	–	–	–

Further analysis of category-specific emissions published by food retailers such as AEON, Ito-Yokado, Lawson, and FamilyMart on the Green Value Chain Platform of the Ministry of the Environment reveals that raw material procurement accounts for the majority of emissions. In contrast, emissions from downstream processes such as transportation and delivery are relatively small. For many companies, CO_2 emissions from the transportation and delivery stages represent less than 1% of total supply chain emissions (Table 7.1).

However, when viewed at the individual company level, transportation-related CO_2 emissions vary significantly depending on business type and operational structure. In particular, companies whose primary activity is transportation tend to have a higher proportion of emissions from logistics, making efficiency improvements and emission reductions critical issues. In 2024, Japan enacted the Comprehensive Logistics Efficiency Act, which promotes the integration of distribution operations—including transportation, storage, cargo handling, and processing—and encourages rationalization through measures such as network consolidation, modal shift, and joint delivery systems.

Despite these developments, the retail sector has yet to establish clear targets for reducing transportation-related CO_2 emissions. This is due in part to the diversity of logistics systems across companies and the wide range of business formats—from food to apparel and daily goods—which complicates standardized measurement. Additionally, the potential impact of carbon pricing and emissions trading schemes on the food industry is being debated in Japan and the EU, and future policy developments are expected to significantly influence decarbonization efforts in the food supply chain.

Given this context, it is urgent to establish comprehensive CO_2 emission management and effective reduction strategies in the food sector based on Scope 1, 2, and 3 classifications. This study aims to quantitatively evaluate CO_2 emissions in the transportation and delivery stages of food retail, identify structural challenges in store-level logistics systems, and propose potential measures for improvement, thereby contributing to the development of a sustainable food system.

2 Methods for Calculating Emissions in the Food Retail Distribution Sector

This section overviews the understandings and operations of nexus in concept and practice. This section calculates the CO_2 emissions generated during the transportation of fresh produce from distribution centers to individual retail stores, based on the *Guidelines for Calculating CO_2 Emissions in the Logistics Sector*. These guidelines were jointly developed by the Ministry of Economy, Trade and Industry and the Ministry of Land, Infrastructure, Transport and Tourism (Fig. 7.2 and Table 7.2) (METI and MLIT 2023) in response to growing concern over greenhouse gas emissions associated with logistics activities.

According to the guidelines, there are four recognized methods for calculating CO_2 emissions from transportation, listed in order of accuracy:

1. Fuel Consumption Method

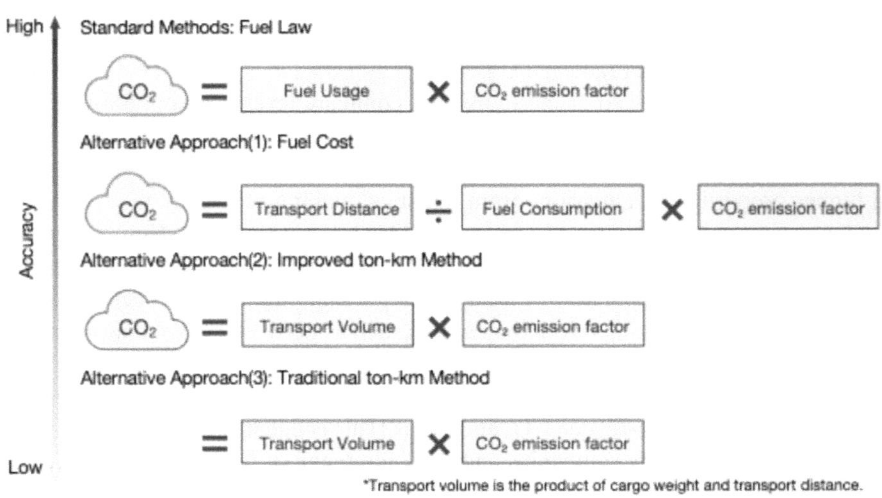

Fig. 7.2 CO_2 emission calculation method for food transportation

Table 7.2 Data requirements for each calculation method

Method	Phase	Fuel usage	Fuel efficiency	Transport distance	Cargo weight (ton)	Cargo volume	Transport volume (ton-km)	Loading rate	Freight cost
Fuel-based method	At time of estimation	○							
	At time of allocation	△		○	△	☆			△
Fuel efficiency method	At time of estimation		○	○					
	At time of allocation	△		○	△	☆			△
Improved ton-km method				○	○		△		
Conventional ton-km method				○	○		☆		

2. Fuel Efficiency Method
3. Improved Ton-Kilometer Method
4. Conventional Ton-Kilometer Method

Each method requires different types of data for calculation. The required data for each method is summarized in Table 7.2.

3 Nexus Principles Case Study: Logistics System of a Food Retailer

This study estimates the CO_2 emissions associated with food transportation for a hypothetical subject, referred to as *Retailer X*, which operates a chain of supermarkets primarily located along railway lines in the Tokyo–Yokohama metropolitan area. As part of a railway company's retail business, *Retailer X* manages 90 stores and leverages its station-front locations to offer highly convenient services. The business is closely aligned with Tokyu Corporation, and its customer base is characterized by a high proportion of commuters and students.

The product range includes fresh produce, prepared foods, and daily necessities, with a strong emphasis on meeting urban consumer demand. The logistics system for food delivery involves a centralized food distribution center located in the coastal area between Tokyo and Kanagawa Prefecture. From this center, goods are transported to individual stores.

To maintain the freshness of perishable items, the distribution center is equipped with a low-temperature management system, and refrigerated or frozen vehicles are used for deliveries. Furthermore, *Retailer X* is actively working to reduce environmental impact by improving the efficiency of truck transportation. This includes optimizing load ratios and delivery routes. In recent years, the company has also introduced electric and hybrid vehicles as part of its efforts to reduce CO_2 emissions.

3.1 Calculation of CO_2 Emissions Associated with Food Transportation

Based on the *Guidelines for Calculating CO_2 Emissions in the Logistics Sector*, this study estimates the CO_2 emissions generated during the transportation of food from the logistics center to individual retail stores. As fuel consumption data is unavailable, the calculation employs the Fuel Efficiency Method, which is the second most accurate among the four methods outlined in the guidelines. This method utilizes data on vehicle fuel efficiency and transportation distance.

Within the Fuel Efficiency Method, transportation distance is categorized into two types:

- *Direct delivery*: Transportation from the logistics center to a single store.
- *Route delivery*: Mixed-load transportation to multiple stores along a delivery route.

$$\begin{aligned}&(\text{CO}_2 \text{ emissions associated with transportation})\\&=(\text{transport distance})\div(\text{fuel efficiency})\times(\text{CO}_2 \text{ emission factor})\times 12\,\text{months}\end{aligned}$$

Using the "Delivery Schedule Table" provided by the logistics department of *Retailer X*, this study calculates the CO_2 emissions for 88 out of 90 stores for which data is available. Since the calculation method varies depending on the delivery type, the data was preprocessed accordingly.

Retailer X employs both direct and route delivery methods. For each store, the number of deliveries and total travel distance per month were compiled. As the delivery schedule reflects daily operations, the number of deliveries per day was multiplied by the number of delivery days in a month to estimate monthly transportation activity. Daytime deliveries, which occur five times per week, were counted over 21 days.

The product of this value and the transportation distance yields the CO_2 emissions for direct delivery. For route delivery, the total transportation distance from the logistics center to multiple stores and back was used. Emissions from shared delivery routes were apportioned to each store based on their respective sales floor area. These apportioned values were then added to the direct delivery emissions to obtain the total CO_2 emissions per store.

3.2 CO_2 Emissions from Food Transportation to Individual Stores

As a result of the analysis, the estimated annual CO_2 emissions from food transportation to the Azamino store amounted to 157.5 tons of CO_2, while the emissions to the Susukino store were 61.7 tons of CO_2. The Azamino store emits approximately 2.5 times more CO_2 than the Susukino store. However, due to differences in sales volume between the two stores, direct comparison is challenging.

To account for this, the CO_2 emissions were normalized by dividing by each store's annual sales revenue. The resulting values were 4.9 t-CO_2/yen/year for the Azamino store and 3.3 t-CO_2/yen/year for the Susukino store. This suggests that the higher number of deliveries to the Azamino store—driven by its larger sales volume—contributes to its greater CO_2 emissions. Additionally, part of the deliveries to the Susukino store are conducted via mixed-load transportation, which may have reduced its transportation-related CO_2 emissions.

A comparative analysis was also conducted between *Retailer X* and other companies in the same industry (see Table 7.3). For this purpose, publicly available data

Table 7.3 Comparison of distribution-based carbon emissions with other companies

	Target X	Ito-Yokado	York-Benimaru
Number of stores (stores)	88	129	237
Carbon emissions (t- CO_2/year)	6754	22,002	18,026
Carbon emissions per store (t- CO_2/(store·year))	76.8	170.6	76.1
Average sales floor area (m^2)	1334.1	9013.7	2299.3
Carbon emissions per sales floor area (kg-CO_2/(m^2·year))	57.6	18.9	33.1

from Seven & I Holdings (Seven & I HLDGS, 2022) was referenced. The results indicate that *Retailer X* emits more CO_2 per store than York Benimaru, a food supermarket under Seven & I Holdings, but less than Ito-Yokado, a general merchandise store.

However, when comparing CO_2 emissions per unit of sales floor area, *Retailer X* was found to emit more than Seven & I Holdings. This finding suggests that there is room for improvement in reducing CO_2 emissions within the logistics operations of *Retailer X*, including considerations such as the location of distribution centers.

4 Nexus in Food Retail Transportation

4.1 Variations in CO_2 Emissions by Store

This study confirmed that CO_2 emissions associated with food transportation vary across individual stores. A comparative analysis was conducted using store area and delivery frequency as key variables.

When store area and delivery frequency are identical, CO_2 emissions were found to be similar. This suggests that stores with the same area tend to receive comparable quantities of goods and delivery frequencies.

When store area and delivery frequency are identical but CO_2 emissions differ, the variation is likely attributable to factors such as the distance from the distribution center to the store and whether mixed-load delivery with other stores is employed.

When store area is similar but delivery frequency and CO_2 emissions differ, this may indicate stores with higher customer traffic and turnover rates. These stores likely adjust their delivery volume to meet increased demand, resulting in higher CO_2 emissions.

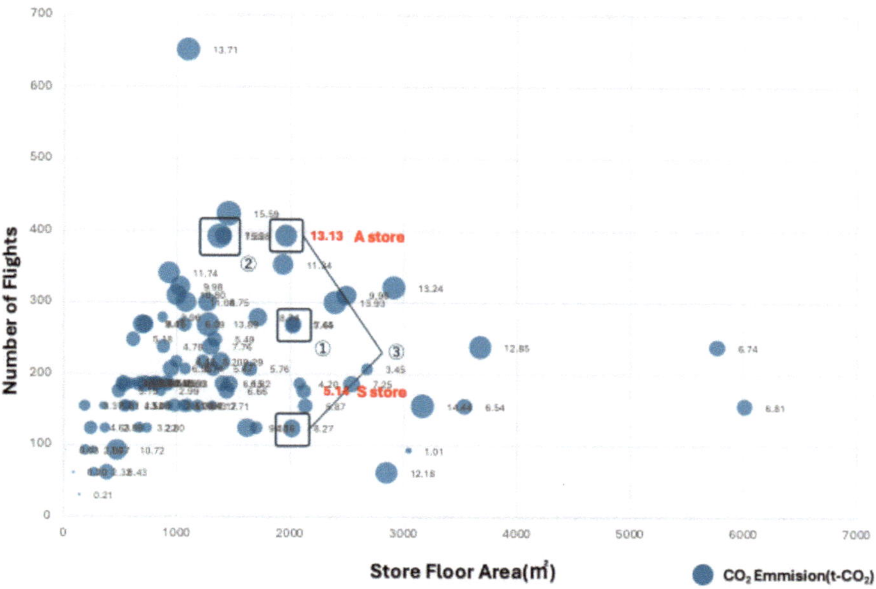

Fig. 7.3 CO_2 emissions in relation to store area and monthly delivery frequency

4.2 Vehicle Size Restrictions for Store Deliveries

Figure 7.3 illustrates the relationship between store area and monthly delivery frequency, showing that smaller stores tend to have higher delivery frequencies, whereas larger stores generally exhibit lower frequencies. This variation in delivery frequency can be attributed, in part, to restrictions on the size of delivery trucks that each store can accommodate.

In particular, stores located directly adjacent to train stations are often limited to smaller vehicles—typically 2-ton trucks, or at most 2.5-ton trucks—due to height restrictions at their delivery bays. As a result, the cargo volume per delivery is constrained, requiring these stores to increase the number of deliveries to meet demand.

Moreover, since these smaller trucks are frequently fully loaded, implementing mixed-load or multi-store route deliveries is challenging, as the limited cargo space restricts flexibility. In contrast, stores with larger loading areas that can accommodate bigger trucks can serve as relay points for nearby stores with stricter vehicle size constraints, enabling the design of more efficient delivery routes.

Incorporating such routing strategies could reduce the total round-trip distance between distribution centers and stores, thereby contributing to a reduction in CO_2 emissions. The trends observed in Fig. 7.3 suggest that the combination of store size, truck restrictions, and delivery logistics plays a significant role in shaping delivery frequency patterns across stores.

4.3 Opportunities for Improvement in Distribution Center Location

From a logistics system perspective, *Retailer X* currently operates its food distribution center in Ogijima, dispatching trucks from this location to its various stores. Depending on store size, delivery configurations vary and may include mixed-load shipments of dry and chilled goods, as well as multi-stop routes. These practices reflect efforts to improve operational efficiency.

However, given the bifurcated store distribution along the Tokyu Denentoshi Line and Tokyu Toyoko Line, the current one-to-many delivery model from Ogijima may result in inefficiencies. As shown in Sect. 3, *Retailer X* exhibits higher CO_2 emissions compared to Seven & I Holdings. One contributing factor may be that Seven & I Holdings operates multiple food distribution centers, enabling shorter transportation distances between centers and stores, and thereby achieving a more efficient logistics system.

For *Retailer X*, a potential improvement scenario involves establishing a new distribution center along National Route 246 in conjunction with the redevelopment of the Seya District in Yokohama (Fig. 7.4). This relocation could shorten the

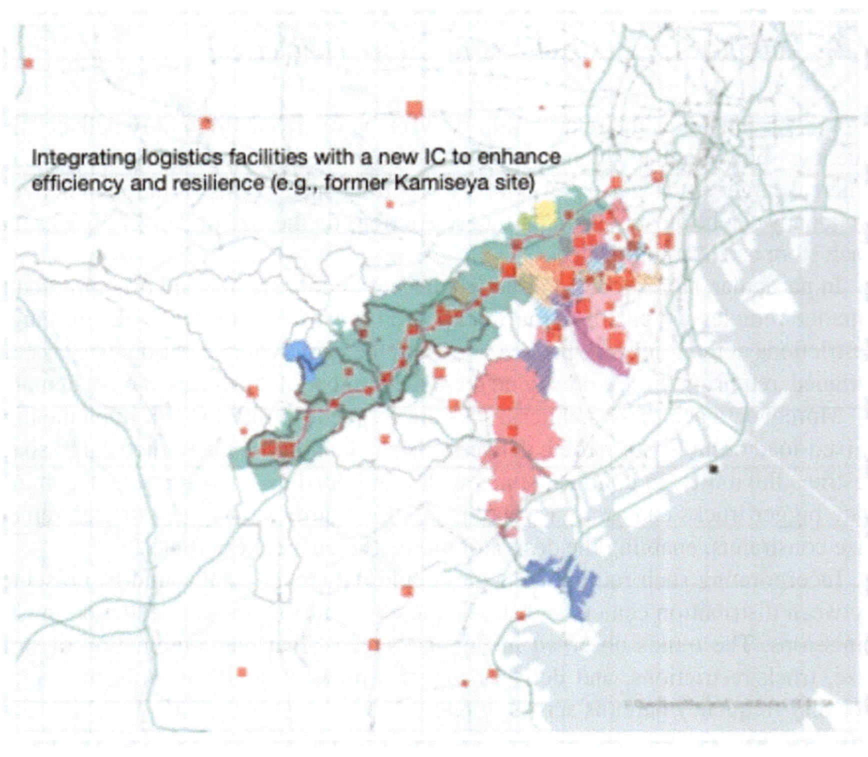

Fig. 7.4 1-to-*N* delivery system from distribution center to store

transportation distance between the distribution center and stores, leading to a reduction in CO_2 emissions associated with food delivery (Koç et al. 2019; Zhang et al. 2023).

5 Discussions

This study estimated the CO_2 emissions associated with food transportation for *Retailer X*, analyzed differences across individual stores, and compared the results with other companies in the same industry. The findings revealed significant variation in CO_2 emissions among stores, primarily influenced by factors such as sales volume, delivery frequency, and transportation efficiency. Furthermore, the location of the distribution center was identified as a potential contributor to increased transportation distances, which may in turn lead to higher overall CO_2 emissions.

It is important to note that in the food retail sector, the proportion of CO_2 emissions from transportation and delivery accounts for only a few percent of the total emissions across the entire food supply chain—from production to disposal. Therefore, for the retail industry to genuinely pursue decarbonization, it is essential to enhance traceability not only in transportation optimization but also in procurement and production methods, including those for fresh produce.

To further reduce CO_2 emissions, a comprehensive approach is required—one that incorporates transportation efficiency, strategic relocation of logistics hubs, adoption of low-impact transportation modes, and consideration of the food–energy–water (FEW) nexus. Specifically, improving delivery efficiency by introducing mixed-load and relay delivery systems between nearby stores, while accounting for vehicle size restrictions, can be effective. Additionally, dynamic route optimization using AI and IoT technologies to reduce empty vehicle rates is a critical strategy.

Establishing new logistics hubs along National Route 246, such as in the redeveloped Seya District of Yokohama, alongside the existing distribution center, could shorten transportation distances and reduce CO_2 emissions. A decentralized logistics system would also enable a more flexible and efficient network.

Moreover, the introduction of electric and fuel-cell trucks can reduce dependence on fossil fuels. For urban store deliveries, implementing small-scale delivery systems using electric vehicles (EVs) is considered effective.

Finally, by comprehensively evaluating energy consumption and water usage in food transportation, and promoting the use of renewable energy and efficient water resource management at logistics hubs, integrated management of food, energy, and water resources can be achieved—further minimizing environmental impact.

Through this study, concrete insights were gained into the current state of CO_2 emissions from food transportation and the potential for their reduction. Moving forward, it will be necessary to conduct simulation analyses using more granular data to develop realistic and effective reduction scenarios, and to formulate policy recommendations.

Acknowledgments This chapter partly draws on the outcomes of the joint research project "Pathways to Decarbonized Community Development along Private Railway Lines," conducted by Tokyu Research Institute, Inc. and Keio University. The author would like to express sincere gratitude to all members who participated in this collaborative research, in particular the advices from the experts of the supermarket stores in the study area.

References

Ballot E, Fontane F (2010) Reducing transportation CO_2 emissions through pooling of supply networks: perspectives from a case study in French retail chains. Prod Plan Control 21(6):640–650

Campbell BM, Beare DJ, Bennett EM, Hall-Spencer JM, Ingram JSI, Jaramillo F, Ortiz R, Ramankutty N, Sayer JA, Shindell D (2017) Agriculture production as a major driver of the Earth system exceeding planetary boundaries. Ecol Soc 22(4)

Falasco S, Ferla G, Caputo P, Mura B, Borrello B, Marson M, León Bravo V, Caniato F (2025) The impacts of food chains: are alternative food networks sustainable? Clean Technol Environ Policy 27:8315

Koç Ç, Erbaş M, Özceylan E (2019) The impact of routing on CO_2 emissions at a retail grocery store chain: a GIS-based solution approach. In: International Series in Operations Research & Management Science. Springer International Publishing, Cham, pp 143–160

MAFF (Ministry of Agriculture, Forestry and Fisheries) (2021) Green Food System Strategy—Realizing Both Increased Productivity and Sustainability in Food, Agriculture, Forestry, and Fisheries Through Innovation (In Japanese)

Marrucci L, Marchi M, Daddi T (2020) Improving the carbon footprint of food and packaging waste management in a supermarket of the Italian retail sector. In: Waste Management, vol 105 (March), New York, pp 594–603

METI and MLIT (Ministry of Economy, Trade and Industry & Ministry of Land, Infrastructure, Transport and Tourism). (2023). Joint Guidelines for CO_2 Emissions Calculation Methods in the Logistics Sector Ver 3.2. https://www.enecho.meti.go.jp/category/saving_and_new/saving/ninushi/pdf/guidelinev3.2.pdf. (in Japanese)

Mladen Krstić, Agnusdei GP, Tadić S, Miglietta PP (2023) Prioritization of e-traceability drivers in the Agri-food supply chains. Agric Food Econ 11:42. https://doi.org/10.1186/s40100-023-00284-5

Oda H, Uchiyam A Y (2008) Analysis of energy and environmental burdens on food sector with input-output tables. Input-Output Analysis 16(2):27–41

Rizet C, Cornélis E, Browne M, Léonardi J (2010) GHG emissions of supply chains from different retail Systems in Europe. Procedia Soc Behav Sci 2(3):6154–6164

Vermeulen SJ, Campbell BM, Ingram JSI (2012) Climate change and food systems. Annu Rev Environ Resour 37:195–222

Yan W, Nakayama S (2021) Redesigning the urban food life through the participatory living lab platform: practices in suburban areas of the Tokyo metropolitan region. In: Roggema R (ed) TransFEWmation: towards design-led food-energy-water systems for future urbanization. Springer, Cham, pp 209–234

Zhang C, Tang L, Zhang J, Gou L (2023) Optimizing distribution routes for chain supermarket considering carbon emission cost. Mathematics, June. https://doi.org/10.3390/math11122734

Chapter 8
Integrated CO_2 Assessment of Water Supply Systems: The Missing Link Between Infrastructure Planning and Urban Planning

Amane Fujita, Shun Nakayama, and Wanglin Yan

Abstract As Japan faces large-scale infrastructure reconstruction due to aging systems and population decline, understanding the environmental implications of water infrastructure decisions becomes critical to avoid carbon lock-in effects. This study presents the first integrated assessment of CO_2 emissions across the entire water supply chain from intake to tap, examining how distribution methods and urban structure interact to determine total emissions. Using a four-step methodology, we analyzed 11 water utilities serving Tokyo and Kanagawa Prefecture, calculating emissions from both utility-managed processes and building-level water supply systems. Results reveal a paradox: while Tokyo achieves lower emission intensity (229.165 g-CO_2/m^3) than Kanagawa (234.545 g-CO_2/m^3) from intake to distribution through systematic gravity flow implementation, this advantage reverses when including water supply to buildings, with Tokyo's per-building emissions (124.448 kg-CO_2/year) exceeding Kanagawa's (111.763 kg-CO_2/year). In Kawasaki City, despite achieving the lowest emission intensity (121.545 g-CO_2/m^3), districts like Musashi-Kosugi with vertical development generate extreme emissions due to energy-intensive pumping requirements. These findings indicate that current approaches optimizing distribution systems independently of urban planning fail to minimize total emissions. Effective decarbonization of water infrastructure requires integrated planning that considers building heights and future urban development patterns alongside technological improvements, ensuring that investments in efficient distribution systems align with the vertical reality of modern cities.

A. Fujita
SFC Research Institute, Keio University, Fujisawa, Kanagawa, Japan

S. Nakayama (✉)
Graduate School of Media and Governance, Keio University, Fujisawa, Kanagawa, Japan
e-mail: shunnkym@keio.jp

W. Yan
Faculty of Environment and Information Studies/Graduate School of Media and Governance, Keio University, Fujisawa, Kanagawa, Japan

W. Yan (ed.), *Urban Nexus Approach for Sustainable Cities*,
https://doi.org/10.1007/978-981-95-6156-8_8

Keywords Water supply systems · Carbon accounting · Urban planning

1 Introduction

Japan is approaching a critical juncture for large-scale infrastructure reconstruction due to the deterioration of infrastructure developed during the high economic growth period and declining profitability of maintenance operations caused by population decline. Once infrastructure systems are constructed, large-scale redevelopment becomes difficult due to cost constraints and other factors, resulting in a certain degree of lock-in of greenhouse gas emissions (Seto et al. 2016). Consequently, there is a recognized need to plan infrastructure reconstruction with comprehensive consideration of not only economic and social costs but also environmental costs.

Water supply systems exemplify this challenge. Since water supply operations account for approximately 0.8% of Japan's total electricity consumption, initiatives to reduce environmental impact are being planned and implemented in conjunction with systematic renewal and sound management of water supply operations (Ministry of Health, Labour and Welfare, Pharmaceutical and Food Safety Bureau, Water Supply Division 2020). In Japan, water utilities maintain and operate five processes: water intake, conveyance, treatment, transmission, and distribution. Environmental impact reduction initiatives have been implemented focusing on these five processes. Research has established that approximately 90% of CO_2 emissions from water intake to distribution result from electricity used by pumps in water treatment and distribution processes (Japan Water Works Association 2022).

To minimize pump usage, conversion from the widely adopted pumped distribution system to a gravity flow system that utilizes natural water flow created by topographical elevation differences is being considered. For example, Kawasaki City reports that by employing a gravity flow system for distribution, CO_2 emissions per cubic meter of distributed water are approximately one-fifth of the average for major urban water utilities (Kawasaki City Waterworks Bureau 2021). Consequently, the Tokyo Metropolitan Government Bureau of Waterworks and the Yokohama Waterworks Bureau are also pursuing conversion to gravity flow systems (Yokohama City Water Bureau 2023; Tokyo Metropolitan Government Bureau of Waterworks 2020).

Beyond utility-managed processes, water supply methods under individual building management also contribute significantly to electricity consumption through pump usage. These methods can be broadly classified into two types with different environmental impacts. The direct water supply system delivers water directly from distribution pipes to consumers' taps using existing water pressure in the distribution network. The water storage tank system stores water from distribution pipes in receiving tanks before pumping it under pressure to supply users.

The direct water supply system offers energy-saving benefits by minimizing pump usage and provides fresh water to consumers. These environmental and sanitary advantages have driven its increasing adoption in recent years. However, this system has limitations including the inability to provide water during service interruptions due to lack of storage capacity and restrictions on building size and height due to its reliance on distribution pipe water pressure. These restrictions are particularly pronounced in areas employing gravity flow systems where water pressure tends to be lower. The water storage tank system, while requiring electricity on the consumer side for pump operation and producing more CO_2 emissions than the direct supply system, offers distinct advantages. It can supply water to high-rise buildings through pump usage and provides access to stored water during service interruptions. The choice between these water supply methods depends on building height, water demand, and the distribution system in place.

Despite the environmental significance of these water supply methods, comprehensive assessment of their CO_2 emissions remains unexplored. This research gap stems from two primary challenges: the complex relationship between water supply methods and urban structure, and the difficulty of accessing data containing sensitive personal information. Consequently, neither current CO_2 emissions from water supply nor their reduction potential have been adequately quantified.

Optimizing water infrastructure for decarbonization requires a comprehensive understanding of total environmental impact across the entire water supply system. This understanding must integrate urban planning perspectives including topographical characteristics, distribution methods, and building heights. However, no existing research has addressed this integrated approach. Therefore, this study aims to estimate CO_2 emissions from water intake to tap and their spatial distribution, taking into account both distribution methods and urban structure.

2 Methodology

2.1 Calculation of CO_2 Emissions from Water Supply

The calculation of CO_2 emissions associated with water supply from intake to final delivery involves four steps. These steps encompass the calculation of annual water demand by building, determination of CO_2 emission intensity from water intake to distribution, calculation of CO_2 emission intensity for water supply, and estimation of total CO_2 emissions from water intake to final delivery.

Step 1: Calculation of annual water demand by building. Calculating water demand for each building requires data on the number of households by household size and corresponding water consumption intensity. Since building-level population data was not available, population data by small area was apportioned based on the floor area of each apartment building. Annual water demand (WD) in cubic meters per year was then calculated by multiplying the average monthly water consumption intensity by household size.

Step 2: The calculation of CO_2 emission intensity from water intake to distribution for each supply area follows a systematic five-phase procedure. The first phase involves collecting and organizing data from environmental reports of each water utility, including water intake volume, water supply volume, CO_2 emissions by water supply process, and environmental initiatives. This data encompasses CO_2 emissions (CE) in kilograms of CO_2 generated from water intake and conveyance, treatment, and transmission and distribution, along with water supply volume (WS) in cubic meters. Environmental initiatives data includes power generation through renewable energy introduction and whether water source forest maintenance and conservation are being conducted.

The second phase addresses water utilities lacking CO_2 emission data from water intake to distribution. For these utilities, the average CO_2 emissions (ACE) from the five processes per cubic meter of water supplied for each target prefecture is calculated from available CO_2 emissions (CE) and water supply volume (WS) data using Eq. (8.1). This average is then multiplied by the water supply volume (WS') of utilities lacking data to calculate their CO_2 emissions (CE') in kilograms of CO_2 per year (kg-CO_2/year) using Eq. (8.2).

$$\mathrm{ACE} = \frac{1}{n}\sum_{k=0}^{n}\left(\frac{\mathrm{CE}}{\mathrm{WS}}\right) \tag{8.1}$$

$$\mathrm{CE}' = \mathrm{WS}' * \mathrm{ACE} \tag{8.2}$$

For water utilities lacking CO_2 emission data from water intake to distribution in the third phase incorporates CO_2 reduction through environmental initiatives alongside emissions from water supply. For renewable energy CO_2 reduction, the actual power generation (EG) in kilowatt-hours (kWh) is multiplied by the CO_2 emission factor (CEF) in kilograms of CO_2 per kilowatt-hour (kg-CO_2/kWh) of the major power company in the target area to calculate CO_2 emission reduction (CER) using Eq. (8.3). CO_2 absorption through water source forest maintenance and conservation utilizes data (CA) from each utility's environmental reports. For utilities that own water source forests but lack CO_2 absorption data, the average CO_2 absorption per square meter of forest (ACA) is calculated from available data using Eq. (8.4), then multiplied by their forest area (*A*') to estimate absorption (CA') through Eq. (8.5).

$$\mathrm{CER} = \mathrm{EG} * \mathrm{CER} \tag{8.3}$$

$$\mathrm{ACA} = \frac{1}{n}\sum_{k=0}^{n}\left(\frac{\mathrm{CA}}{A}\right)_{\mathrm{k}} \tag{8.4}$$

$$\mathrm{CA}' = A' * \mathrm{ACA} \tag{8.5}$$

The fourth phase calculates the annual CO_2 emissions (YCE) from water intake to distribution for each water utility. This calculation, shown in Eq. (8.6), incorporates the CO_2 emissions (CE or CE') from the second phase and the CO_2 reduction (CER) and absorption (CA or CA') from the third phase.

$$\mathrm{YCE} = \mathrm{CE}\left(\mathrm{or\,CE'}\right) - \left(\mathrm{CER} + \mathrm{CA}\left(\mathrm{or\,CA'}\right)\right) \tag{8.6}$$

The fifth phase determines the CO_2 emission intensity (WCEI) in kilograms of CO_2 per cubic meter (kg-CO_2/m^3) for water intake to distribution. This is calculated using Eq. (8.7) by dividing the annual CO_2 emissions (YCE) by the water supply volume (WS) for each utility.

$$\mathrm{WCEI} = \frac{\mathrm{YCE}}{\mathrm{WS}} \tag{8.7}$$

Step 3: Calculation of CO_2 emission intensity for water supply based on the estimation formula by Mashiko et al. (2010). CO_2 emission intensity is calculated for both direct water supply systems (DCEI) and water storage tank systems (SCEI). Although direct water supply systems typically lack storage tanks due to their direct delivery nature, daily water usage is considered as overflow capacity for estimation purposes. For water storage tank systems, since tank capacity design standards specify half the daily water usage, this study adopts half the daily water usage per building as the storage tank capacity (SC) following Matsuu et al. (2020). These calculations are performed using Eqs. (8.8) and (8.9), respectively.

$$\mathrm{DCEI} = 0.0004 * \mathrm{SC} + 0.0866 \tag{8.8}$$

$$\mathrm{SCEI} = 0.027\mathrm{Ln}\left(\mathrm{SC}\right) - 0.014 \tag{8.9}$$

Step 4: Estimation of total CO_2 emissions from water intake to final delivery. The total annual CO_2 emissions (BCE) for each building are calculated using Eq. (8.10). This involves summing the CO_2 emission intensity from water intake to distribution (WCEI) and the CO_2 emission intensity for water supply (DCEI or SCEI) for each building, then multiplying by the building's annual water demand (WD).

$$\mathrm{BCEI} = \left(\mathrm{WCEI} + \mathrm{DCEI}\left(\mathrm{or\ SCEI}\right)\right) * \mathrm{WD} \tag{8.10}$$

2.2 Data Sources

For the calculation of annual water demand by building in Step 1, this study focuses on the residential sector. Water demand estimation for the residential sector has been extensively conducted using statistical resources such as actual condition surveys by the Ministry of Health, Labour and Welfare and household expenditure surveys (Fukuyo 2011). These studies have clarified correlations between consumption and housing types as well as household composition (Hasegawa and Inoue 2004). However, statistical data on water demand is extremely limited compared to energy demand data. Therefore, this study utilizes average monthly water consumption data by household size from the 2021 Tokyo Metropolitan Government Bureau of Waterworks' water and sewerage rate tables, combined with data on general households by housing type and average household size by small area from the 2015 national census.

For the calculation of CO_2 emission intensity from water intake to distribution in Step 2, data from 2020 environmental reports of each water utility is employed. This includes CO_2 emissions from the five processes, water supply volume, water source forest area and CO_2 absorption, and renewable energy generation. The CO_2 emission reduction intensity for renewable energy uses the 2020 Tokyo Electric Power Company CO_2 emission factor of 0.441 kg-CO_2/kWh.

The calculation of CO_2 emission intensity for water supply in Step 3 incorporates implementation conditions for direct water supply systems from 2020 water utility websites, the estimation formula by Mashiko et al. (2010), implementation conditions for direct water supply systems from each water utility, and building height data from the 2015 Basic Survey of City Planning.

3 Case Study

3.1 Target Area

This study calculates CO_2 emissions from water intake to water supply for residential buildings throughout Tokyo Metropolis and Kanagawa Prefecture. These two metropolitan areas share a water distribution relationship, drawing from common water sources. Tokyo's water supply relies predominantly on river water, with 80% sourced from the Tone River and Ara River systems. Until the 1950s, water sources depended heavily on the Tama River system, but increasing demand necessitated greater reliance on the Tone River system as water resource development progressed in that watershed. Kanagawa Prefecture, in contrast, depends primarily on the Sagami River flowing through central regions and the Sakawa River in western areas.

3.2 Characteristics of Water Utilities in the Target Area

While the Water Supply Act stipulates that municipalities should fundamentally operate water services, challenges in securing water sources and funding, as well as maintaining technical expertise, often lead to cross-municipal operations spanning broader areas. In Tokyo's Tama region, water services were initially operated by individual municipalities. However, the Tokyo Metropolitan Government Bureau of Waterworks now operates water services for 25 of the 29 municipalities in the Tama region, excluding Musashino City, Akishima City, Hamura City, and Hinohara Village (hereafter referred to as nonintegrated municipalities). As of fiscal year 2020, the Tokyo Metropolitan Government Bureau of Waterworks serves approximately 13.63 million residents across an area of about 1239 square kilometers. The nonintegrated municipalities blend provisional water distributions from the Tokyo Metropolitan Government Bureau of Waterworks with groundwater for their supply.

In Kanagawa Prefecture, water is supplied primarily through 18 water utilities. Among these, four major utilities account for approximately 91% of the prefecture's total annual water supply: Kanagawa Prefectural Enterprise Agency serving 12 cities and 6 towns, Yokohama Waterworks Bureau, Kawasaki City Waterworks and Sewerage Bureau, and Yokosuka City Waterworks and Sewerage Bureau (Kanagawa Prefecture Water Utility Review Committee 2010). These four utilities jointly established the Kanagawa Prefectural Public Corporation for Water Supply (hereinafter referred to as the Corporation), a special local public entity, to promote effective regional utilization of water resources. The Corporation supplies water to these four utilities, accounting for approximately half of their total water supply as of fiscal year 2021. Additionally, the entire Kawasaki City area and parts of Yokohama City employ gravity flow distribution systems. This distribution method utilizes natural water flow energy created by topographical elevation differences and produces lower CO_2 emissions compared to the more commonly adopted pumped distribution systems.

This study considers 11 water utilities while accounting for differences in supply areas and distribution methods: Tokyo Metropolitan Government Bureau of Waterworks, the four nonintegrated municipalities of Musashino City, Akishima City, Hamura City, and Hinohara Village, Kanagawa Prefectural Public Corporation for Water Supply, Kanagawa Prefectural Water Supply, Yokohama City, Kawasaki City, Yokosuka City water utilities, and other municipal water utilities within Kanagawa Prefecture.

3.3 Implementation Conditions for Water Supply Methods by Each Water Utility

The choice between direct water supply systems and water storage tank systems is determined by building height conditions arising from each utility's distribution method. This study selected water supply methods for individual buildings based on the implementation conditions shown in Table 8.1.

However, direct water supply systems and water storage tank systems can be further subdivided, with implementation conditions varying accordingly. For example, direct water supply systems can be divided into direct pressure water supply systems and direct boosted water supply systems. In Musashino City, direct pressure water supply systems can be implemented up to the third floor, while direct boosted water supply systems are available from the third to tenth floors. However, such detailed information is not publicly available and could not be obtained for utilities other than Musashino City. Therefore, this study does not differentiate between direct pressure and boosted water supply systems when calculating CO_2 emissions.

3.4 Evaluation of CO_2 Emission Intensity by Water Utility

Table 8.2 presents the CO_2 emission intensity for the entire water supply system from water intake to distribution. The data reveals that within the Tokyo Metropolitan Government Bureau of Waterworks, the Asaka Water Purification Plant exhibits the

Table 8.1 Water supply system types and application conditions

Water authority	Direct water supply system application conditions	Tank water supply system application conditions
Tokyo Metropolitan Government Bureau of Waterworks	Up to 200 units	–
Musashino City	Up to 10 floors	10 floors and above
Kanagawa Prefecture	Maximum daily water usage 143 m^3 or less	–
Yokohama City (Gravity-fed system)	Up to 5 floors	–
Yokohama City (Pump distribution system)	Up to 15 floors	High-rise condominiums, etc.
Kawasaki City	Mid-to-high-rise buildings (up to 5 floors)	High-rise buildings
Yokosuka City	Up to approximately 265 units	–
Others	Small-to-medium scale condominiums/buildings (up to 5 floors)	Large condominiums and critical facilities, etc.

Table 8.2 CO_2 emission factors by water authority/treatment plant

Prefecture	Water authority/treatment plant	CO_2 emission factor (g-CO_2/m^3)
Tokyo	Hamura City Water Bureau	258.9
Tokyo	Musashino City Water Bureau	258.3
Tokyo	Akishima City Water Bureau	255.7
Tokyo	Himura Water Treatment Plant	244.2
Tokyo	Takatsuki Water Treatment Plant	244.2
Tokyo	Ogouchi Water Treatment Plant	244.2
Tokyo	Ozaku Water Treatment Plant	244.2
Tokyo	Nagasawa Water Treatment Plant	244.2
Tokyo	Kinuta-shimo Water Treatment Plant	244.2
Tokyo	Misato Water Treatment Plant	244.1
Tokyo	Kanamachi Water Treatment Plant	243.9
Tokyo	Misono Water Treatment Plant	241.7
Tokyo	Higashimurayama Water Treatment Plant	239.9
Tokyo	Asaka Water Treatment Plant	215.7
Tokyo	Hinohara Village Water Bureau	−286
Tokyo	Others	242.8
Kanagawa	Yokohama City Water Bureau (Pump distribution)	335
Kanagawa	Yokosuka City Water Bureau	299.7
Kanagawa	Kanagawa Prefectural Water Bureau	286.1
Kanagawa	Minamiashigara City Water Bureau	237.8
Kanagawa	Yokohama City Water Bureau (Gravity-fed)	131.5
Kanagawa	Others	258.9

lowest CO_2 emission intensity, while other facilities demonstrate relatively uniform values. Among the nonintegrated municipalities, Musashino City and Hamura City show high CO_2 emission intensity, whereas Hinohara Village records negative values.

Focusing on Kanagawa Prefecture, among the four water utilities that jointly established the Corporation, Kawasaki City Waterworks Bureau's supply area demonstrates the lowest CO_2 emission intensity at 121.545 grams of CO_2 per cubic meter (g-CO_2/m^3), followed by Yokohama City's gravity flow area at 131.475 g-CO_2/m^3. In contrast, Yokohama City's pumped distribution area and Yokosuka City Waterworks Bureau show significantly higher values around 300 g-CO_2/m^3. Among utilities outside the four major operators, Matsuda Town, which has implemented numerous environmentally conscious technologies, achieved negative values.

Comparing Tokyo and Kanagawa Prefecture, Tokyo's average CO_2 emission intensity is 229.165 g-CO_2/m^3, while Kanagawa Prefecture's average is 234.545 g-CO_2/m^3, indicating that Kanagawa Prefecture has higher CO_2 emission intensity.

3.5 Distribution of CO_2 Emissions per Building

Figure 8.1 spatially visualizes annual CO_2 emissions from the entire process spanning water intake to distribution and water supply. The average CO_2 emissions per building in Tokyo are 124.448 kilograms of CO_2 per year (kg-CO_2/year), while Kanagawa Prefecture's average is 111.763 kg-CO_2/year, indicating higher emissions per building in Tokyo. This result contrasts with the earlier finding that Tokyo's average CO_2 emission intensity is lower than Kanagawa Prefecture's, demonstrating an inverse relationship between emission intensity and total emissions.

Within Tokyo, CO_2 emissions are generally higher within the 23 special wards compared to outer areas, though Sumida Ward, Taito Ward, and Arakawa Ward show relatively lower emissions. In Kanagawa Prefecture, several buildings along the Yokohama coastal area and in Fujisawa City exceed 180 kg-CO_2/year. Areas with lower CO_2 emissions include the vicinity of Noborito in Kawasaki City and the Aoba Ward area of Yokohama City, which show lower values compared to other municipalities. The highest average CO_2 emissions per building in Kanagawa Prefecture occur in Tsuzuki Ward at 151.961 kg-CO_2.

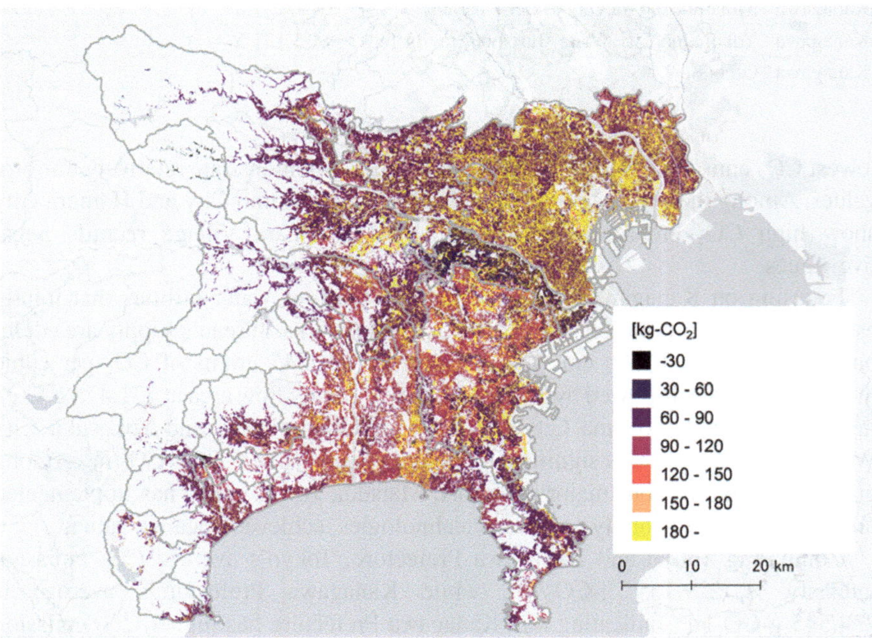

Fig. 8.1 Spatial distribution of CO_2 emissions from water supply and demand by individual building

4 Discussion

The results reveal two distinct yet interconnected findings regarding CO_2 emissions from water supply systems in Tokyo and Kanagawa Prefecture, highlighting the complex relationship between distribution systems, urban structure, and environmental impact.

4.1 Regional Differences in CO_2 Emission Intensity and the Paradox of Total Emissions

First, a paradoxical pattern emerges when examining emission intensity across different phases of water supply. Tokyo demonstrates lower average CO_2 emission intensity from water intake to distribution at 229.165 g-CO_2/m^3 compared to Kanagawa Prefecture's 234.545 g-CO_2/m^3. However, when water supply to buildings is included, Tokyo's average emissions per building reach 124.448 kg-CO_2/year, exceeding Kanagawa Prefecture's 111.763 kg-CO_2/year.

Tokyo's lower emission intensity in the intake-to-distribution phase can be attributed to its systematic pursuit of gravity flow system conversion. The Tokyo Metropolitan Government Bureau of Waterworks, which manages water supply for 25 of 29 municipalities in the Tama region and serves approximately 13.63 million residents, has implemented unified strategies for transitioning to gravity flow systems. This centralized management enables coordinated infrastructure investments and consistent implementation of energy-efficient technologies. In contrast, Kanagawa Prefecture's water supply involves 18 different utilities, each operating under distinct natural environments, historical development patterns, and current financial constraints. These diverse circumstances result in varying levels of commitment and capability for gravity flow system conversion. For instance, while Kawasaki City has achieved extensive gravity flow implementation with emission intensity as low as 121.545 g-CO_2/m^3, other utilities such as Yokosuka City record values around 300 g-CO_2/m^3.

The reversal in total emissions when including water supply to buildings likely stems from differences in urban density and building characteristics between the two regions. Within Tokyo, areas such as Sumida Ward, Taito Ward, and Arakawa Ward show relatively lower emissions despite being within the 23 special wards. These areas benefit from both low emission intensity water sources such as the Asaka and Kanamachi Water Purification Plants and lower per-building demand due to higher proportions of single-person households. While this analysis identifies building density and height as contributing factors, quantitative verification of their specific impacts remains a subject for future research.

4.2 Trade-offs Between Gravity Flow Systems and Vertical Urban Development

Second, the results illuminate critical trade-offs inherent in gravity flow systems, particularly evident in Kawasaki City's experience. As the municipality with the most extensive gravity flow distribution implementation, Kawasaki City demonstrates both the benefits and limitations of this approach. The city's strategic positioning of distribution reservoirs and water purification plants at higher elevations minimizes pump usage during distribution, contributing to its low emission intensity. This advantage is amplified by the city's building stock, where less than 10% of structures exceed six stories, allowing most buildings to utilize energy-efficient direct water supply systems without requiring additional pumping.

However, this environmental advantage faces challenges from changing urban development patterns. The Musashi-Kosugi area exemplifies this tension, where former industrial sites have been transformed into clusters of high-rise residential towers. Despite benefiting from the city's efficient gravity flow distribution system, these areas now generate some of the highest CO_2 emissions per building in the region. The extreme water demand from these vertical communities effectively eliminates the emission reduction benefits of the gravity flow system, as water must be pumped to upper floors using energy-intensive water storage tank systems.

This pattern extends beyond Musashi-Kosugi to other developing areas in both prefectures. In Kanagawa Prefecture, coastal areas of Yokohama and parts of Fujisawa City show buildings exceeding 180 kg-CO_2/year, often corresponding to areas with recent high-rise development. The highest average emissions per building in Kanagawa Prefecture occur in Tsuzuki Ward at 151.961 kg-CO_2, an area characterized by large-scale residential complexes.

While gravity flow systems represent optimal technology for minimizing emissions from water intake to distribution, their effectiveness diminishes in areas with vertical urban development. This suggests that water infrastructure planning must evolve beyond optimizing individual system components to consider the entire water supply chain from intake to final delivery at the tap. Future infrastructure decisions should integrate projections of urban development patterns, building heights, and population density to ensure that investments in gravity flow systems align with the actual built environment they will serve. Without such integrated planning, the environmental benefits of advanced distribution systems may be undermined by the energy demands of water supply in an increasingly vertical urban landscape.

5 Conclusion

This study aimed to estimate CO_2 emissions from water intake to tap and their spatial distribution while considering both distribution methods and urban planning, addressing the lack of comprehensive environmental assessment in water infrastructure planning. Despite recognition that water supply methods under individual building management significantly impact electricity consumption, previous research had not examined these emissions in relation to urban planning due to data accessibility challenges and the complexity of integrating multiple system components.

To address this gap, we developed a four-step methodology calculating emissions across the entire water supply chain for residential buildings in Tokyo and Kanagawa Prefecture. This approach integrated data from 11 water utilities' environmental reports with building-level characteristics, enabling the first comprehensive quantification of emissions from both utility-managed processes (intake through distribution) and building-level water supply systems. The spatial analysis incorporated building height data and water supply method selection criteria specific to each utility's distribution system.

The results revealed two key findings. First, Tokyo's lower average emission intensity from intake to distribution (229.165 g-CO_2/m^3) compared to Kanagawa (234.545 g-CO_2/m^3) reversed when including water supply, with Tokyo's per-building emissions (124.448 kg-CO_2/year) exceeding Kanagawa's (111.763 kg-CO_2/year). Second, gravity flow systems showed highly variable effectiveness depending on urban context, with Kawasaki City achieving 121.545 g-CO_2/m^3 in low-rise areas but generating extreme emissions in high-rise districts like Musashi-Kosugi.

These findings demonstrate that optimizing individual water system components without considering urban structure fails to minimize total emissions. As Japan undertakes large-scale infrastructure reconstruction, water utilities must adopt integrated planning approaches that consider future building patterns alongside technological improvements. The effectiveness of gravity flow systems, currently being implemented by major utilities, depends critically on the building heights they serve.

Acknowledgments This study is done as part of The Moveable NEXUS: Design-Led Urban Food, Water and Energy Management Innovation in New Boundary Conditions of Change (No. 11314551), a Belmont Forum grant project under the Collaborative Research Area on Sustainable Urbanization Global Initiative (SUGI)/Food–Water–Energy Nexus.

References

Fukuyo K (2011) Residential energy and water consumption in the 2000s in the coastal cities of the Seto Inland Sea. Trans Soc Heating Air-Condit Sanit Eng Jpn 36(171):9–22

Hasegawa Y, Inoue T (2004) Energy consumption in housing on the basis of National Scale Questionnaire: study on influence of residential characteristic and dispersion of energy consumption part 1. J Environ Eng (Trans AIJ) 69(583):23–28

Japan Water Works Association (2022) Water Supply Statistics, Fiscal Year 2021 (No. 104)

Kanagawa Prefecture Water Utility Review Committee (2010) Report of the Kanagawa Prefecture water utility review committee. https://www.pref.kanagawa.jp/documents/78259/51295.pdf

Kawasaki City Waterworks Bureau (2021) Annual environmental plan report: fiscal year 2020 final accounts version

Mashiko A, Matsuzawa M, Ashida H, Tamura S, Koizumi A (2010) Study on the reduction of CO2 emission by replacing receiving tank system to directly connected water supply system. Environ Eng Res 47:329–337

Matsuu S, Hideyo N, Hitoshi A (2020) Actual Survey of water contamination by bacteria in water storage tank part 4: measurement investigation in 2019. Techinical Papers of Annual Meeting the Society of Heating,Air-Conditioning and Sanitary Engineers of Japan 1:9–12

Ministry of Health, Labour and Welfare, Pharmaceutical and Food Safety Bureau, Water Supply Division (2020) Report on comprehensive survey for the development of carbon-neutral water supply systems

Seto KC, Davis SJ, Mitchell RB, Stokes EC, Unruh G, Ürge-Vorsatz D (2016) Carbon lock-in: types, causes, and policy implications. Annu Rev Environ Resour. https://doi.org/10.1146/annurev-environ-110615-085934

Tokyo Metropolitan Government Bureau of Waterworks (2020) Tokyo Metropolitan Government Bureau of Waterworks Environmental Five-Year Plan 2020–2024

Yokohama City Water Bureau (2023) Toward environmentally friendly water supply. J Jpn Water Works Assoc 92(4):7–10

Chapter 9
Building–Urban Integrated Decarbonization Strategy Based on Resource-Use Intensity Database

Shun Nakayama and Wanglin Yan

Abstract The building sector's decarbonization requires comprehensive approaches beyond conventional energy-focused methods. This chapter presents two integrated methodologies: Service Point-Based Carbon Accounting (SPBCA) and BEMS-Driven Energy Audit Method (BDEAM). SPBCA employs GIS to quantify building-level CO_2 emissions across energy, water, and food sectors, covering approximately 70% of global emissions compared to 50% in traditional methods. BDEAM leverages BEMS data and industry standards to enable precise performance evaluation by space use and equipment type, identifying specific inefficiencies invisible to conventional whole-building audits. The integration of both methods through a shared resource-use intensity database enables detailed building diagnostics to inform city-wide carbon reduction strategies. However, this requires transitioning from fragmented BEMS data storage to centralized, systematically managed systems. Only through such infrastructure can evidence-based decarbonization strategies effectively bridge building and urban scales.

Keywords Carbon accounting · Energy audit · Resource-use intensity

S. Nakayama (✉)
Graduate School of Media and Governance, Keio University, Minato, Tokyo, Japan

School of Network and Information, Senshu University, Chiyoda, Tokyo, Japan
e-mail: shunnkym@keio.jp

W. Yan
Faculty of Environment and Information Studies/Graduate School of Media and Governance, Keio University, Fujisawa, Kanagawa, Japan
e-mail: w_yan@keio.jp

W. Yan (ed.), *Urban Nexus Approach for Sustainable Cities*,
https://doi.org/10.1007/978-981-95-6156-8_9

1 Carbon Emissions and Decarbonization in the Building Sector

The building sector represents a major source of carbon emissions, accounting for over 40% of global energy consumption and one-third of greenhouse gas emissions (U.S. Energy Information Administration 2019). According to Figs. 9.1 and 9.2, energy consumption in Japan's building sector increased by 8% compared to 1990 levels and expanded to account for 30.1% of total energy use in 2022 (Agency for Natural Resources and Energy 2022). Consequently, the Agency for Natural Resources and Energy's plan aims to achieve 47% of Japan's total energy savings through the building sector (Ministry of Economy, Trade and Industry 2015). According to this document, energy-saving measures through urban redevelopment and urban rehabilitation—including improvements in energy efficiency for new buildings, retrofitting of existing buildings, enhancement of energy-consuming equipment, thorough energy management through BEMS (Building Energy Management System), and introduction of high-efficiency lighting and water heaters—account for 95% of the energy efficiency plan for the building sector (Ministry of Economy, Trade and Industry 2015).

Meanwhile, with the transition from a low-carbon to a decarbonized era, approaches to capturing carbon emissions in the building sector are also evolving. Fig. 9.3 shows the proportion of sectors in CO_2 emissions. While conventional assessments focusing on the energy sector alone capture only about 50% of global CO_2 emissions, including water and food in addition to energy increases this proportion to approximately 70% (Ramaswami et al. 2016). This trend is consistent

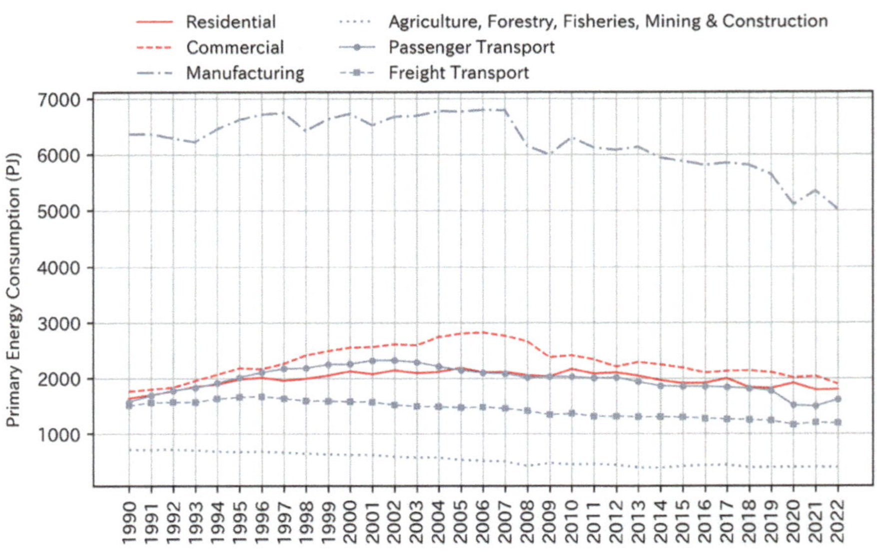

Fig. 9.1 Energy consumption trends in Japan

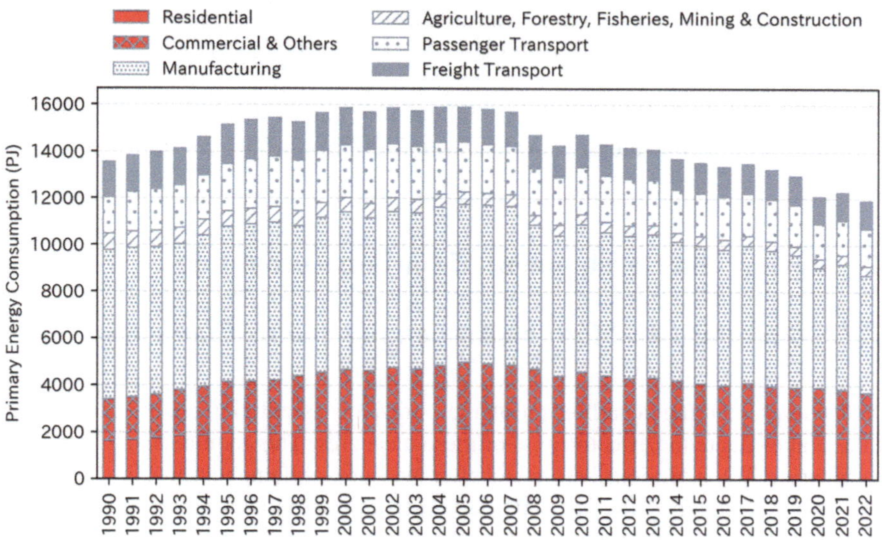

Fig. 9.2 Trends in building sector share of total energy consumption in Japan

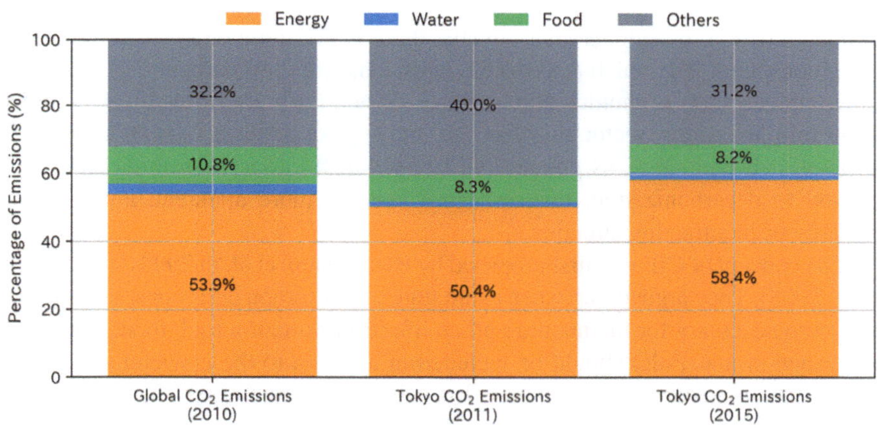

Fig. 9.3 Sectoral breakdown of CO_2 emissions

with previous studies using Tokyo as a case study (Hu and Yan 2024). Although the food sector has rarely been discussed alongside energy and water, it accounts for 19–29% of anthropogenic greenhouse gas emissions (Campbell et al. 2017; Vermeulen et al. 2012) and approximately 10% of global CO_2 emissions and 7% of Japan's CO_2 emissions (Oda and Uchiyam A 2008), and is increasingly being recognized as a major decarbonization sector (Jones and Kammen 2011; Song et al. 2021). Energy, water, and food constitute urban core infrastructure and are positioned as essential resources for urban life. These resources are closely interrelated in the process of being delivered as services (Keairns et al. 2016; Ramaswami et al.

2017), and it has been pointed out that considering their interactions presents potential for improving resource consumption efficiency (Hoff 2011; Yan and Roggema 2019). In the building sector as well, the potential for improving resource consumption efficiency through leveraging the interactions among these three types of infrastructure is attracting attention (Valencia et al. 2022).

2 Review and Development of Carbon Accounting

2.1 Building-Level Carbon Accounting and Its Challenges

When implementing decarbonization measures, it is essential to quantitatively measure and evaluate the effectiveness of these measures. This requires calculating and comparing CO_2 emissions before and after measure implementation using appropriate carbon accounting methods. While numerous carbon accounting methods have been developed (Chen et al. 2019), these can be broadly classified into three approaches (Seto et al. 2021). First, there is the purely in-boundary source-based footprint (IB), which targets direct combustion within the subject area. Second, there is the consumption-based footprint (CBF), which is based on consumption by residents within the subject area. Finally, there is the community-wide infrastructure footprint (CIF), which targets CO_2 emissions from infrastructure across entire communities such as municipalities. Since IB and CBF are limited to CO_2 emissions from the energy sector, they face the challenge of covering only about 50–60% of total CO_2 emissions, as pointed out in Sect. 1. As awareness shifts from low-carbon to decarbonization, this issue has become more apparent, and CIF has recently been attracting attention.

The concept of CIF was first proposed by Ramaswami et al. in 2008 (Ramaswami et al. 2008). They argued that when estimating urban carbon emissions, major infrastructure necessary for maintaining urban life (transportation fuel, food, water, and construction materials) should be included in addition to the conventional energy sector. The boundary of "community" in CIF has predominantly been defined as municipalities in most previous studies. For example, Chavez et al. analyzed temporal changes in CIF for Delhi, revealing how increased infrastructure demand accompanying urbanization drives up greenhouse gas (GHG) emissions (Chavez et al. 2012). Additionally, Lin et al. evaluated the ripple effects of emissions through urban supply chains in Xiamen, China, by combining CIF with input–output analysis (Lin et al. 2015). CIF has also been applied to multiple US cities (Ramaswami and Chavez 2013), major Chinese cities (Chen et al. 2020), and Delhi (Ramaswami et al. 2017), contributing to a more comprehensive understanding of urban GHG emission structures.

However, methods for implementing CIF in a bottom-up manner have not yet been established (Chavez and Ramaswami 2013). The primary reason is the need to integrate data with different purposes, sources, and resolutions for carbon accounting. In a study comparing six global cities, the authors point out that the granularity of intensity data available for carbon accounting differs among energy, water, and food sectors, with particularly coarse data granularity available for the food sector (ten Caat et al. 2022). Traditionally, CO_2 emissions from energy, water, and food have often been calculated separately, and the minimum spatial units that can be statistically calculated also differ by sector. For example, the Building Stock Energy Model (BSEM) represents a typical example of conventional bottom-up carbon accounting (Langevin et al. 2020; Yamaguchi et al. 2022). In bottom-up CBF targeting the energy sector, such as BSEM, using a single database enabled evaluation of carbon emissions without omissions or duplications. The accuracy of this model depends on databases where resource consumption intensities are uniformly and comprehensively organized, and several representative databases are known in the energy sector, including CBECS in the United States, ODYSSEE in the EU, and DECC in Japan (Takaguchi et al. 2012). However, there are no examples of single databases like those used in BSEM that comprehensively organize resource consumption intensities for energy, water, and food. Due to these challenges, carbon accounting targeting multiple resources has been limited to top-down estimation methods at the municipal level or coarser spatial scales (Cohen and Ramaswami 2014; Hillman and Ramaswami 2010; Feng et al. 2013). To solve this problem, it is necessary to effectively integrate multiple databases for each resource prepared at different levels—such as census-based, building use-based, and tenant-based—and establish complementary relationships.

2.2 *Service Point-Based Carbon Accounting (SPBCA)*

To address the challenges identified through the review of previous studies, the authors developed SPBCA (Service Point-Based Carbon Accounting), a bottom-up carbon accounting method capable of estimating carbon emissions from each sector at the building level (Nakayama and Yan 2025). While SPBCA follows the community-wide infrastructure footprint (CIF) methodology, it achieves finer spatial resolution. The distinctive feature of this method lies in utilizing Geographic Information Systems (GIS) as the foundational technology to integrate diverse spatial data at the building level and develop and apply resource demand intensities and CO_2 emission factors consistent with this integration.

SPBCA consists of four stages. The first stage, "Spatial Data Inventory," involves collecting multiple spatial data layers including buildings, POI (Point of Interest), and census tracts, and integrating them using building data as the foundation. This process also integrates the attribute information from each layer. For example, even when a building use is classified only as "mixed-use facility," if the POI layer reveals the presence of a supermarket and apartments, the building use can be identified with greater accuracy. Additionally, population attributes can be added to buildings identified as having residential use by proportionally allocating census population data.

The second stage, “Intensity Data Inventory,” involves collecting and organizing resource consumption intensities and carbon emission factors for energy, water, and food. This stage must be implemented iteratively with the first stage. This is necessary to collect appropriate intensities according to spatial data attributes while organizing spatial data attributes to match available intensities. The authors’ previous report demonstrates that each resource demand can be calculated by multiplying social features such as population or physical features such as floor area by their respective use intensities (Nakayama et al. 2024). This iterative process ensures consistent integration of spatial data and intensities.

The third stage, “Record Linkage,” connects spatial data, resource consumption intensity data, and carbon emission factor data. First, social feature data such as census data are joined to building data, which are physical features, gradually integrating data at different spatial scales including buildings, tenants, and land. Next, resource consumption intensities are joined to the integrated building data. Starting with exact matches of attributes such as building use, construction year, and floor area, and gradually relaxing conditions while assigning intensities, this approach addresses data incompleteness and ambiguity while achieving accurate intensity assignment. This approach enables the integrated treatment of energy, water, and food sectors at the common spatial unit of building level.

The fourth stage, “Spatial Assessment of Carbon Emissions,” first calculates the consumption of each resource using resource consumption intensities assigned to each building, then quantifies carbon emissions at the building level by multiplying these by the corresponding CO_2 emission factors for each resource. Furthermore, by aggregating these, the total CO_2 emissions for the entire target area can be determined. This progressively refined carbon accounting enables spatial assessment of the distribution and density of carbon emissions.

SPBCA has made it possible to determine carbon emissions at the building level, which was difficult with conventional top-down methods at coarse spatial scales such as municipalities. Moreover, it has realized integrated analysis including not only the energy sector but also water and food sectors, enabling comprehensive carbon accounting that covers approximately 70% of global CO_2 emissions.

3 Review and Development of Energy Audit

3.1 Verification of Building Decarbonization Potential and Its Challenges

All buildings must not only achieve high environmental performance at the design stage but also continuously verify and update their performance during the operational phase. For this purpose, energy auditing has gained attention. Energy auditing refers to the process of diagnosing energy consumption patterns in buildings in detail and quantitatively evaluating energy-saving potential through operational improvements and equipment retrofitting. Today, it is estimated that over 100 building energy audit methods exist worldwide (Schleich and Fleiter 2019). Among the most widely referenced guidelines are those developed by ASHRAE (American Society of Heating, Refrigerating and Air-Conditioning Engineers), the Energy Efficiency Directive (EED) established in the EU, and ISO 50001 (Merabtine et al. 2018; Atiba et al. 2019). While these energy audits differ in details, they typically consist of standard processes: (1) energy consumption profiling, (2) benchmarking, and (3) equipment diagnostics (Deru et al. 2011; Kluczek and Olszewski 2017; Hong et al. 2014).

Buildings function not merely as physical structures but as comprehensive systems where various elements—including building structure, equipment, operational methods, and occupant behavior—interact in complex ways. For example, a building's insulation performance is closely related to the efficiency of its air conditioning system. Buildings with high insulation performance reduce heat loss, thereby reducing the load on air conditioning systems and decreasing energy consumption (Kumar et al. 2020). Furthermore, building operational methods and occupant behavior significantly impact building energy consumption. For instance, even in identical buildings, energy consumption varies greatly depending on factors such as room temperature settings, lighting hours, and equipment usage patterns (Masoso and Grobler 2010). Additionally, building shape and placement affect the surrounding wind environment, influencing the effectiveness of natural ventilation (Ramponi et al. 2014). In essence, energy audits tailored to individual building characteristics are inherently required.

However, the reality of existing energy audits fails to meet these requirements. For example, regarding energy consumption profiling in the first of the four standard phases, while there are cases where energy consumption and efficiency have been organized according to space use and building equipment, these use and equipment classifications were not designed with benchmarking in mind and cannot be compared (Liddiard 2014). Furthermore, in the benchmarking phase, only total building energy consumption is considered, and energy efficiency is evaluated between buildings with entirely different characteristics based solely on energy consumption per unit floor area. As a result, it is believed that feasible measures cannot be appropriately selected, leading to errors in energy audit results.

3.2 BEMS-Driven Energy Audit Method (BDEAM)

In parallel with SPBCA, the authors developed a new energy audit method, BDEAM (BEMS-Driven Energy Audit Method), which enables more precise building performance evaluation by utilizing BEMS data (Nakayama, Yan, and Fujita 2024). Conventional energy audits did not adequately consider the unique characteristics of buildings, necessitating a new approach to clarify the decarbonization potential of buildings.

The limitations of conventional energy audits stem from the fact that many existing buildings measure energy consumption only based on utility bills, consequently requiring buildings to be treated as single entities. As a result, energy consumption profiling and benchmarking could only capture total building energy consumption. Even if energy consumption could be determined by space use or building equipment, industry standards for these classifications did not exist, and similar information for other properties used in benchmarking had not been collected.

However, today, many buildings are equipped with advanced BEMS, and industry standards are being developed across the building sector. BEMS covers not only energy consumption but also water consumption, enabling measurement by space use and building equipment at least on an hourly basis. This has made it possible to understand energy and water consumption by space use and building equipment, and to collect data at similar granularity for other buildings.

The standardization of industry standards is also eliminating barriers in benchmarking. Energy use intensity (EUI) and water use intensity (WUI) are used for evaluating energy and water consumption efficiency in benchmarking (Ma and Cheng 2016; Ahn et al. 2016; Hadjikakou et al. 2015). While areas commonly defined across different buildings other than gross floor area were previously limited, industry standards representing more detailed space uses are being developed in recent years. In Japan, for example, the Building Energy Efficiency Act defines uses such as offices, hotels, hospitals, department stores, schools, restaurants, assembly halls, and factories. If energy and water consumption data are associated with each space use under these de facto industry standards in both BEMS and reference resource consumption intensity databases, EUI and WUI can be calculated based on floor area for each space use, enabling more accurate comparisons between buildings.

BDEAM enables diagnostics limited to equipment with poor energy or water consumption performance and allows for appropriate measure selection by referencing industry standard databases. Furthermore, the economic costs of measures are also an important concern and are increasingly being made available on the internet. As a result, it has become possible to conduct feasibility studies to determine whether measures are economically rational. The proliferation of BEMS and establishment of industry standards have realized high-precision benchmarking capable of identifying building equipment and space uses with poor energy or water consumption performance, making it possible to link this to equipment diagnostics.

4 Database as the Core of the Methodologies

The building-level carbon accounting method SPBCA and the energy audit method BDEAM, which enables more precise building performance evaluation, may appear to be separate methods at first glance, but they are fundamentally interconnected. At the core of both methods is a database of resource-use intensity. This resource-use intensity functions as resource consumption intensity in SPBCA and as a benchmarking comparison reference in BDEAM. Through this database, carbon accounting and energy auditing are systematically integrated.

Integrating SPBCA and BDEAM with the resource-use intensity database at their core enables two important developments in urban decarbonization efforts. First, it establishes a pathway for directly feeding energy audit calculation results back into carbon accounting. For example, if a sample survey of buildings demonstrates an average 10% reduction in energy consumption through air conditioning system improvements, this effect can be extrapolated to the entire district, enabling quantitative evaluation of energy consumption reduction and associated CO_2 emission reductions. This realizes the linkage of micro-to-macro analysis, where insights from individual building diagnostics are utilized to evaluate the decarbonization potential of entire districts. Second, the cross-sectional approach to buildings and cities enables the development of BDEAM, currently targeting individual buildings, into an energy audit method that can also consider area-wide energy systems such as self-consignment and district heating and cooling. To accurately evaluate the effects of such energy interchange between buildings and shared infrastructure, an analytical framework that comprehensively addresses buildings and cities is essential, and this becomes feasible through the common database of resource-use intensity.

However, constraints currently exist in the connectivity between these two methods. In carbon accounting, resource-use intensity primarily uses databases organized by building or industry type. This is necessary because calculating CO_2 emissions by building for entire districts is required. Meanwhile, energy auditing uses resource-use intensity databases organized by equipment or space use. This is because the objective is to analyze individual buildings in detail and derive specific improvement measures, making building-level resource-use intensity insufficient in granularity. This difference in resolution presents a barrier to integrated analysis across buildings and cities.

To overcome this constraint, comprehensive development of high-granularity resource-use intensity databases is essential. Currently, BEMS data are stored individually in each building's servers, but rather than leaving these data unutilized, a mechanism is required to centrally aggregate and analyze them, and feed the insights back to society as a whole. In Japan's current context, while developers might assume this data aggregation role, individual operators face limitations in the number of data they can collect, making it difficult to construct statistically reliable, high-granularity resource-use intensity databases. Therefore, establishing a systematic data collection and management system by an organization with broader data access and public character is necessary. Through the development of such public data infrastructure, the effectiveness of decarbonization strategies encompassing building to city levels can be truly ensured.

5 Conclusion

Decarbonization of the building sector requires a comprehensive approach that encompasses not only the energy sector but also the water and food sectors. To address this requirement, this chapter has developed SPBCA, a building-level carbon accounting method, and BDEAM, an energy audit method utilizing BEMS data, and presented a framework for a new decarbonization strategy through their integration.

SPBCA overcomes the limitations of conventional top-down carbon accounting at the municipal level and enables integrated calculation of carbon emissions from energy, water, and food at the building level using GIS. Through a four-stage process—spatial data inventory, intensity data inventory, record linkage, and spatial assessment of carbon emissions—it achieves comprehensive carbon accounting covering approximately 70% of global CO_2 emissions.

BDEAM utilizes detailed energy and water consumption data made available through the proliferation of BEMS and industry standards established through regulations such as the Building Energy Efficiency Act, enabling precise benchmarking and diagnostics by space use and building equipment. This clarifies inefficient equipment and spaces that could not be identified through conventional energy audits treating buildings as single entities, supporting the selection of effective improvement measures.

The core of both methods lies in their systematic integration through the resource-use intensity database. This database functions as resource consumption intensity in SPBCA and as benchmarking criteria in BDEAM, enabling a cyclical process where detailed diagnostic results from building samples are reflected in estimates of carbon reduction effects for entire cities. Furthermore, expansion to evaluate area-wide energy systems such as self-consignment and district heating and cooling comes into view.

However, to fully realize this integration, establishing a system to centrally aggregate and analyze BEMS data currently distributed across individual servers is essential. Since individual developers face limitations in data collection capacity, development of high-granularity resource-use intensity databases by organizations with public character is required. Only when this data infrastructure is established will it become possible to develop and implement evidence-based decarbonization strategies that span from buildings to cities.

Acknowledgments This study is done as part of The Moveable NEXUS: Design-Led Urban Food, Water and Energy Management Innovation in New Boundary Conditions of Change (No. 11314551), a Belmont Forum grant project under the Collaborative Research Area on Sustainable Urbanization Global Initiative (SUGI)/Food–Water–Energy Nexus.

References

Ahn J, Cho S, Chung DH (2016) Development of a statistical analysis model to benchmark the energy use intensity of subway stations. Appl Energy 179:488–496

Agency for Natural Resources and Energy (2022) General energy statistics. https://www.e-stat.go.jp/en/statistics/00551010

Atiba OE, Efemwenkiekie UK, Olatunji RO, Ohunakin OS, Adelekan DS, Oyeleke OA (2019) Walk-through energy audit of an institutional building. J Phys Conf Ser 1378:032051

Campbell BM, Beare DJ, Bennett EM, Hall-Spencer JM, Ingram JSI, Jaramillo F, Ortiz R, Ramankutty N, Sayer JA, Shindell D (2017) Agriculture production as a major driver of the earth system exceeding planetary boundaries. Ecol Soc 22(4). https://www.jstor.org/stable

Chavez A, Ramaswami A (2013) Articulating a trans-boundary infrastructure supply chain greenhouse gas emission footprint for cities: mathematical relationships and policy relevance. Energy Policy 54(March):376–384

Chavez A, Ramaswami A, Nath D, Guru R, Kumar E (2012) Implementing trans-boundary infrastructure-based greenhouse gas accounting for Delhi, India. J Ind Ecol 16(6):814–828

Chen G, Shan Y, Yuanchao H, Tong K, Wiedmann T, Ramaswami A, Guan D, Shi L, Wang Y (2019) Review on city-level carbon accounting. Environ Sci Technol 53(10):5545–5558

Chen S, Long H, Chen B, Feng K, Hubacek K (2020) Urban carbon footprints across scale: important considerations for choosing system boundaries. Appl Energy 259(February):114201

Cohen E, Ramaswami A (2014) The water withdrawal footprint of energy supply to cities. J Ind Ecol 18(1):26–39

Deru MP, Deru MP, Kelsey J (2011) Procedures for commercial building energy audits. American Society of Heating, Refrigerating, and Air-Conditioning Engineers

Feng K, Davis SJ, Sun L, Li X, Guan D, Liu W, Liu Z, Hubacek K (2013) Outsourcing CO2 within China. Proc Natl Acad Sci USA 110(28):11654–11659

Fujita A, Nakayama S, Yan W (2022) Estimation of spatial distribution of CO_2 emissions in urban water utilities: a case study of Tokyo and Kanagawa prefecture

Hadjikakou M, Miller G, Chenoweth J, Druckman A, Zoumides C (2015) A comprehensive framework for comparing water use intensity across different tourist types. J Sustain Tour 23:1445–1467

Hillman T, Ramaswami A (2010) Greenhouse gas emission footprints and energy use benchmarks for eight U.S. cities. Environ Sci Technol 44(6):1902–1910

Hong T, Yang L, Hill D, Feng W (2014) Data and analytics to inform energy retrofit of high performance buildings. Appl Energy 126:90–106

Hoff H (2011) Understanding the Nexus

Hu X, Yan W (2024) Quantifying the impact of earthquakes on urban energy consumption and carbon emissions in Tokyo from a Nexus perspective. Environ Res Commun 6(7):075020

Jones CM, Kammen DM (2011) Quantifying carbon footprint reduction opportunities for U.S. households and communities. Environ Sci Technol 45(9):4088–4095

Keairns DL, Darton RC, Irabien A (2016) The energy-water-food Nexus. Annu Rev Chem Biomol Eng 7(June):239–262

Kluczek A, Olszewski P (2017) Energy audits in industrial processes. J Clean Prod 142:3437–3453

Kumar D, Alam M, Zou PXW, Sanjayan JG, Memon RA (2020) Comparative analysis of building insulation material properties and performance. Renew Sustain Energy Rev 131:110038

Langevin J, Reyna JL, Ebrahimigharehbaghi S, Sandberg N, Fennell P, Nägeli C, Laverge J et al (2020) Developing a common approach for classifying building stock energy models. Renew Sust Energ Rev 133(November):110276

Liddiard R (2014) Room-scale profiles of space use and electricity consumption in non-domestic buildings. Build Res Inf 42:72–94

Lin J, Hu Y, Cui S, Kang J, Ramaswami A (2015) Tracking urban carbon footprints from production and consumption perspectives. Environ Res Lett 10(5):054001

Ma J, Cheng JCP (2016) Estimation of the building energy use intensity in the urban scale by integrating GIS and big data technology. Appl Energy 183:182–192

Masoso OT, Grobler LJ (2010) The dark side of occupants' behaviour on building energy use. Energy Build 42:173–177

Merabtine A, Maalouf C, Al Waheed Hawila A, Martaj N, Polidori G (2018) Building energy audit, thermal comfort, and IAQ assessment of a school building: a case study. Build Environ 145:62–76

Ministry of Economy, Trade and Industry (2015) Long-term energy supply and demand outlook

Nakayama S, Yan W, Fujita A (2024) Developing an energy audit methodology for assessing decarbonization potential in high performance buildings. Energy Convers Manag: X 24

Nakayama S, Yan W (2025) Unravelling food carbon footprint heterogeneity in metropolitan areas using Tokyo as a case study. Sustain Cities Soc 106236

Oda H, Uchiyam A Y (2008) Analysis of energy and environmental burdens on food sector with input-output tables. Input-Output Anal 16(2):27–41

Ramaswami A, Chavez A (2013) What metrics best reflect the energy and carbon intensity of cities? Insights from theory and modeling of 20 US cities. Environ Res Lett 8(3):035011

Ramaswami A, Hillman T, Janson B, Reiner M, Thomas G (2008) A demand-centered, hybrid life-cycle methodology for city-scale greenhouse gas inventories. Environ Sci Technol 42(17):6455–6461

Ramaswami A, Russell AG, Culligan PJ, Sharma KR, Kumar E (2016) Meta-principles for developing smart, sustainable, and healthy cities. Science 352(6288):940–943

Ramaswami A, Boyer D, Nagpure AS, Fang A, Bogra S, Bakshi B, Cohen E, Rao-Ghorpade A (2017) An Urban Systems framework to assess the trans-boundary food-energy-water Nexus: implementation in Delhi, India. Environ Res Lett ERL [Web Site] 12(2):025008

Ramponi R, Angelotti A, Blocken B (2014) Energy saving potential of night ventilation: sensitivity to pressure coefficients for different European climates. Appl Energy 123:185–195

Schleich J, Fleiter T (2019) Effectiveness of energy audits in small business organizations. Res Energy Econ 56:59–70

Seto KC, Churkina G, Hsu A, Keller M, Newman PWG, Qin B, Ramaswami A (2021) From low- to net-zero carbon cities: the next global agenda. Annu Rev Environ Resour 46(1):377–415

Song L, Cai H, Zhu T (2021) Large-scale microanalysis of U.S. household food carbon footprints and reduction potentials. Environ Sci Technol 55(22):15323–15332

Takaguchi H, Izutsu S, Washiya S, Kametani S, Hanzawa H, Yoshino H, Asano Y et al (2012) Development and analysis of DECC (data-base for energy consumption of commercial building). J Environ Eng 77(678):699–705

ten Caat PN, Tenpierik MJ, Sanyal T, Tillie NMJD, van den Dobbelsteen AAJF, Thün G, Cullen S, Nakayama S, Karanisa T, Monti S (2022) Towards fossil free cities—emission assessment of food and resources consumption with the FEWprint carbon accounting platform. Clean Environ Syst 4(March):100074

U.S. Energy Information Administration (2019) Annual Energy Outlook 2014 with projections to 2040

Valencia A, Hossain MU, Chang N-B (2022) Building energy retrofit simulation for exploring decarbonization pathways in a community-scale food-energy-water-waste nexus. Sustain Cities Soc 87(December):104173

Vermeulen SJ, Campbell BM, Ingram JSI (2012) Climate change and food systems. Annu Rev Environ Resour 37:195–222

Yamaguchi Y, Kim B, Kitamura T, Akizawa K, Chen H, Shimoda Y (2022) Building stock energy modeling considering building system composition and Long-term change for climate change mitigation of commercial building stocks. Appl Energy 306(January):117907

Yan W, Roggema R (2019) Developing a design-led approach for the food-energy-water nexus in cities. Urban Plan 4(1):123–138

Chapter 10
Nexus of Urban Mobility and Food Access in Ageing Society

Takahiro Kanamori, Ryoko Okumura, and Wanglin Yan

Abstract Ensuring physical access to groceries is a prerequisite for residents to continue living in a given area. This issue is especially critical for people without a driver's license, such as many older adults, whose mobility options are often limited. In Japan's super-aged society, such problems are now emerging even in suburban areas near Tokyo—some of which are characterized by steep topography. This chapter proposes a method for evaluating physical accessibility to grocery stores, taking into account differences in age, geographical conditions, and mobility. Using this approach, we examine food-access challenges in the suburban Tokyo region, where demographic aging is rapidly advancing. The discussion contributes to the achievement of the Sustainable Development Goals (SDGs), which aim to "leave no one behind."

Keywords Food access · Ageing society · Walking accessibility · Ensure access for all

T. Kanamori (✉)
SFC Research Institute, Keio University/PASCO Corporation, Fujisawa, Kanagawa, Japan
e-mail: tiarko7548@pasco.co.jp

R. Okumura
Tokyu Research Institute, Shibuya, Tokyo, Japan
e-mail: ryoko-o@triinc.co.jp

W. Yan
Faculty of Environment and Information Studies/Graduate School of Media and Governance, Keio University, Fujisawa, Kanagawa, Japan
e-mail: w_yan@keio.jp

W. Yan (ed.), *Urban Nexus Approach for Sustainable Cities*,
https://doi.org/10.1007/978-981-95-6156-8_10

1 Introduction

Our daily lives are supported by access to a wide range of services. These include infrastructure services such as gas, electricity, and water-and-sewerage services; services that provide daily necessities such as groceries and clothing; educational services such as elementary and junior high schools; and medical services such as hospitals and pharmacies. In Japan, for example, the water-supply coverage rate is 98% (Ministry of Land, Infrastructure, Transport and Tourism 2023), the electricity coverage rate is 100% (World Bank Open Data 2023), and the enrollment rate for compulsory education is 99.9% (Ministry of Education, Culture, Sports, Science and Technology 2024), ensuring that virtually everyone can access these infrastructure and educational services. The ways in which these services are provided vary: some, such as infrastructure utilities, are supplied directly to the home, while others, such as groceries, require visiting a retail outlet. Public-transport services—buses and railways—are also provided so that people can reach these services physically.

As a result, our lives are sustained only when physical access to these services is secured. In other words, no matter how excellent a service may be, its benefits cannot be realized if people cannot reach it. In particular, if access to fundamental services such as food, water, and energy becomes difficult, daily life itself may be jeopardized. A study of hilly and mountainous areas has likewise shown that the adequacy of shopping services and the availability of an automobile are decisive factors in residents' willingness to remain in their community (Arikawa et al. 2009).

In short, access is the key that activates the spatial functioning of the mutually dependent relationship among essential resources—food, water, and energy (the FEW nexus). Nexus thinking calls for viewing the production, supply, and consumption processes cross-sectoral manner and evaluating both trade-offs and synergies so as to raise the efficiency of the entire system. The quality of access that links supply with consumption therefore plays a pivotal role in determining residents' quality of life, which is the very core of the nexus. Measuring and improving access with precision is thus indispensable if the FEW nexus is to be designed and managed in a sustainable way.

Meanwhile, cities are home to diverse populations, and constraints related to physical ability or available transportation can create significant disparities in service access. Many older adults, for example, find it difficult to drive, making it harder for them to reach everyday services; this can threaten the sustainability of their daily lives and reduce their quality of life (Hasegawa and Suzuki 2011). Notably, frequent access to grocery stores is indispensable for maintaining day-to-day living. Moreover, even for those who are able to drive, promoting automobile use poses challenges from an environmental perspective. Compared with travel modes such as walking or public transportation, automobiles produce significantly higher per-capita CO_2 emissions, making access an important factor in the pursuit of decarbonization.

To achieve the Sustainable Development Goals' pledge to "leave no one behind," it is essential to address the access issues faced by older adults and secure accessibility, the ease with which necessary services can be reached. In this chapter, we focus on physical access and present a method for evaluating accessibility to grocery stores that takes into account differences in age and geographical conditions. We also address CO_2 emissions resulting from variations in mobility as one of the challenges arising in areas where access is limited. Finally, we examine the issue of food access in the suburban Tokyo area in the context of an aging society.

1.1 Urban Growth and Suburbanization of Housing

Japan is an island nation ringed by the sea and studded with mountains; about 67% of its land area is covered by forest (Forestry Agency 2023). Because of this topography, human activities have long been concentrated in limited pockets of flat land, where settlements and urban districts took shape. Many prefectural capitals originated before the Edo period (1603–) as castle towns, agricultural market towns, temple-and-shrine towns, or post towns and then continued to develop (National Institute for Land and Infrastructure Management 2013). These historical cities were relatively compact, with homes and shops located within easy walking distance.

From the start of the Meiji era (1868–), however, urban landscapes changed dramatically. Fires and wartime bombing destroyed wide swathes of city fabric, while the haibutsu-kishaku movement led to the demolition of castles and temple-shrine complexes. Rapid population growth further reshaped cities. Tokyo's population, for example, jumped from 1.12 million in 1886 to 2.35 million in 1918, and Osaka's from 0.36 million to 1.64 million over the same period (Toyo Keizai Shinpo-sha 1927). In Yokohama, a horse-carriage road opened in 1867, followed in 1872 by the first rail link to Tokyo (City Planning Institute of Japan 2009). Transportation thus shifted from a pedestrian focus to include horse-drawn vehicles and railways.

The 1950s ushered in rapid economic expansion, nationwide population growth, and a pronounced drift toward the big cities (Fig. 10.1). In Tokyo and other major metropolitan areas, soaring demand for housing led to the conversion of farmland and hilly terrain on the urban fringe into residential districts capable of absorbing large numbers of new households. Residential development proceeded mainly through (1) small-scale private subdivisions, (2) large-scale new-town projects led by national or local governments, (3) trackside developments by private railway companies, and (4) land-readjustment schemes and projects by public housing agencies. Although some of these projects were carefully planned to integrate housing with basic services, many areas experienced unregulated urban sprawl.

During the same period, worldwide motorization took off, and Japan saw a surge in private car ownership. Automobiles accelerated suburbanization, encouraged the spread of large-scale commercial facilities into the suburbs, and widened people's daily activity spheres. While dense rail and bus networks were established in core areas such as Tokyo and Yokohama, gaps in public-transport coverage appeared on the periphery, leaving many suburban residents dependent on cars to reach everyday services.

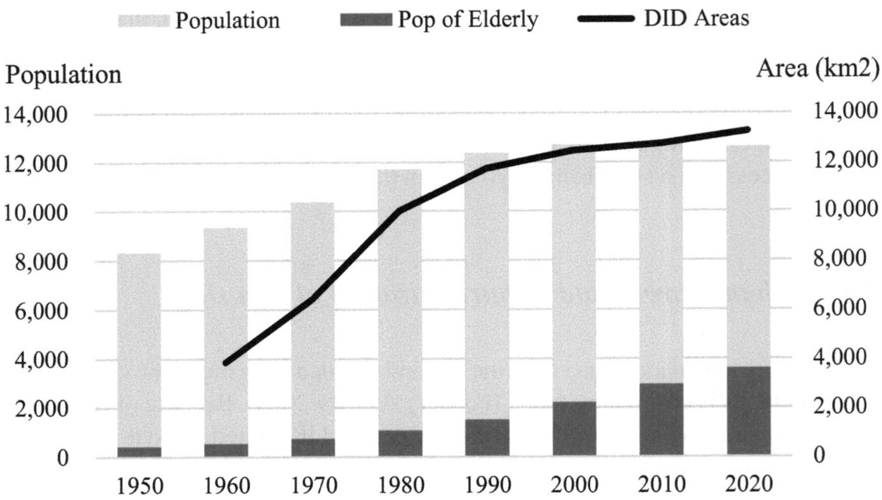

Fig. 10.1 Trends in Japan's population and the area of densely inhabited districts (DIDs), based on the national census

1.2 Urban Structure in the Tokyo Suburban Region

In the suburban belt around Tokyo, roads and rail lines radiate outward from the city center, and urban development is strongly concentrated around railway stations. These station-centered spaces can be divided into five zones—Station-Interior, Station-Front, Near-Station area, Far-Station area, and Remote-Station area (Fig. 10.2)—each with its own characteristic functions.

A. *Station-Interior (Eki-naka)*—Spaces inside the ticket gates or directly integrated with the station structure. Most facilities are developed by the railway operators and include convenience stores, grocery stores, and other outlets for convenience and comparison goods; many stations also host full-scale shopping complexes.
B. *Station-Front (Eki-mae)*—The interface between the railway operator's premises and the surrounding community. It typically contains bus terminals, commercial buildings, everyday-service facilities, and offices. Land prices here are often the highest within the station's sphere of influence.
C. *Near-Station area (Eki-chika)*—Neighborhoods within comfortable walking distance of the station, where a relatively large number of commercial and daily-life services are clustered.
D. *Far-Station area (Eki-en)*—Areas that remain within the station catchment but are hard to reach on foot; they are predominantly residential. Small outlets for convenience goods are scattered throughout the neighborhood, and shops selling comparison goods sometimes line the major roads.

Fig. 10.2 Typical urban structure in the Tokyo suburban region

E. *Remote-Station area (Eki-oku)*—The outer fringe of the station sphere, located farthest from the station. In addition to housing, these districts often contain woodlands or farmland.

Because essential services cluster around stations, the Far-Station and Remote-Station areas often lack adequate service provision. Route buses therefore connect these outlying residential areas to the Station-Front, forming a daily-mobility ecosystem centered on the station.

1.3 The Onset of Population Decline and Super-Aged Society

Societal changes—population growth, economic development, and motorization—fueled rapid urban expansion and pushed housing ever farther into the suburbs. This pattern is common worldwide, but every wave of growth eventually hits a limit. In Japan, the limits have become clear: the country has entered an era of population decline and a super-aged society.

Japan's population peaked at 128.08 million in 2008 (Statistics Bureau, Ministry of Internal Affairs and Communications 2019). Since then, it has fallen steadily, reaching 123.79 million in October 2024 (Statistics Bureau 2024). While some municipalities in the major metropolitan areas still grow through net in-migration, many regional cities and rural communities are shrinking. The chief driver of this decline is a natural decrease caused by population aging. The share of people aged 65 and older was 22.1% in October 2008 but had risen to 29.3% by October 2024 (Statistics Bureau 2008, 2024), signaling Japan's entry into an unprecedented super-aged society.

1.4 The Emergence of Food Deserts and the Need for Independent Living

A falling regional population means total demand for food declines. At the same time, motorization pushes large-scale commercial facilities to the suburbs and draws shoppers after them. Together, these forces make it more likely that grocers will close their downtown stores.

Population loss and motorization also undercut access to food. Fewer residents and more private cars translate into lower public-transport ridership and wider "service gaps." Yet even as depopulation advances, many people—often older adults in this super-aged society—remain in those underserved areas. Some seniors still drive, but many surrender their licenses as their physical ability wanes, and a good number of Tokyo-area suburban households never own a car at all (National Police Agency 2023). Single-person and elderly-only households without nearby family support are also common, leaving some residents with no one to help them shop.

For people who must rely solely on walking, distance itself becomes a major hurdle. Parts of the Tokyo suburbs are hilly, and walking those slopes is especially taxing for older adults, whose physical capacity is generally lower than that of younger people. Under such conditions, Far-Station and Remote-Station areas that lack both transit service and grocery stores can turn into food deserts—places where buying food is difficult and residents' quality of life declines.

2 Method for Evaluating Walking Accessibility

Understanding walking accessibility for older adults is essential for creating communities where no one is left behind. The issue is how to quantify that accessibility accurately. Japan's *Handbook for Evaluating Urban Structure* (Ministry of Land, Infrastructure, Transport and Tourism 2014) recommends the walkable catchment—a simple circle (e.g., 400 m or 800 m) drawn around a facility—to gauge everyday convenience. In dense city centers with well-connected street networks, this radial "buffer" serves as a quick, practical indicator.

A circular buffer, however, assumes travel "as the crow flies." Block geometry, cul-de-sacs, and other barriers lengthen actual walking paths. A more realistic measure is path distance, which traces the street network that pedestrians are expected to follow. Path distance refines the catchment boundary, yet distance alone still fails to capture two key factors: (1) the added effort of hills, and (2) the differing stamina of younger and older walkers.

To incorporate those factors, this chapter adopts metabolic distance—an index that converts walking effort into energy expenditure—developed by Hara et al. (2009). Energy consumption E is given by.

$$E = \left(\text{RMR} + 1.2\right) \times \text{BMR} \times W \times T \tag{10.1}$$

where RMR is the relative metabolic rate (multiples of basal metabolism), BMR is basal metabolic rate, W is body weight, and T is walking time. When walking speed v = 80 m/min[1], RMR varies with slope s (rise ÷ run) as.

$$\begin{aligned} &\text{when}\, v = 80 \\ &\text{RMR} = 10.0 && (s \le -0.25) \\ &\text{RMR} = -58.07s - 4.52 && (-0.25 \le s \le -0.11) \\ &\text{RMR} = 3.113e^{4.614s} && (-0.11 \le s \le 0.25) \\ &\text{RMR} = 10.0 && (0.25 \le s) \end{aligned} \tag{10.2}$$

Steeper slopes raise *RMR*, while downgrades lower it. Because *BMR* declines with age and older adults usually walk more slowly, the same route requires more energy from a person aged 75 or older than from a younger adult aged 20–29. Parameter settings and map legends are listed in Tables 10.1 and 10.2.

Table 10.1 Parameters

Age group	v (m/min)	BMR (kcal/kg/min)	W (kg)
20–29	66.7	0.01601	56.6
75+	41.8	0.01370	56.32

Table 10.2 Abbreviations and meaning

Variable	Description	Units
E	Energy consumption	kcal
RMR	Relative metabolic rate	–
BMR	Basal metabolic rate per unit weight and time	kcal/kg/min
W	Weight	kg
T	Walking time ($T = l/v$)	min
l	Road distance to the destination facility	m
v	Walking speed	m/min
s	Slope	%

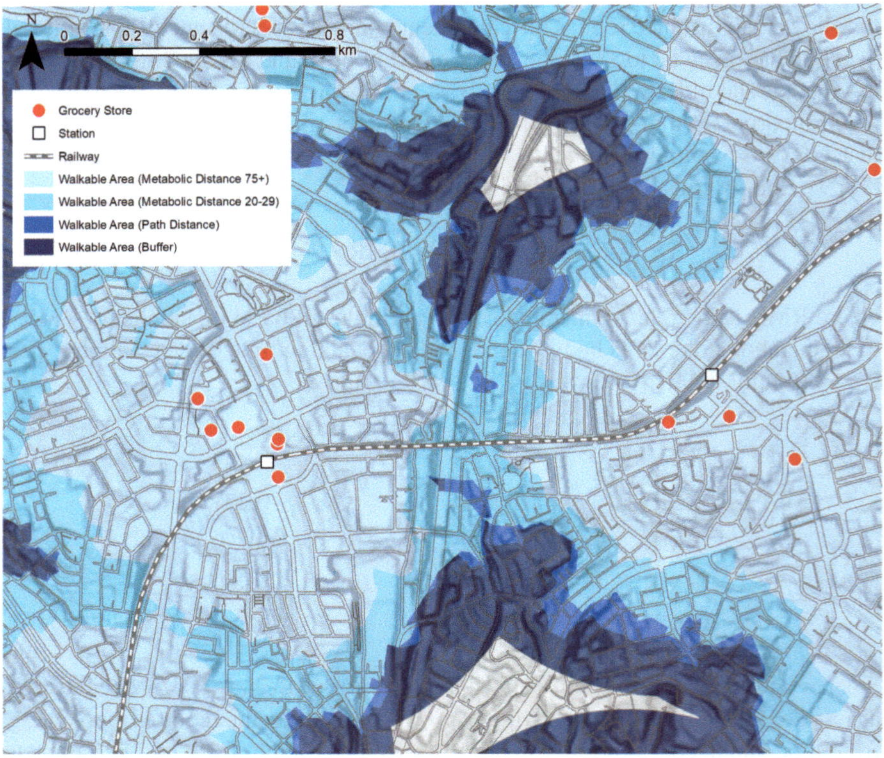

Fig. 10.3 Comparison of 800 m walkable catchment areas by distance metric (buffer, path, metabolic) and age group (younger adults 20–29 years, later-stage older adults 75+ years)
• Digital National Land Information (Railway data), Ministry of Land, Infrastructure, Transport and Tourism, processed by author
• Basic Geospatial Information (Road Edge data; Digital Elevation Model), Geospatial Information Authority of Japan (GSI), processed by author

Figure 10.3 compares three 800-m catchments around a suburban rail station—buffer distance, path distance, and metabolic distance—for younger adults (20–29 years) and later-stage older adults (75 years and over). In districts with sparse streets or steep slopes, the path-distance catchment is narrower than the buffer, and the metabolic-distance catchment is narrower still. These results highlight how terrain and age must be considered when evaluating walkability, especially in hillside suburbs where many older residents already face mobility constraints.

3 Overview of Yokohama's Aoba Ward

In this chapter, Yokohama's Aoba Ward (hereafter "Aoba Ward") is taken as a representative suburb adjoining Tokyo. Aoba Ward covers 35.1 km^2 and has a population of 309,399 (Yokohama City Aoba Ward 2023); 22.9% of residents are aged 65 or older. Much of the ward lies on hilly terrain. Serving largely as a bedroom community for Tokyo, it is traversed by the Tokyu Den-en-toshi Line, which runs straight to central Tokyo, and by the Yokohama Municipal Subway, which provides direct access to Yokohama Station.

Many housing developments date from the opening of the Tokyu Den-en-toshi Line, which triggered hillside construction around each station; as the population grew, extensive residential areas also spread into Far-Station and Remote-Station zones.

The ward's busiest stop is Azamino Station, which handled roughly 190,000 passengers per day in FY 2022 (Tokyu Corporation 2024; Yokohama City 2023). Inside the station (Eki-naka) are eateries and boxed-lunch shops, while the Station-Front (Eki-mae) area has shopping facilities and a bus terminal. Route buses link outlying residential complexes and shopping centers to the station. These buses mainly run along arterial roads or through large housing estates with high day-and-night populations, and stops are spaced about 100–200 m apart.

Grocery stores—a common source of daily food—are heavily concentrated in Station-Front and Near-Station areas, though a few are scattered in Far-Station neighborhoods. Their numbers, however, are far lower than in the station core (Fig. 10.4).

In Aoba Ward, aging ratios are especially high in Far-Station districts, in some cases exceeding 30% (Fig. 10.4). Such neighborhoods tend to have high shares of single or two-person elderly households, many of whom must manage daily life on their own despite living far from grocery stores.

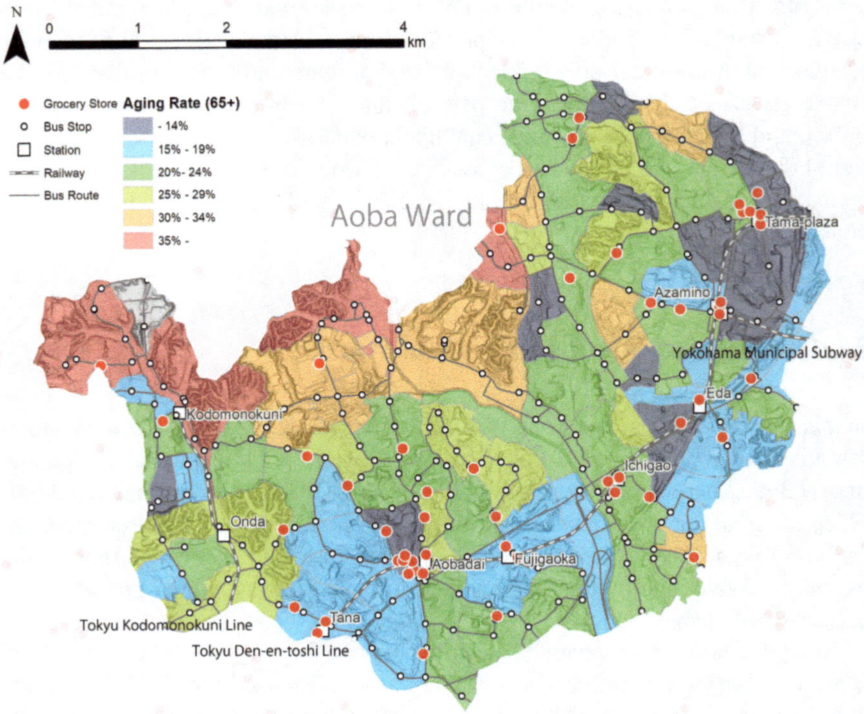

Fig. 10.4 Distribution of public transit stops, grocery stores, and aging rates in Aoba Ward, Yokohama
• Digital National Land Information (Railway, Bus Stop, and Bus Route data), Ministry of Land, Infrastructure, Transport and Tourism, processed by author
• Basic Geospatial Information (Digital Elevation Model), Geospatial Information Authority of Japan (GSI), processed by author
• Government Statistics Portal (e-Stat), Statistics Bureau of Japan, processed by author

4 Who Can—And Cannot—Access Grocery Stores

Building on the above, this section uses hilly Aoba Ward as a case study to reveal—via metabolic distance—the latent hurdles later-stage older adults face when obtaining food on foot. Travel to grocery stores is assumed to occur on foot, while publicly accessible transit is considered as an alternative. Residents who live inside a grocery store's walkable catchment (800 m) are assumed to walk; those outside that catchment but within 800 m of a rail station or 300 m of a bus stop are assumed to use transit.

Figure 10.5 maps walkable areas for younger adults, the reference group. Colored regions denote zones where at least one travel mode (walking, bus, or rail) leads to a grocery store, and each color corresponds to a specific set of mobility options. Dark-pink areas mark locations where the station, bus stop, and grocery store all lie

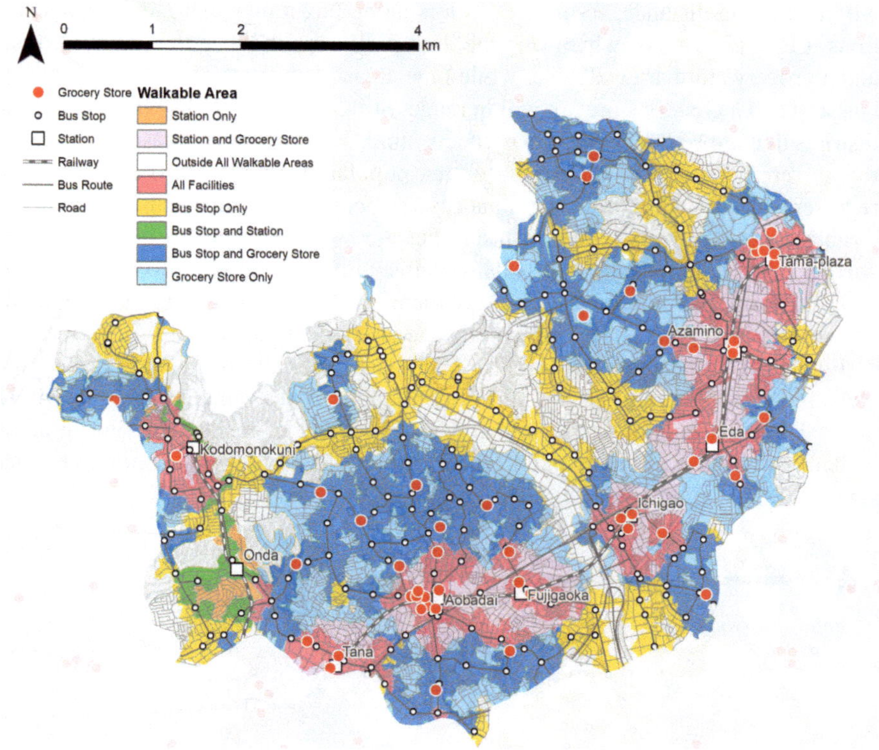

Fig. 10.5 Walkable catchment areas by mobility type for younger adults (20–29 years)
• Digital National Land Information (Railway, Bus Stop, and Bus Route data), Ministry of Land, Infrastructure, Transport and Tourism, processed by author.
• Basic Geospatial Information (Road Edge data; Digital Elevation Model), Geospatial Information Authority of Japan (GSI), processed by author.

Table 10.3 Area and estimated population of walkable catchments by mobility type (younger adults)

Types of walkable area	Area (km^2)	Estimated population	
		Persons	Ratio (%)
Bus stop and grocery store	9.46	10,115	33.4
Outside all walkable areas	7.05	2755	9.1
Grocery store only	5.43	5114	16.9
Bus stop only	5.40	3126	10.3
All facilities	4.42	4950	16.4
Station and grocery store	2.66	3713	12.3
Bus stop and station	0.55	310	1.0
Station only	0.25	176	0.6
Total	35.22	30,259	100.0

within walking distance, forming corridors along bus routes that radiate from stations. Light-pink areas, which ring the dark-pink core, indicate that both a station and a grocery store are walkable, while blue areas identify neighborhoods where a bus stop and a grocery store are within range. Still farther out, light-blue areas show districts that are walkable only to a grocery store. Taken together, these four categories cover roughly 79% of the study area population (Table 10.3) and therefore represent places where transit is unnecessary for grocery shopping. By contrast, sizable orange patches reveal districts that are walkable only to bus stops, and smaller pockets show neighborhoods walkable only to a station or to both a station and a bus stop. Uncolored zones, most of them in Far-Station districts, have no facility within walking distance of any mode and are home to about 9.1% of residents.

Figure 10.6 repeats the analysis for later-stage older adults. In every respect the walkable areas contract. Near stations, contiguous zones remain, but even within about 1 km inaccessible gaps now appear. The share of residents living in areas walkable to both a bus stop and a grocery store falls from 33.4% among younger

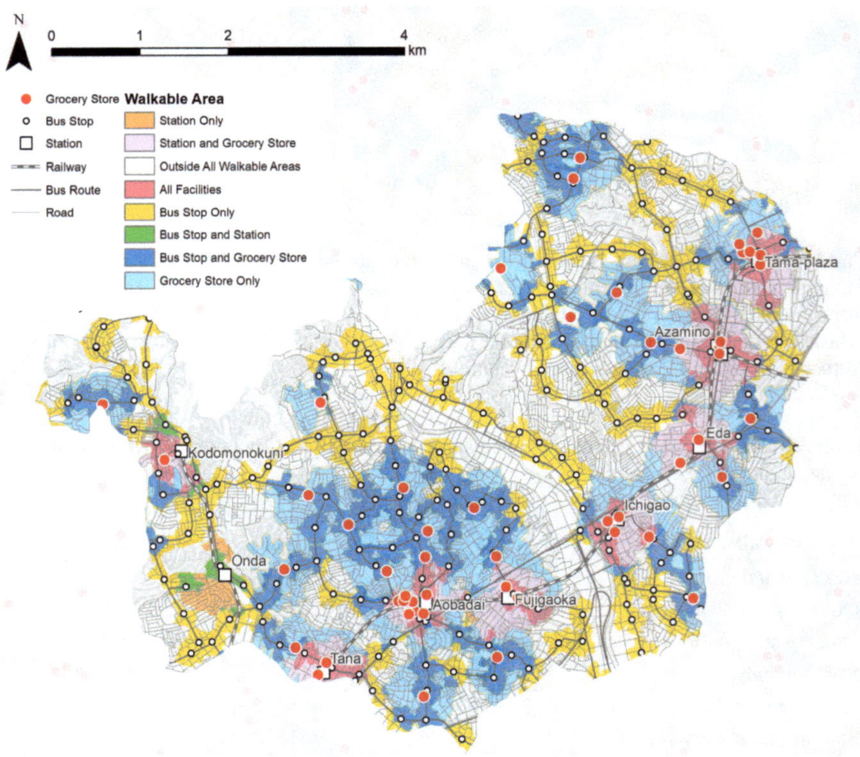

Fig. 10.6 Walkable catchment areas by mobility type for later-stage older adults (75+ years)
• Digital National Land Information (Railway, Bus Stop, and Bus Route data), Ministry of Land, Infrastructure, Transport and Tourism, processed by author.
• Basic Geospatial Information (Road Edge data; Digital Elevation Model), Geospatial Information Authority of Japan (GSI), processed by author.

Table 10.4 Area and estimated population of walkable catchments by mobility type (later-stage older adults)

Types of walkable area	Area (km^2)	Estimated population	
		Persons	Ratio (%)
Outside all walkable areas	13.67	8657	27.5
Bus stop only	6.27	5235	16.6
Grocery store only	5.90	6760	21.5
Bus stop and grocery store	5.20	6783	21.5
Station and grocery store	2.01	2342	7.4
All facilities	1.64	1398	4.4
Station only	0.27	210	0.7
Bus stop and station	0.25	101	0.3
Total	35.22	31,485	100.0

adults to 21.5% among older adults, an 11.9-point drop (Table 10.4). Conversely, the proportion of people living in "no-facility" areas rises to 27.5%, 18.4 points higher than among younger adults. Although overall walkability has shrunk, the fractions of residents who can walk only to a grocery store or only to a bus stop both grow, suggesting that modal choices become more constrained with age. In districts remote from stations, limited walking and transit options are therefore likely to increase reliance on automobiles for everyday shopping.

5 CO_2 Emissions in Areas with Limited Access

The previous section showed that accessibility to grocery stores varies greatly with both age and terrain, with walkable access areas contracting significantly for later-stage older adults. In particular, compared with younger residents, these older adults are more likely to live in districts where reaching a grocery store on foot is difficult. In such areas, where walking or using public transit is impractical, even older adults who can still drive are assumed to rely on private cars. Automobiles are advantageous in that they allow easy travel from remote neighborhoods or across steep topography to reach grocery stores. However, per-capita CO_2 emissions from cars are substantially higher than those from public transport (Ministry of the Environment 2025).

To quantify this difference, we focused on the vicinity of Azamino Station—one of the busiest stations in Aoba Ward—and calculated one-way, per-capita CO_2 emissions for accessing grocery stores, based on the accessibility framework introduced in the previous section. Emission factors were set at 10.4 g CO_2/km for walking, 51 g/km for buses, and 147 g/km for private cars (Forest Agency 2023, 2025). Figure 10.7 presents the results. In neighborhoods close to grocery stores, where walking is possible, CO_2 emissions remain under 5 g per trip, with such areas concentrated relatively near the station. By contrast, areas reachable only by bus show

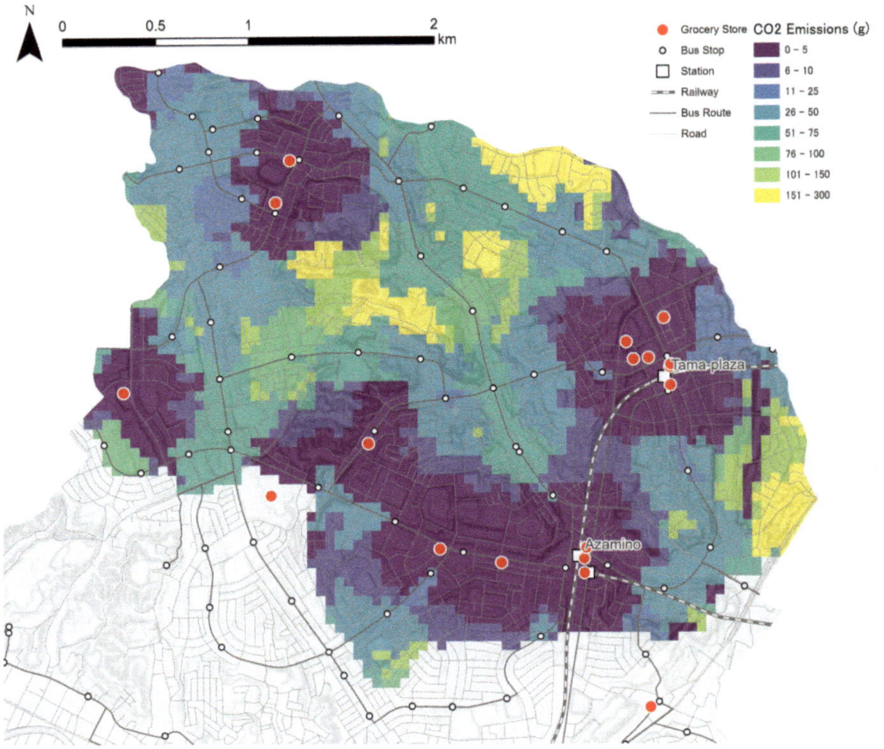

Fig. 10.7 One-way, per-capita CO_2 emissions resulting from access to grocery stores
• Digital National Land Information (Railway, Bus Stop, and Bus Route data), Ministry of Land, Infrastructure, Transport and Tourism, processed by author.
• Basic Geospatial Information (Road Edge data; Digital Elevation Model), Geospatial Information Authority of Japan (GSI), processed by author.

higher emissions, and the most distant bus stops from grocery stores record even greater figures. In districts far from both grocery stores and bus stops—where car travel is assumed—one-way emissions can exceed 150 g.

Assuming a shopping frequency of three round trips per week, total annual emissions would amount to 1.56 kg for walking (10 g per round trip), 15.6 kg for buses (100 g per round trip), and 46.8 kg for cars (300 g per round trip). A single cedar tree absorbs approximately 8.8 kg of CO_2 per year (Forestry Agency 2023, 2025). On this basis, the emissions from 1 year of shopping trips would require the absorption capacity of 0.18 cedar trees for walking, 1.77 trees for bus travel, and 5.32 trees for car use. In other words, the difference in annual CO_2 emissions between living in a walkable location and living in a car-dependent one can be on the order of 30-fold—solely from grocery shopping.

6 Conclusions

In this chapter, we proposed a method for evaluating physical accessibility to grocery stores that explicitly accounts for age, terrain, and available modes of mobility. Using Yokohama's Aoba Ward as a case study, we clarified the food-access challenges facing the suburban Tokyo region in an aging society. Our analysis shows that walkable catchments shrink sharply with steeper topography and advancing age and that mobility options likewise narrow. Consequently, a nontrivial share of residents—particularly those in Far-Station districts—may be unable to reach a nearby grocery store solely by walking or by public transit.

Far-Station and Remote-Station neighborhoods often appeal as housing choices because land prices are lower and the surroundings quieter than in Station-Front or Near-Station areas. From a food-access standpoint, however, longer travel distances increase physical strain and frequently force residents to rely on alternative modes of travel. In low-density districts, maintaining mass-transit services such as buses is difficult, so residents are often pushed toward private cars. Although cars are convenient for people with sufficient physical and financial resources, heavier reliance on them raises per-capita environmental burdens, and driving is not feasible for individuals with physical or economic constraints. As a result, vulnerable groups—especially older adults, whose walkable range is further curtailed by age and hilly terrain—face elevated risks of food insecurity in Far-Station and Remote-Station areas. Food-access problems of this kind have already been flagged in smaller regional cities that are losing population, and they are likely to become more visible in the Tokyo metropolitan area and its suburbs as depopulation and aging deepen.

The national government warns that, under population decline and super-aging, urban sprawl into Far- and Remote-Station zones undermines the sustainable provision of core urban services such as public transit and daily-life facilities. It therefore promotes the "Compact + Network" strategy, which steers housing and services toward core areas—such as Near-Station districts—while connecting those nodes by public transport. Achieving a society in which no one is left behind requires not only residential consolidation and stronger transit networks but also mechanisms that let people already living in Far- and Remote-Station areas, including older adults, obtain food services safely and reliably. To fulfill the Bonn Nexus Conference's call to "ensure access for all" (Hoff 2011), solutions must work within existing settlement patterns. The metabolic-distance approach presented here is a useful tool for visualizing where the gaps lie.

Going forward, coordinated action among local governments, transit operators, community groups, and retail and logistics firms will be essential to expand mobility assistance, attract grocery outlets, and enhance delivery services tailored to local conditions. Such initiatives should yield effective measures for resolving food-access problems in a super-aged society.

Acknowledgments This chapter draws in part on the outcomes of the joint research project "Pathways to Decarbonized Community Development along Private Railway Lines" conducted by Tokyu Research Institute, Inc. and Keio University. The author would like to express sincere gratitude to all members involved in this collaborative research.

References

Arikawa T, Tsukai M, Kuwano M, Fujiyama K, Yamada K (2009) An analysis on the willingness of settling by the convenience of daily activity in middle-mountainous area. Infrastruct Plan Rev 26:383–391

City Planning Institute of Japan (2009) History of the modern city planning system

Forestry Agency (2023) Forest rate and artificial forest rate by prefecture in 2022

Forestry Agency (2025) How much CO_2 do forests absorb?. https://www.rinya.maff.go.jp/j/sin_riyou/ondanka/20141113_topics2_2.html. Accessed 31 July 2025

Hara T, Ishizaka K, Ohashi Y (2009) Evaluation of residential area from a point of elderly people's walking accessibility in local center city. J Architect Plan 74(635):129–135

Hasegawa N, Suzuki H (2011) A study on the situation of living consideration in detached housing estates by aging and location. J Architect Plan 76(663):965–971

Hoff H (2011) Understanding the Nexus: background paper for the Bonn2011 Nexus Conference. Stockholm Environment Institute. https://www.sei.org/publications/understanding-the-nexus/. Accessed 31 July 2025

Ministry of Education, Culture, Sports, Science and Technology (2024) Statistical abstract of education, culture, sports, science and technology (FY2023)

Ministry of Internal Affairs and Communications, Statistics Bureau (2008) Estimated population as of October 1, 2008

Ministry of Internal Affairs and Communications, Statistics Bureau (2019) Statistical Topics No.119: Statistics narrate the path of the Heisei era

Ministry of Internal Affairs and Communications, Statistics Bureau (2024) Population estimates, October 2024

Ministry of Land, Infrastructure, Transport and Tourism. (2014). Handbook on urban structure evaluation

Ministry of Land, Infrastructure, Transport and Tourism. (2023). Basic statistics of water supply—FY2022 water supply population and coverage rate

Ministry of the Environment (2025) Data on current emissions and related data in the transport sector. https://www.env.go.jp/content/900445318.pdf. Accessed 31 July 2025

National Institute for Land and Infrastructure Management (2013) Guidelines for historic town development (Draft)

National Police Agency (2023) Driver's license statistics for FY2022

Tokyu Corporation (2024) Number of passengers in FY2023

Toyo Keizai Shinpo-sha (ed) (1927) Comprehensive national census overview: Meiji and Taisho eras

World Bank (2023) Access to electricity. World Bank Open Data. https://data.worldbank.org/indicator/EG.ELC.ACCS.ZS

Yokohama City (2023) Yokohama City statistical yearbook

Yokohama City Aoba Ward (2023). Naruhodo Aoba

Part IV
Nexus Applications in Urban Planning and Environmental Actions

Chapter 11
Urban Nexus Approach to TOD: Enhancing Resource Integration and Compact City Development

Weiyao Yang and Wanglin Yan

Abstract Transit-oriented development (TOD) has become a key pathway toward sustainable urban transformation by enhancing resource efficiency and promoting compact city design. High-density development around transit hubs not only reduces automobile dependence and optimizes land and energy use but also facilitates the coordinated management of essential resources—such as water, electricity, and land—within the framework of the Urban Nexus, thereby alleviating ecological stress and reducing carbon emissions. Furthermore, TOD increasingly reflects an integrated environmental dimension that bridges the ecological (natural system resilience) and carbon (human-induced pressure) aspects into a unified "environment" perspective. This chapter examines the interconnections among compact cities, TOD, and urban ecology by proposing two analytical frameworks—the Node–Place–Ecology (NPE) and Node–Place–Carbon (NPC) models—and further extends them to an environment-centered evaluation paradigm. Using empirical evidence from the Tokyo Metropolitan Area, the study reveals multi-scalar sustainability transitions across urban cores and suburban contexts. The results demonstrate that incorporating the environment dimension helps achieve a dynamic balance between transportation, land use, and resource systems, offering new analytical insights and policy implications for developing compact, low-carbon, and resilient metropolitan regions.

Keywords Compact cities · Transit-oriented development · Node–Place–Ecology model · Metropolitan areas

---

W. Yang (✉)
School of Urban Planning and Design, Peking University, Shenzhen, China
e-mail: weiyao.yang@pku.edu.cn

W. Yan
Faculty of Environment and Information Studies/Graduate School of Media and Governance, Keio University, Fujisawa, Kanagawa, Japan
e-mail: w_yan@keio.jp

W. Yan (ed.), *Urban Nexus Approach for Sustainable Cities*,
https://doi.org/10.1007/978-981-95-6156-8_11

1 Introduction

1.1 What Is Compact Cities?

The concept of the compact city emerged in the mid-to-late twentieth century as a response to the growing problems of urban sprawl, which produced inefficient land use, social segregation, and environmental degradation. During the mid-twentieth century, suburbanization was often celebrated as a symbol of modern living, yet it also led to fragmented communities and the erosion of urban vitality. In her seminal work *The Death and Life of Great American Cities* (originally published in 1961), Jane Jacobs (1992) offered a profound critique of this trend. She condemned the "anti-city" ideology promoted by Garden City proponents and Decentrists, who encouraged residents to relocate to low-density suburban areas. According to Jacobs, such planning ideals—though intended to improve living conditions—undermined the complexity, diversity, and human interactions that sustain urban life. She argued that thriving cities depend on density, mixed land uses, and the coexistence of different activities within walkable neighborhoods rather than on spatial separation between living and working zones. Her arguments directly inspired later generations of planners and scholars to reexamine urban density and to explore whether compact urban forms could serve as a more sustainable alternative to uncontrolled urban expansion. Since the beginning of the twenty-first century, the compact city theory has evolved from a conceptual reaction against urban sprawl into a multidimensional framework for sustainable urban transformation. Scholars have progressively shifted from normative advocacy toward empirical evaluations that examine the social, environmental, and economic outcomes of compact urban forms. The framework is now widely regarded as a methodological foundation for measuring urban sustainability and guiding city redevelopment policies worldwide.

The distinctive features of compact cities lie in their spatial, functional, and social integration. They promote the close concentration of residential, commercial, and public facilities to create efficient urban systems that maximize the use of limited land and energy resources. Compact cities are characterized by convenient public transport networks that encourage walking and cycling, reducing automobile dependence, congestion, and greenhouse gas emissions. At the same time, they retain Howard's (2006) "Garden City" vision by establishing green belts and ecological corridors to contain urban growth and preserve agricultural and natural areas. Compared with sprawling urban forms, compact cities emphasize intensive land use, low-carbon mobility, and mixed-use development while fostering social diversity, inclusiveness, and neighborhood cohesion (Mouratidis 2018; Kjærås 2021).

Contemporary research on compact cities has increasingly focused on operationalizing the concept through quantitative assessment and data-driven analysis. With rapid urbanization and global sustainability agendas advancing, studies now explore the relationships between compact urban structures and community participation, mixed land-use efficiency, transport network performance, carbon-reduction strategies, and the emergence of sharing-economy practices. These efforts have transformed the compact city from an abstract planning philosophy into a practical and policy-oriented analytical framework, positioning it at the forefront of modern urban planning and sustainable development research (Lee et al. 2015; Kikuchi et al. 2022; Haarstad et al. 2023; Jiang et al. 2025; Xu et al. 2025; Yang et al. 2025).

1.2 Urban Nexus in TOD: A Means of Resource Integration

A key milestone in compact city theory was the empirical exploration of the link between urban density and transportation behavior. Newman and Kenworthy (1989) showed that across cities in North America, Europe, Asia, and Australia, higher population density strongly correlates with lower car dependence and per capita fuel use, confirming that compact forms improve energy efficiency and reduce emissions. Bertaud and Richardson (2004) reinforced this by comparing Barcelona and Atlanta, illustrating how low-density sprawl increases travel distance, fuel consumption, and carbon output, while compact, transit-oriented structures enhance connectivity and sustainability. Collectively, these studies revealed the fundamental density–transport–energy coupling mechanism, laying the empirical groundwork for subsequent sustainable urban planning theories. Calthorpe (1993) proposed the concept of transit-oriented development (TOD), framing transportation systems as the central mechanism for structuring urban form and allocating resources. TOD not only inherits the compact city's emphasis on density, mixed land uses, and walkability but also advances it by positioning public transit as the backbone of urban growth. Over time, TOD evolved beyond the compact city framework, emerging as an independent theoretical and practical paradigm that redefined sustainable urban development through the integration of transportation and spatial planning.

As illustrated in Fig. 11.1, transit-oriented development (TOD) is typically defined as a mixed-use community located within an average walking distance of approximately 2000 feet (around 600 m) from a transit station and its core commercial area. In such compact environments, residential, commercial, office, public, and open-space functions are integrated to promote walking and provide convenient access by public transit, cycling, or car. The core zone should concentrate major commercial and employment activities around the station, supported by adjacent public spaces that maintain neighborhood vitality. Surrounding residential zones are developed with progressively lower densities, forming a smooth urban gradient. Beyond the core, secondary TOD areas may emerge, accommodating low-density housing, parks, schools, and other community facilities. The street network in these peripheral zones should ensure direct, efficient connections to the core area and incorporate park-and-ride facilities to enhance overall accessibility.

Although TOD, while deriving from the compact city theory, is also inspired by earlier urban planning concepts, particularly the Garden City. Howard (2006) attempted to reconcile the harmony between rural and urban areas by proposing a combination of rural environmental quality and comfort with urban work

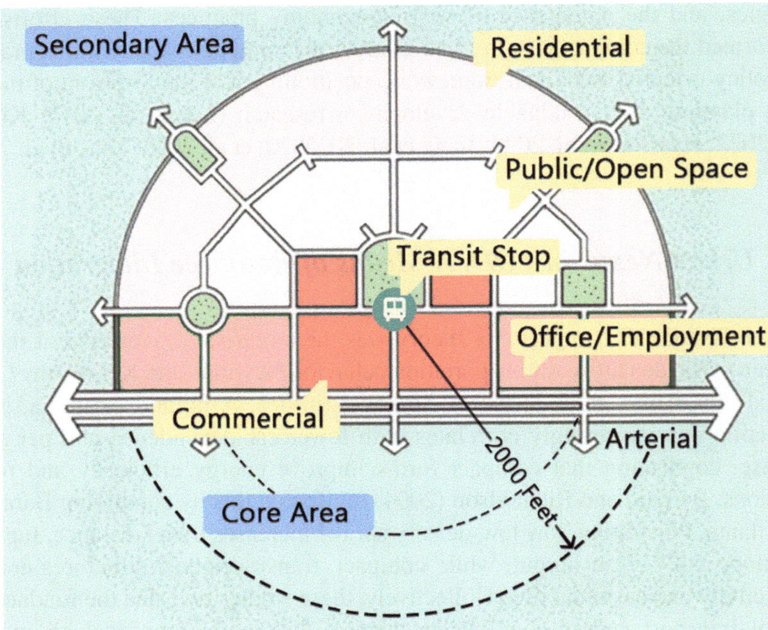

Fig. 11.1 According to the TOD schematic diagram redrawn by Calthorpe (1993)

opportunities and income levels. TOD incorporates this aspect into its concept, reflected in the core area with open spaces (typically referring to urban parks with extensive greenery). This makes the core area different from the bustling, crowded, and chaotic city centers of traditional urban planning, while still being vibrant, lively, and functional. Additionally, the TOD concept's advocacy for dense, compact, and walkable residential areas aligns with the principles of the Garden City concept. However, these two concepts differ in terms of spatial layout for resource integration. First, the primary employment locations in Garden City are situated on the outskirts of urban clusters, while in TOD, employment is concentrated in central areas. This difference stems from the distinct historical contexts in which the two models were developed: the garden city concept was introduced during the late Second Industrial Revolution in Europe, a period of rapid urbanization when large populations were engaged in heavy industry, which needed to be kept at a certain distance from residential areas. And TOD was proposed in North America at the turn of the century, when the tertiary sector was already booming and heavy industry had almost disappeared in many developed cities, especially as we enter the twenty-first century. This makes TOD less of a consideration for the location of heavy industry. Second, in garden cities, life revolves around a central square, typically housing medical, administrative, and cultural facilities, with boulevards radiating outward. In contrast, in TOD, the core is a public transit hub, highlighting the importance of transportation infrastructure in contemporary urban life. This also indicates that TOD, as an efficient means of resource integration, not only involves the dense and well-connected arrangement of residential, commercial, and public facilities around transit hubs but also considers the impact of ecological construction. This layout not only enhances the efficiency of land use, including green spaces, but also promotes the intensification of urban functions, reducing reliance on private vehicles, thereby lowering energy consumption and environmental pollution (Liu et al. 2022; Yang et al. 2024b).

In this context, the concept of Urban Nexus is increasingly embodied within TOD, emphasizing the synergistic integration of key urban systems such as transport, energy, water, and land use. By efficiently clustering housing, employment, public services, and ecological spaces around transit hubs, TOD fosters interconnections among sectors that were traditionally planned in isolation. This integrated approach not only enhances resource-use efficiency but also strengthens the resilience and sustainability of urban systems. The Urban Nexus perspective, which underscores the interdependence of systems, thus serves as a critical framework for TOD to support urban functionality and facilitate low-carbon transitions.

1.3 The Necessity of Environmental Considerations in the TOD

Ecology initially emerged from the study of natural and seminatural environments such as forests, grasslands, farmlands, dunes, lakes, and oceans. In contrast, urban ecology originated from the built environment, which differs fundamentally from natural ecosystems. Early studies often borrowed the analytical framework of natural ecology, paying limited attention to the social and spatial complexity unique to cities. With technological progress and intensified human activities, urban ecology has gradually developed into a distinct branch of ecological science, increasingly emphasizing social attributes (Chan et al. 2017). Today, it not only includes natural elements within cities but also examines land use, built structures, anthropogenic flows, and human decisions and behaviors. This marks a shift from viewing cities as ecological exceptions to recognizing them as integrated socio-ecological systems. The rise of the eco-city and low-carbon city paradigms further links urban ecology with global sustainability goals. These concepts associate ecological thinking with low-carbon lifestyles, energy conservation, emission reduction, biodiversity protection, and climate regulation (de Jong et al. 2015). Thus, contemporary urban ecology bridges environmental and social dimensions, providing a framework for understanding how cities can evolve toward ecological resilience and low-carbon sustainability.

TOD is intrinsically linked to urban ecology, serving as a key mechanism that promotes sustainable resource management and carbon reduction. The concept of Green TOD highlights the integration between TOD and ecological urbanism, emphasizing how transit-oriented planning can foster environmentally friendly urban regeneration and green urbanization. The environmental and social benefits of Green TOD are multifaceted. It enhances public transit use and reduces automobile dependence, improves energy efficiency in heating and cooling systems, encourages waste recycling and resource recovery, and promotes mixed land use and nonmotorized transport, enabling multifunctional urban spaces to operate efficiently. In addition, increasing open spaces and community greenery mitigates the heat-island effect and enhances urban livability. Green TOD also supports the use of renewable energy, for example, by installing photovoltaic systems on station roofs or public buildings to power nearby facilities, or by utilizing renewable energy generated within the built environment to support transportation systems. Despite attempts by subsequent researchers to evaluate Green TOD quantitatively, most analyses still focus predominantly on the physical built environment, overlooking its essential nature as an organic integration of transport, land use, and ecological systems (Cervero and Sullivan 2011).

The objective of this chapter is to highlight the necessity of embedding environmental considerations into TOD within the broader framework of the Urban Nexus. By systematically linking transportation, land use, and the ecological–carbon dimension, the chapter emphasizes how TOD can function as a nexus that coordinates urban resource flows while enhancing resilience and sustainability. Specifically, it seeks to extend the classical Node–Place model with an environmental axis, demonstrate how this nexus perspective addresses the limitations of traditional TOD assessments that focus mainly on transport and land diversity, and provide a framework for evaluating TOD in both urban cores and suburban contexts. This nexus-based approach sets the foundation for the subsequent discussion of the Node–Place–Ecology and Node–Place–Carbon models, as well as their application in case studies of the Tokyo Metropolitan Area.

2 A Framework Based on the Node–Place Model

2.1 Node–Place Model

The Node–Place (NP) model, proposed by Bertolini (1996, 1999) in the 1990s alongside the rise of the transit-oriented development (TOD) concept, is a theoretical framework designed to reveal the interactive relationship between urban transport systems and spatial structures. By linking transport nodes with the diversity of surrounding land uses, the model innovatively explains the dynamic coupling between mobility, land-use functions, and urban form. It quickly gained attention within the academic community and has since become a widely adopted tool for evaluating the development potential and spatial coordination of transit hubs and their catchment areas. Initially, the NP model served primarily as a conceptual paradigm to interpret the balance between transport accessibility and land-use intensity within urban networks. However, with the advancement of data availability and analytical techniques, it has demonstrated remarkable extensibility—integrating multidimensional statistical methods, GIS-based spatial analysis, and multisource data modeling. In recent years, its integration with machine learning, spatial clustering, and transport visualization analytics has further enhanced its quantitative and predictive capabilities. Today, the NP model has evolved from a theoretical construct into a comprehensive analytical instrument used to identify TOD development stages, evaluate Node–Place imbalances, and provide scientific support for optimizing sustainable urban spatial structures.

The NP model is based on a Cartesian coordinate system designed to reveal the coordination between urban transport nodes and place functions. The vertical axis (*Y*-axis) represents the node function, referring to the accessibility and connectivity of a station or transport hub; the horizontal axis (*X*-axis) represents the place function, which reflects the intensity of land use and the diversity of population activities within the station's catchment area. By comparing the relative strengths of these two dimensions, the model intuitively captures the development balance of each station area and classifies them into five typical categories: (1) *Balance Zone*—located along the diagonal, where node and place functions mutually reinforce one another, forming areas with efficient land use, convenient transport, and coordinated development—an ideal state of TOD; (2) *Stress Zone*—situated above the diagonal to the right, typically corresponding to urban core areas where both functions are strong but compete for limited space and resources, often leading to "overconcentration"; (3) *Dependency Zone*—found below and to the left of the diagonal, where both node and place functions are weak, and development relies on external policy intervention or infrastructure investment; (4) *Unsustained Node Zone*—located above the diagonal, where node functions are advanced but place development lags behind, representing areas with "sufficient mobility but insufficient vibrancy"; (5) *Unsustained Place Zone*—located below the diagonal, where place functions dominate but node capacity is limited, resulting in poor accessibility and transportation bottlenecks.

In theory, these categories are not static but can transform dynamically through adjustments in node and place functions. By enhancing public transport accessibility, optimizing land use, or strengthening regional connectivity, different areas can gradually evolve toward the "Balance Zone." The NP model thus serves as a key theoretical framework for assessing the integrated development potential of station areas, guiding spatial renewal strategies, and supporting the phased evolution of TOD.

2.2 *Introducing the Concept of Eco-Cities into the Node–Place Model*

In its initial theoretical framework, transit-oriented development (TOD) and ecology share common interests—both seek to optimize spatial organization and resource utilization to achieve cities that are more sustainable, low-carbon, and livable. TOD was originally regarded as an important pathway to ecological urbanization: through compact, high-density, and mixed-use forms centered on public transport, it reduces energy consumption and carbon emissions while improving urban environmental quality and residents' well-being. However, in global TOD research and practice, mainstream attention has focused primarily on transport efficiency and land-use diversity, with insufficient consideration of the ecological characteristics and ecosystem services of station areas and their surroundings. Even the emerging notions of "Green TO" and "Sustainable TOD" often remain superficial, emphasizing environmental friendliness at the design level. Their indicator systems tend to be homogeneous and fail to capture the complex interactions among ecological resilience, environmental capacity, and social sustainability. Against this backdrop, the concept of the eco-city provides new theoretical insight and methodological support for TOD. Both frameworks share the goal of developing compact, efficient, low-carbon, and livable urban systems that satisfy daily mobility and living needs while minimizing environmental pressure. Unlike traditional TOD, the eco-city perspective highlights the coevolution of natural and human systems, advocating the use of green infrastructure, ecological networks, and renewable energy to achieve balance among social, economic, and environmental dimensions. Incorporating the eco-city concept into the TOD framework thus extends its sustainability vision and enables a more integrated form of urban governance that coordinates transport, land use, and ecological environments (Cervero and Sullivan 2011; Tiwari et al. 2023; Wang et al. 2023).

With the growing awareness of the profound impact of urbanization and industrialization on the environment, and the increasing demand for sustainable urban development models, the concept of the eco-city has gradually emerged and become a central topic in global urban planning. The "Garden City" idea proposed by the British planner Howard (2006) in *Garden Cities of Tomorrow* is often regarded as an early precursor to the eco-city vision. While Howard's proposal primarily sought to integrate the benefits of urban and rural life, it laid the conceptual foundation for achieving social, economic, and environmental balance within modern industrial societies. The notion of the eco-city in its contemporary form took shape in the mid-1970s, following the establishment of the Urban Ecology Group, which aimed to reconstruct cities in harmony with natural systems. In *Ecocity Berkeley: Building Cities for a Healthy Future*, Register (1987) further articulated that an eco-city should operate within environmental limits, ensuring that all urban elements remain within the ecological carrying capacity of the planet. Since then, as challenges such as climate change, excessive energy use, uncontrolled land expansion, and ecological degradation have intensified, an increasing number of scholars have come to view the eco-city as a mainstream pathway toward environmental sustainability and urban transformation (Lepczyk et al. 2017; Wu et al. 2020). More importantly, the eco-city is not merely a planning model but a systemic governance paradigm. It seeks to achieve integrated optimization across multiple urban subsystems—energy, water, land use, and transportation—shifting cities from resource-intensive to ecologically coordinated structures. Its ultimate goal is to establish a "nature-based" urban development mechanism, fostering synergy among ecological restoration, green infrastructure, and social participation, and thereby enabling genuine coexistence and long-term resilience between humans and nature.

Developing an eco-city is an inherently complex and multidisciplinary process that requires a comprehensive understanding of economic, social, cultural, technological, and environmental dimensions. The assessment of an eco-city's performance should therefore adopt a diverse and context-sensitive set of indicators, reflecting both its spatial characteristics and developmental priorities. These indicators may include quantitative measures such as greenhouse gas emissions, renewable energy adoption rates, and the extent of green infrastructure and open space networks, as well as qualitative dimensions related to community participation and institutional capacity. Importantly, no single evaluation framework can be universally applied. Each eco-city must establish a set of locally adapted metrics that correspond to its ecological conditions, governance structures, and policy goals. Such contextualization ensures that eco-city development remains both scientifically measurable and socially meaningful, supporting continuous improvement toward urban sustainability and resilience (Yang et al. 2024c).

2.3 *Expanding the Concept of Sustainability: Inclusion of Low-Carbon Cities*

Yang et al. (2024b) incorporated the concept of eco-cities into the classical Node–Place (NP) model by introducing an additional and independent third dimension—ecology—as a means to mitigate the unsustainable trends resulting from the overheated development of node and place functions. This modification expanded the analytical scope of the NP framework, offering a novel approach to balancing ecological systems and transit-oriented growth, particularly in suburban contexts. However, the model still presents a notable limitation: the added "ecology" dimension was primarily designed for peripheral and suburban areas of large metropolitan regions, raising concerns about its applicability in dense urban cores. In such central districts, the limited availability of ecological resources—such as marine areas, arable land, or forests—combined with high population density and intensive land use, constrains the operational relevance of ecological indicators.

In this regard, the concept of low-carbon cities provides a broader and more adaptable framework for advancing sustainable urban development. Compared with eco-cities, low-carbon cities possess a stronger theoretical foundation and wider applicability, capable of addressing both metropolitan centers and suburban environments. Their primary objective lies in reducing greenhouse gas emissions and energy consumption, thus facilitating cross-scalar sustainability transitions (Xu et al. 2021, 2023; Li et al. 2022; Zhai et al. 2023, 2025). From a governance perspective, the low-carbon city approach emphasizes multi-sectoral coordination and systemic energy planning, achieved by restructuring land-use patterns, energy supply systems, transportation modes, and residential lifestyles. In doing so, it aligns closely with the core vision of TOD—achieving compact, accessible, and energy-efficient urban forms. Furthermore, low-carbon city frameworks have moved beyond theory into practice, with numerous quantifiable and comparable indicator systems developed globally to evaluate urban energy efficiency, carbon reduction, and policy effectiveness (Tan et al. 2017).

Building upon this logic, Yang et al. (2024d) integrated the low-carbon city paradigm into the NP framework, establishing the Node–Place–Carbon (NPC) model within a broader sustainability context. Compared with the earlier Node–Place–Ecology (NPE) model, the NPC framework demonstrates greater methodological adaptability and cross-scalar relevance. It enables systematic evaluation of spatial structure, transport accessibility, and carbon efficiency from city centers to suburban areas, offering a robust analytical tool to examine the dynamic coupling among accessibility, land-use intensity, and carbon performance—ultimately providing new insights for low-carbon transition and spatial governance in metropolitan regions.

2.4 *Node–Place–Ecology (NPE) Model and Node–Place–Carbon (NPC) Model*

While the NP model offers a flexible and widely applicable analytical framework, relying solely on its two traditional dimensions—node and place—has become inadequate for capturing the growing complexity of contemporary urban development. To address this limitation, recent studies have expanded the framework by introducing an additional dimension, giving rise to the Node–Place–"X" paradigm, where "X" represents a newly incorporated variable that enriches spatial interpretation and analytical depth (Yang et al. 2024a). Building on this extension, Yang et al. (2024c, d) proposed two advanced three-dimensional models—the Node–Place–Ecology (NPE) model and the Node–Place–Carbon (NPC) model—each tailored to different research contexts and development phases, as illustrated in Fig. 11.2.

3 Case Studies of the NPE–NPC Model

The NPE–NPC model, as two newly developed TOD evaluation model, is currently applied primarily in the suburban areas of mega-metropolitan areas (with Odawara City as an example) and across research areas spanning from urban centers to suburbs within mega-metropolitan areas (with Tokyo and Kanagawa Prefectures as an example). The two study areas have distinctly different focal points. The study area in suburban cities of mega-metropolitan regions focuses more on the interaction between natural ecological resources and TOD. In contrast, the study area spanning from urban centers to suburbs emphasizes the transition from urban center TOD to

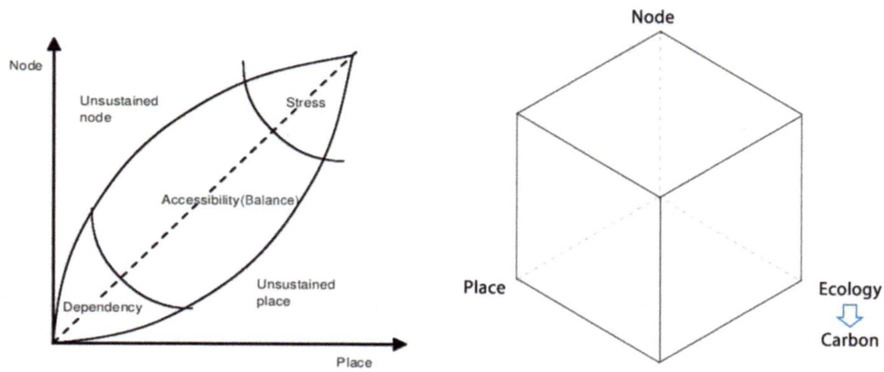

Fig. 11.2 Schematic diagram of NP model and NPE–NPC model. (Yang et al. 2024c, d)

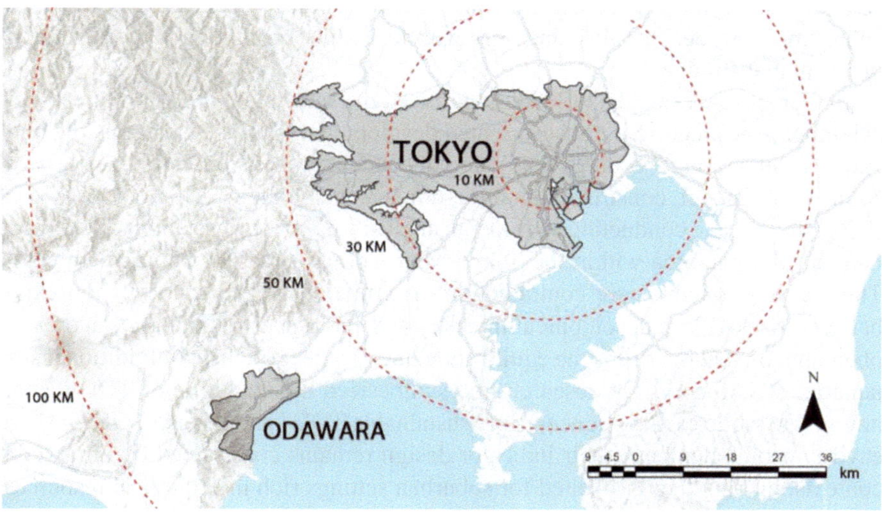

Fig. 11.3 Location of Tokyo and Odawara. (Yang et al. 2024c)

suburban TOD. More critically, Yang et al. (2024d) introduced, for the first time, the concept of a "timeline-based" TOD assessment, highlighting whether the transition from urban center TOD to suburban TOD becomes more sustainable over time.

3.1 *Metropolitan Suburb: A Case in Odawara*

As shown in Fig. 11.3, Odawara City is located approximately 100 km southwest of Tokyo and maintains excellent connectivity to the metropolitan area through an extensive rail and highway network. Due to its convenient accessibility, many residents commute daily to the capital for work, and Odawara has gradually evolved into a typical bedroom community for Tokyo. At the same time, the city benefits from a diverse and favorable natural environment, situated between the mountains and the sea. Its mild climate and abundant rainfall foster rich biodiversity and provide ideal conditions for a wide range of ecological resources to flourish. According to the Odawara City Location Optimization Plan released in 2019, the city is currently facing several structural challenges, including a rapidly aging population, a declining birth rate, and the resulting stagnation of regional industries and decline in urban vitality. Continuous suburban expansion—such as the spread of residential and commercial areas—has led to urban sprawl and low-density development, making it increasingly difficult to provide public services efficiently under constrained financial conditions. In response, city planners proposed a vision for a "compact and networked city," aiming to ensure that residents can conveniently access medical, welfare, and commercial facilities via public transportation, rather than relying heavily on private vehicles. This urban restructuring strategy seeks to concentrate essential daily services and administrative functions near residential areas, thereby enhancing both accessibility and sustainability while promoting a more resilient urban spatial structure.

In their study, Yang et al. (2024c) addressed the systematic omission of ecological dimensions in current transit-oriented development (TOD) evaluations by innovatively integrating "ecology" as an independent dimension into the classical Node–Place model, constructing a three-dimensional Node–Place–Ecology Model (NPE Model), and conducting an in-depth analysis of 18 railway stations in Odawara City, a suburban area within the Tokyo Metropolitan Area, as empirical subjects. The study revealed a core contradiction: a significant negative correlation exists between Node–Place development intensity and ecological value in station areas, a phenomenon that overturns the equilibrium development premise of traditional sustainable TOD theory. The research first synthesized the evolution of TOD theory, noting that while existing concepts of "sustainable TOD" and "Green TOD" involve environmental elements, their indicator design remains constrained by urban-core contexts, making them ill-suited for suburban settings rich in ecological resources. To address this, Yang et al. (2024c) distilled quantifiable indicators from "eco-city" theory (e.g., open green space area, biodiversity support capacity, stormwater infiltration potential), combined them with node dimension metrics (number of metro service directions, daily service frequency, public transport accessibility) and place dimension metrics (population density, job distribution, commercial service density), established an evaluation system comprising 24 indicators, and engaged 15 interdisciplinary experts (spanning landscape planning, environmental economics, ecological science, etc.) to determine weights using an AHP-entropy weight-integrated weighting method, ultimately completing the three-dimensional model construction.

The study indicates that suburban TOD development must transcend the traditional "balanced growth" paradigm and establish dynamic compensation mechanisms—restricting Node–Place expansion in ecologically sensitive zones (e.g., NEB, KOZ) while offsetting development via ecological service functions, and implementing ecological restoration (e.g., vertical greening, ecological corridors) in core development areas like ODA. The model simultaneously provides precise targets for policy intervention: for instance, NEB Station requires enhanced transport connectivity and job creation, while ODA Station urgently needs improved green space connectivity. Three key limitations are noted: data constraints from reliance on open-source databases excluded critical ecological indicators such as noise pollution and building energy consumption; the small sample size (18 stations) hindered capturing micro-fluctuations in ecological parameters; and the absence of a temporal dimension prevented tracking dynamic effects of policy interventions. Future research will expand to multi-suburb comparisons across the Tokyo Metropolitan Area and explore constructing a four-dimensional "Node–Place–Ecology–Time" model. The model's innovation lies in its pioneering deep integration of the ecological dimension with TOD assessment, providing a scientific tool that balances efficiency and ecological resilience for the sustainable development of suburban towns worldwide.

3.2 *From City Center to Suburbs: The Case of the Odakyu Line*

As shown in Fig. 11.4, Tokyo Metropolitan Area (including surrounding areas such as Chiba Prefecture, Kanagawa Prefecture, Saitama Prefecture, and others) is one of the most populous and developed metropolitan regions in the world. It is precisely

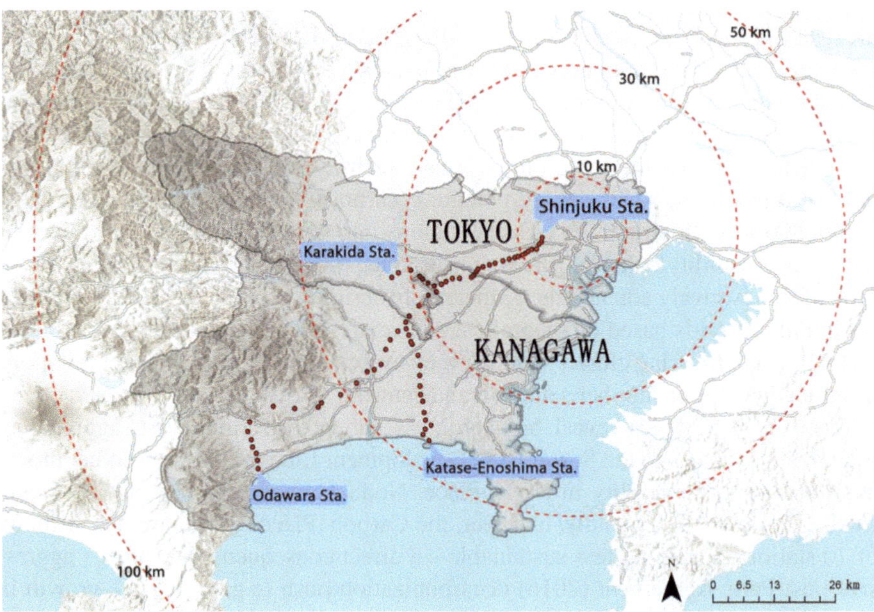

Fig. 11.4 Location map of the 70-station Odakyu Line. (Yang et al. 2024a, b, c, d)

because it has the complexity of a metropolis that it has value as a study area for Yang et al. (2024d). However, at the same time, the complexity of the Tokyo Metropolitan Area is also evident in data collection and processing. Therefore, Yang et al. (2024d) hope to make an attempt starting from a micro-level perspective. To reflect their research purpose, Yang et al. (2024d) will direct their attention to the urban center of Tokyo and the suburbs of Kanagawa Prefecture. Kanagawa and Tokyo are both economic centers in Japan, with many large enterprises, financial institutions, and commercial centers concentrated in these two areas. Thus, there is a close economic connection between them, including business cooperation and talent mobility.

While TOD is globally embraced for promoting sustainable urban growth through integrated transport–land use planning, existing assessments predominantly capture static snapshots, failing to track how station areas evolve toward or away from sustainability over time—particularly under pressures like urban shrinkage. Yang et al. (2024d) work responds to this by developing a novel timeline-based Node–Place–Carbon model, which expands Bertolini's foundational Node–Place framework to incorporate a "carbon" dimension rooted in low-carbon city principles. This triad—node (transport accessibility), place (land use intensity/diversity), and carbon (environmental performance)—forms a 3D analytical axis where sustainability is defined by dynamic equilibrium among dimensions, operationalized through the Sustainable Development Line (SDL), expressed as $x = y = z$ in coordinate space. Deviations from this line signal imbalances, such as over-prioritizing transport nodes at the expense of carbon efficiency.

Yang et al. (2024d) applied this model to 70 stations along Tokyo's Odakyu Line, spanning from the metropolitan core (e.g., Shinjuku) to peripheral Kanagawa suburbs, across an 8-year timeline (2011–2019). This corridor epitomizes Japan's urban trajectory: early TOD adoption, intense reliance on private railways, and contemporary shrinkage driven by aging and population decline. Data were drawn from open-source repositories (e.g., e-Stat, national land databases) and field surveys, covering 18 indicators across the three dimensions—such as transit frequency (node), job density (place), and solar energy potential (carbon)—normalized and aggregated within 800-m station radii. To classify stations into six sustainability zones (dependency, accessibility/balance, stress, unsustained node/place/carbon), Yang et al. (2024d) innovatively adapted the K-means clustering algorithm. By predetermining six initial centroids based on theoretical zone extremes (e.g., unsustained carbon at [0,1,0]) and optimizing cluster stability, they overcame K-means' typical limitations in subjective cluster-number selection and centroid sensitivity.

Yang et al. (2024d) reveal both progress and polarization. First, aggregating Euclidean distances to the Sustainable Development Line (SDL) showed net movement toward sustainability in Node–Place, Node–Carbon, and the full 3D model from 2011 to 2019. Crucially, however, the Carbon–Place dyad regressed, with 35 of 70 stations becoming less sustainable—a direct consequence of Japan's aggressive post-Paris Agreement (2016) decarbonization push (e.g., explosive growth in EV charging stations [C1] and building energy retrofits [C5]), coinciding with Place–Dimension erosion in shrinking suburbs, where population and jobs declined.

Second, K-means results exposed intensifying monopolization by central hubs: Shinjuku Station (the sole persistent "stress" zone occupant) dramatically advanced in all three dimensions, leveraging policy focus and investment to amplify its transit dominance, commercial density, and carbon tech deployment. Conversely, stations such as Shimokitazawa and Machida slipped from "stress" to "accessibility" zones. Shimokitazawa's transport gains (node ↑) via Setagaya City's redevelopment plans were undermined by lagging place/carbon metrics; Machida's place stability could not offset declines in node/carbon amid slow implementation of its 2016 urban renewal strategy. Third, 18 stations remained stubbornly "unsustained carbon" both years. These were primarily suburban areas with abundant, expanding green spaces—a paradoxical artifact of shrinkage reducing human pressure but failing to translate ecological presence into holistic low-carbon performance (e.g., integrating renewables or walkability).

These dynamics underscore urban shrinkage's double-edged nature: while resource concentration in cores like Shinjuku can curb sprawl inefficiencies, it risks marginalizing suburbs where Place–Dimension decay outpaces carbon interventions. Policy alignment is thus critical—Yang et al. (2024d) model empirically links Japan's top-down carbon targets to uneven local outcomes, revealing where integrated interventions (e.g., coupling EV infrastructure with job retention initiatives) are urgent. Methodologically, Yang et al. (2024d) proved the Node–Place-Carbon model's adaptability across the metropolitan gradient, resolving prior ecological–dimension limitations that hindered urban-core applicability. By embedding timeline analysis and SDL quantification, they offer planners a tool to diagnose station-specific sustainability pathways, prioritize corrective actions, and simulate policy impacts dynamically.

Limitations persist. Data scarcity confined our timeline to 2011–2019; longer spans might reveal different trends. Focusing solely on Odakyu Line stations, though illustrative, limits generalizability across Tokyo's complex rail network. Some indicators (e.g., granular energy-use data) were omitted due to accessibility gaps, slightly constraining Carbon–Dimension richness. Nevertheless, this study establishes a replicable framework for global cities navigating transit-oriented sustainability amid growth or decline. Future work will expand spatial coverage, integrate real-time data streams, and refine carbon metrics—empowering cities to balance node, place, and carbon not as competing priorities, but as synergistic pillars of enduring urban resilience.

4 Discussion

The Tokyo Metropolitan Area TOD dynamic assessment study (Yang et al. 2024d) represents a paradigmatic breakthrough beyond the Odawara suburban static model (Yang et al. 2024c) as illustrated in Fig. 11.5. This evolution manifests first through dimensional advancement in model architecture: While the Odawara study pioneered the three-dimensional "Node–Place–Ecology" framework by introducing

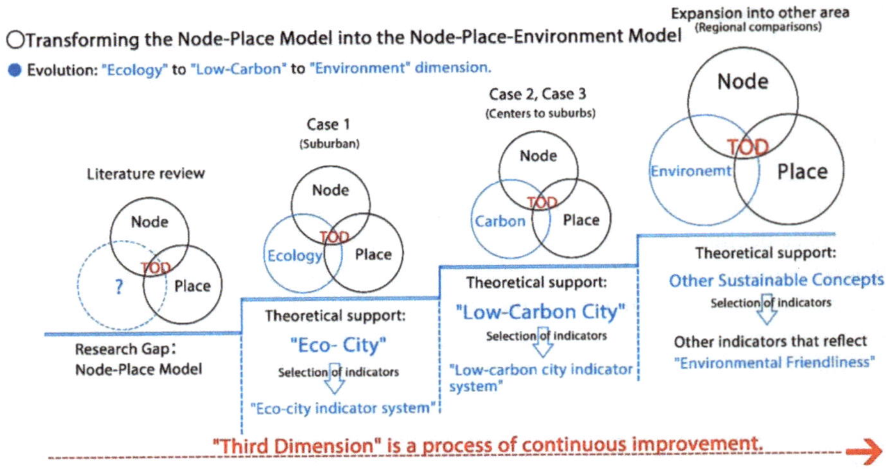

Fig. 11.5 Evolution process diagram: "Ecology" to "Low-Carbon" to "Environment"

ecological considerations, its static nature could only capture spatial characteristics at a single timepoint (e.g., the negative correlation between ecological resources and development intensity). Moreover, its ecology indicators (green spaces, biodiversity, etc.) proved difficult to adapt to high-density urban cores. The Tokyo research achieved a dual innovation to construct a four-dimensional dynamic assessment system: it transformed the geographically dependent "ecology" dimension into a universally applicable "carbon" dimension (encompassing quantifiable metrics such as building energy consumption and renewable energy infrastructure), while simultaneously embedding a "timeline" to form a Node–Place–Carbon–Time model. By tracking 70 stations from 2011 to 2019, it revealed the dynamic trajectories of TOD sustainability for the first time.

The methodological evolution signifies a deeper level of advancement. The Odawara model relied on expert-weighted calculations of static three-dimensional values, capable of diagnosing spatial conflicts such as "ecological displacement" but unable to explain their evolutionary mechanisms. The Tokyo study developed two dynamic analytical tools: First, it proposed the Sustainable Development Line (SDL), defining the ideal equilibrium state as $x = y = z$ in three-dimensional space. Through Euclidean distance deviation values, it precisely quantified annual changes. Second, it refined the K-means clustering algorithm by predefining six spatiotemporal centroids, visualizing cross-zone migration patterns of stations (e.g., a 27% increase in dependency zone stations over 8 years). This empirically demonstrated the dynamics of central polarization during urban contraction—a spatiotemporal interaction effect only discernible through the temporal dimension.

This dimensional upgrade directly catalyzed a transformation in policy intervention paradigms. The Odawara study, while exposing planning misalignments, could only offer corrective suggestions for the status quo with its static model. The Tokyo research, leveraging the temporal dimension, established a spatiotemporal control mechanism: By correlating policy milestones with station displacement trajectories (e.g., Machida Station's cluster downgrade due to delayed implementation of its 2016 redevelopment plan), it identified sensitive windows for policy implementation. Furthermore, based on SDL deviation convergence patterns, it distilled highly effective policy combinations, providing a replicable dynamic intervention template for global cities experiencing shrinkage. Ultimately, embedding the temporal dimension transformed TOD assessment from spatial diagnosis to temporal navigation. The Odawara model functioned like a static CT scanner, locating ecological pathologies in suburban development. In contrast, the Tokyo study evolved into a real-time MRI system, dynamically visualizing pathology evolution and treatment responses. When stations are conceptualized as observation points within a space–time continuum, the essence of sustainable development is redefined—it ceases to be a static equilibrium in spatial dimensions but becomes a continuous process of converging toward dynamic equilibrium over time.

This dimensional evolution reveals a fundamental theoretical convergence: the "ecology" dimension (emphasizing natural system integrity in Odawara) and the "carbon" dimension (prioritizing anthropogenic environmental pressure in Tokyo) collectively constitute a comprehensive "environment" dimension. The Odawara model's ecology metrics—such as biodiversity support and green space conservation—represent the capacity of ecosystems to sustain development, while the Tokyo model's carbon metrics—including energy consumption and renewable infrastructure—quantify the impact of development on planetary boundaries. Together, they encapsulate the dual imperatives of environmental sustainability: preservation of ecological functions and mitigation of human-induced stressors. Integrating these into a singular "environment" axis would not only bridge urban–rural assessment paradigms but also align TOD evaluation with planetary-scale sustainability frameworks such as the UN Sustainable Development Goals (SDGs) and the Planetary Boundaries model. Such synthesis would empower planners to simultaneously address habitat protection in ecologically sensitive suburbs (Odawara's lesson) and carbon neutrality in high-density cores (Tokyo's focus), transforming TOD into a holistic practice of environmental stewardship across the urban continuum.

5 Conclusion

Add closing remarks by summarizing the three evolutions and respond to the objective about Urban Nexus in TOD. This chapter has traced the progressive evolution of transit-oriented development (TOD) assessment models through three critical stages. The first evolution expanded Bertolini's classical Node–Place framework by introducing the ecological dimension, forming the Node–Place–Ecology (NPE) model. This innovation highlighted the need to integrate biodiversity, green space, and ecological resilience into TOD evaluations, particularly in suburban contexts such as Odawara, where ecological resources play a decisive role in shaping sustainable development.

The second evolution transformed the ecological axis into a carbon dimension, giving rise to the Node–Place–Carbon (NPC) model. By embedding quantifiable low-carbon indicators such as energy use and renewable infrastructure, this approach overcame the limitations of applying ecological metrics to dense urban cores. Furthermore, by introducing a temporal perspective, the NPC model advanced TOD assessment from a static spatial snapshot to a dynamic spatiotemporal analysis. This allowed researchers to capture sustainability trajectories across the Tokyo Metropolitan Area, identify imbalances between transport, land, and carbon functions, and align policy interventions with sensitive temporal windows.

The third evolution synthesized ecology and carbon into a broader environmental dimension. This convergence recognizes that sustainable TOD requires addressing both the preservation of ecological functions and the mitigation of anthropogenic pressures. By reframing TOD as part of the Urban Nexus—where transportation, land use, and environment are systematically integrated—TOD is positioned not only as a planning strategy but also as a nexus mechanism for coordinating urban resource flows and guiding compact, low-carbon, and resilient city development.

In sum, this chapter has demonstrated that the future of TOD lies in moving beyond transport–land coordination toward a nexus-based approach that incorporates environmental sustainability at its core. By embedding environmental considerations through the NPE and NPC models and situating them within the broader Urban Nexus framework, TOD can serve as a robust instrument for balancing efficiency with ecological and carbon imperatives. This synthesis directly responds to the chapter's objective: to establish TOD as a key integrative framework that links nodes, places, and the environment, enabling cities to pursue compact, resilient, and sustainable pathways for the twenty-first century.

Acknowledgments The authors sincerely acknowledge all members of the Yan (Eco-GIS) Laboratory at Keio University for their invaluable contributions. The collaborative environment and cutting-edge computational resources of the laboratory proved fundamental to this study.

References

Bertaud A, Richardson HW (2004) Transit and density: Atlanta, the United States and western Europe. In: Urban sprawl in Western Europe and the United States. Ashgate, London, pp 293–310

Bertolini L (1999) Spatial development patterns and public transport: the application of an analytical model in The Netherlands. Plan Pract Res 14:199–210. https://doi.org/10.1080/02697459915724

Bertolini L (1996) Nodes and places: complexities of railway station redevelopment. Eur Plan Stud 4:331–345. https://doi.org/10.1080/09654319608720349

Calthorpe P (1993) The next American metropolis: ecology, community, and the American dream. Princeton Architectural Press

Cervero R, Sullivan C (2011) Green TODs: marrying transit-oriented development and green urbanism. Int J Sustain Dev World Ecol 18:210–218. https://doi.org/10.1080/13504509.2011.570801

Chan EHW, Conejos S, Wang M (2017) Low Carbon Urban Design: Potentials and Opportunities
de Jong M, Joss S, Schraven D et al (2015) Sustainable-smart-resilient-low carbon-eco-knowledge cities; making sense of a multitude of concepts promoting sustainable urbanization. J Clean Prod 109:25–38. https://doi.org/10.1016/j.jclepro.2015.02.004
Haarstad H, Kjærås K, Røe PG, Tveiten K (2023) Diversifying the compact city: a renewed agenda for geographical research. Dialogues Human Geogr 13:5–24. https://doi.org/10.1177/20438206221102949
Howard E (2006) Garden cities of to-morrow. Routledge, London
Jacobs J (1992) The death and life of great American cities; reissue edition. Vintage, New York. Google Scholar
Jiang Y, Palazzo E, Raval S (2025) A framework for mapping socio-ecological dynamics to support sustainable post-mining regional transitions. IET Smart Cities 7:e70016. https://doi.org/10.1049/smc2.70016
Kikuchi H, Emberger G, Ishida H et al (2022) Dynamic simulations of compact city development to counter future population decline. Cities 127:103753. https://doi.org/10.1016/j.cities.2022.103753
Kjærås K (2021) Towards a relational conception of the compact city. Urban Stud 58:1176–1192. https://doi.org/10.1177/0042098020907281
Lee J, Kurisu K, An K, Hanaki K (2015) Development of the compact city index and its application to Japanese cities. Urban Stud 52:1054–1070. https://doi.org/10.1177/0042098014536786
Lepczyk CA, Aronson MFJ, Evans KL et al (2017) Biodiversity in the city: fundamental questions for understanding the ecology of urban green spaces for biodiversity conservation. Bioscience 67:799–807. https://doi.org/10.1093/biosci/bix079
Li S, Xu Q, Liu J et al (2022) Experience learning from low-carbon pilot provinces in China: pathways towards carbon neutrality. Energy Strategy Rev 42:100888. https://doi.org/10.1016/j.esr.2022.100888
Liu Y, Nath N, Murayama A, Manabe R (2022) Transit-oriented development with urban sprawl? Four phases of urban growth and policy intervention in Tokyo. Land Use Pol 112:105854. https://doi.org/10.1016/j.landusepol.2021.105854
Mouratidis K (2018) Is compact city livable? The impact of compact versus sprawled neighbourhoods on neighbourhood satisfaction. Urban Stud 55:2408–2430. https://doi.org/10.1177/0042098017729109
Newman PG, Kenworthy JR (1989) Cities and automobile dependence: an international sourcebook
Register R (1987) Ecocity Berkeley: building cities for a healthy future. North Atlantic Books
Tan S, Yang J, Yan J et al (2017) A holistic low carbon city indicator framework for sustainable development. Appl Energy 185:1919–1930. https://doi.org/10.1016/j.apenergy.2016.03.041
Tiwari R, Nigro A, Bondada MV (2023) Analysing urban form on transit oriented development (TOD) principles comparison of two areas in Naples, Italy. Int Rev Spatial Plan Sustain Dev 11:141–157
Wang R, Wu W, Yao Y, Tan W (2023) "Green transit-oriented development": exploring the association between TOD and visible green space provision using street view data. J Environ Manag 344:118093. https://doi.org/10.1016/j.jenvman.2023.118093
Wu Y, Swain RE, Jiang N et al (2020) Design with nature and eco-city design. Ecosyst Health Sustain 6:1781549. https://doi.org/10.1080/20964129.2020.1781549
Xu Q, Chen C, Huang X, Yang W (2025) Assessment of residential renewable energy investment under dynamic market environment: aspect from household benefits. J Environ Manag 374:124060. https://doi.org/10.1016/j.jenvman.2025.124060
Xu Q, Hwang B-G, Choo RQ et al (2023) Comparison of construction project risks before and during COVID-19 in Singapore: criticality and management strategies. Constr Manag Econ 41:875–891. https://doi.org/10.1080/01446193.2023.2211179
Xu Q, Hwang B-G, Lu Y (2021) Influencing paths of the behavior-driven household energy-saving exploring the influencing paths of the behavior-driven household energy-saving

intervention-household energy-saving option. Sust Cities Soc 71:102951. https://doi.org/10.1016/j.scs.2021.102951

Yang W, Tian S, Zhai M et al (2025) Can sustainable policies drive TOD effectively? Insights from multi-scenario simulations. J Environ Manag 380:125067. https://doi.org/10.1016/j.jenvman.2025.125067

Yang W, Xu Q, Yi S et al (2024a) Enhancing transit-oriented development sustainability through the integrated node-place-ecology (NPE) model. Transp Res Part D Transp Environ 136:104456. https://doi.org/10.1016/j.trd.2024.104456

Yang W, Xu Q, Zhai M et al (2024b) Are different TOD circles oriented towards sustainability amidst urban shrinkage? Evidence from urban areas to suburbs in the Tokyo metropolitan area. J Environ Manag 372:123274. https://doi.org/10.1016/j.jenvman.2024.123274

Yang W, Yan W, Chen L et al (2024c) Developing a TOD assessment model based on node-place-ecology for suburban areas of metropolitan cities: a case in Odawara. Environ Plan B Urban Analytics City Sci 51:839–853. https://doi.org/10.1177/23998083231202880

Yang W, Yan W, Chen L, Li H (2024d) From city center to suburbs: developing a timeline-based TOD assessment model to explore the dynamic changes in station areas of Tokyo metropolitan area. Environ Plan B Urban Analytics City Sci. https://doi.org/10.1177/23998083241258240

Zhai H, Gu B, Wang Y (2023) Evaluation of policies and actions for nature-based solutions in nationally determined contributions. Land Use Pol 131:106710. https://doi.org/10.1016/j.landusepol.2023.106710

Zhai M, Tian X, Liu Z et al (2025) Advancing just transition: the role of biomass co-firing in emission reductions and employment for coal regions. Sustain Energy Technol Assess 75:104246. https://doi.org/10.1016/j.seta.2025.104246

Chapter 12
Impact of Compact City Policy on Urban Access

Haichen Wei and Wanglin Yan

Abstract Japan's compact city policy, a rational response to demographic decline, is built on a core assumption: improving accessibility in urban cores will attract population. This chapter challenges this foundational premise through a case study of Odawara City. Our analysis reveals a critical paradox: areas with the highest accessibility are, counterintuitively, experiencing the sharpest population decline. This trend highlights a critical misalignment between existing facility distribution and evolving demographic patterns, challenging the foundational objectives of the government's Location Optimization Plan. Our analysis over the past 5 years reveals significant discrepancies between policy intent and reality. Despite approximately 80% of the population residing within a nominal 15-minute living sphere, overall accessibility to essential services is declining. Crucially, the anticipated accessibility improvements around major railway stations have failed to materialize, signaling an inability to effectively intensify core urban functions. Through a detailed examination of development patterns around key nodes, this research assesses the policy's limited effectiveness and illuminates the divergent, often contradictory, development trajectories unfolding across the urban landscape.

Keywords Compact city · Accessibility · Population distribution · Population decline

H. Wei
Data Intelligence Team, JTB Corporation, Tokyo, Japan

W. Yan (✉)
Faculty of Environment and Information Studies/Graduate School of Media and Governance, Keio University, Fujisawa, Kanagawa, Japan
e-mail: w_yan@keio.jp

W. Yan (ed.), *Urban Nexus Approach for Sustainable Cities*,
https://doi.org/10.1007/978-981-95-6156-8_12

1 Introduction

1.1 The Dilemma of Suburban Compactness

The compact city concept, while having global intellectual roots, finds a unique and urgent expression in Japan. Unlike its Western counterparts, which primarily sought to make urban growth more sustainable, the Japanese model is fundamentally a strategy for managing a planned, dignified urban contraction in the face of systemic population decline and hyper-aging (Kaido 2007). The sprawling suburbs and extensive infrastructure developed during the postwar economic boom have become fiscally unsustainable and functionally inefficient for a nation with a smaller, older population. This demographic reality catalyzed a decisive policy shift, culminating in the 2014 revision of the Urban Regeneration Special Measures Law. This legislation formally institutionalized Japan's unique interpretation of the compact city through a powerful instrument: the Location Optimization Plan (MLIT 2020). At its core is a dual-pronged strategy designating Urban Function Induction Areas (UFIA) to concentrate essential services around transit hubs, and surrounding Residential Inducement Areas (RIA) to maintain sufficient population density. The synergistic goal is to proactively restructure the urban form into a "Compact City Plus Network" of self-sufficient, walkable clusters (MLIT 2015).

However, this national strategy confronts a significant dilemma when applied to the suburban cities of the Tokyo Metropolitan Area. These cities, once symbols of growth, are now caught in a double bind: they must maintain aging, growth-era infrastructure while simultaneously grappling with population outflow to central Tokyo and the pressures of a rapidly aging local populace. For these areas, the compact city model is not an abstract ideal but a high-stakes test of managed adaptation.

Odawara City serves as a quintessential case embodying this dilemma. Blessed with a rich transportation network of six railway lines and 18 stations, it possesses a strong foundation for transit-oriented development. Nevertheless, it simultaneously struggles with entrenched car dependency and steady population decline. It is precisely in such a context that one of the policy's greatest challenges becomes apparent: its nonmandatory nature. Local governments are encouraged, but not required, to implement the Location Optimization Plan, creating a critical gap between planning ambitions and market-driven realities. The risk is that this voluntary framework will be insufficient to steer development, leading to unintended consequences such as the exacerbation of urban–suburban disparities and a decline in service access for the most vulnerable residents in peripheral areas. This chapter, therefore, investigates the central question arising from this dilemma: Does the policy's attempt to enhance accessibility in suburban nodes succeed, or does it falter against the persistent and powerful logic of the existing urban fabric and prevailing market forces?

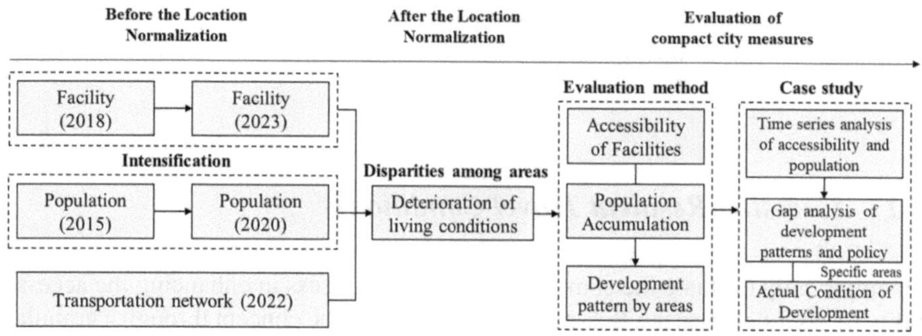

Fig. 12.1 The research framework

1.2 Research Objectives and Framework

This research, therefore, moves beyond a simple evaluation of policy implementation. It aims to dissect the on-the-ground reality of Odawara's compact city strategy by asking a critical question: Has the Location Optimization Plan successfully steered population and services toward transit nodes, or has it inadvertently created new forms of spatial inequality and inefficiency?

To answer this central question, the study undertakes a rigorous spatial–temporal analysis of facility accessibility and population shifts. It critically evaluates the delicate balance between service provision and population needs within the compact city framework, investigating whether public transportation networks have functioned as intended conduits of accessibility or as lines of exclusion, particularly in areas experiencing population decline.

The research seeks to uncover the fundamental misalignments between demographic change and the availability of services, thereby assessing the Location Optimization Plan's practical effectiveness in aligning population dynamics with planned urban development. By doing so, this study aims to illuminate the policy's systemic shortcomings and offer data-driven insights for a more rational and equitable urban future. The overall research framework is illustrated in Fig. 12.1.

Through this evaluation, this research provides a crucial empirical touchstone for the broader discourse on urban sustainability and resilience in an era of demographic contraction. Ultimately, the findings offer a framework for interrogating the real-world outcomes of compact city policies, ensuring they remain adaptable and genuinely responsive to the evolving needs of Japan's urban and suburban populations.

2 A Framework for Diagnosing the Paradox: Measuring Accessibility and Population Shifts

To capture the nuanced and often contradictory impacts of the compact city policy, a simple, city-wide analysis is insufficient. We therefore developed a multistep, GIS-based methodology designed to move from macro-level accessibility trends to the micro-dynamics of neighborhood change. This process allows us to systematically quantify the complex interplay between facility provision, resident mobility, and demographic shifts.

2.1 Assessing Resident Travel Conditions

To empirically evaluate the compact city policy's success in enhancing the accessibility of essential services, this study operationalizes the concept through a granular, GIS-based analysis of resident travel times. Our methodology calculates the time required for residents to reach key life-support facilities from their homes, employing a definition of travel time aligned with the proposed framework (Akimoto et al. 2014). The analysis utilizes the National Census 1/4 Area Mesh Data to define residential origins, integrating this with detailed road network and public transportation data to model realistic travel routes and times. Considering Odawara City's function as a commuter suburb in the Tokyo Metropolitan Area, and in alignment with the station-centric goals of its Location Optimization Plan, this study focuses on railway and walking modes. This choice is supported by data from the Tokyo Metropolitan Area Person Trip (PT) Survey, which identifies these as primary modes of public transit-related travel. The core of this analysis is a multimodal travel time calculation for each residential mesh, represented by the formula (Wang et al. 2022):

$$T = T1 + T2 + T3 \tag{12.1}$$

where $T1$ represents the walking time from a residence to the nearest transit hub (e.g., railway station), $T2$ is the travel time on the public transport network, including transfers, and $T3$ is the final walking time from the transit hub to the destination facility. This approach accurately reflects the varied mobility patterns within the city, capturing both pedestrian-only journeys and the more complex, multimodal trips that are central to the "Compact City Plus Network" concept.

However, a simple travel time to a single type of facility is an insufficient measure of overall urban livability. Daily life requires access to a variety of services with differing levels of importance and frequency of use. To create a more holistic and realistic metric, we developed a weighted average travel time for each residential mesh. The calculation for this composite score is defined by the following weighted average formula:

$$Ti = \frac{\sum_{j=1}^{n}\left(T_{ij} \times M_j\right)}{\sum_{j=1}^{n} M_j} \tag{12.2}$$

where T_i is the weighted average travel time for the i-th mesh, T_{ij} represents the travel time from mesh i to facility type j, and M_j is the weight assigned to facility j, reflecting its importance to residents. In applying this formula, we assigned weights based on the centrality of each service to residents' daily needs. Reflecting their role as the primary source for daily necessities, supermarkets were assigned the highest weight at 35%, followed by the high-frequency access provided by convenience stores at 25%. Essential health services, including both clinics and drugstores (pharmacies), were each assigned a weight of 20%, recognizing their critical importance. This weighting system, grounded in typical usage frequency and the critical nature of these services, ensures that the resulting accessibility score accurately reflects the lived experience of convenience.

2.2 *Accessibility and Population Distribution Analysis*

Once travel times are calculated, the next step in the analysis is to assess accessibility and examine how it correlates with population distribution. This involves a two-part process: first, converting the calculated travel times into a standardized accessibility score, and second, employing the Population Accumulation Index (PAI) to measure spatial shifts in population (Hoshi et al. 2019).

To translate raw travel times into a more intuitive metric, we developed a quantitative accessibility score. Recognizing that accessibility is inversely related to travel time, we first convert the weighted average travel time for each mesh into an accessibility score using the formula:

$$S_i = \frac{K}{T_i + C} \tag{12.3}$$

where S_i is the accessibility score for the i-th mesh, T_i is its weighted average travel time, and K and C are constants used for scaling and stabilization, respectively. This ensures that areas with shorter travel times receive higher scores. To allow for meaningful comparisons across different areas, these scores are then normalized to a uniform scale from 0 to 1 using the formula:

$$S_i' = \frac{S_i - \min(S)}{\max(S) - \min(S)} \tag{12.4}$$

where S'_i is the final normalized score for the i-th mesh, and min(S) and max(S) are the minimum and maximum scores in the study area. This normalization yields a consistent metric where a score of 0 indicates the least accessible areas and 1 represents the most accessible.

To understand how these varying levels of accessibility affect population distribution, the study employs the Population Accumulation Index (PAI). This index measures population concentration relative to the land area of each mesh, offering a more nuanced view than simple density. The PAI is calculated as:

$$\mathrm{PAI}_i = \frac{P_i \,/\, P_{\mathrm{total}}}{A_i \,/\, A_{\mathrm{total}}} \tag{12.5}$$

where P_i and A_i are the population and land area of the i-th mesh, and P_total and A_total represent the total population and land area of the study area. A PAI value of 100 indicates an evenly distributed population, while higher values signify a greater concentration. By comparing changes in PAI values with our calculated accessibility scores, the analysis can reveal the critical relationship between service accessibility and demographic shifts, identifying where population growth is—or is not—aligning with the policy's spatial objectives.

2.3 *Classifying Regional Trajectories: Five Patterns of Urban Development*

By integrating the temporal changes in our normalized Accessibility Score and the Population Accumulation Index (PAI) for each urban mesh, we can move beyond city-wide averages to classify distinct regional development patterns. This classification reveals the uneven impact of the compact city policy, showing how different areas have responded to planning interventions. It provides a diagnostic framework for understanding where the strategy has succeeded in fostering sustainable, transit-oriented development, and where it has failed, resulting in population decline or infrastructural strain.

The spatial analysis of integrated accessibility and population metrics uncovers distinct development patterns that have emerged in response to the compact city policy. By tracking the shifts in PAI and Accessibility Score over the study period, we classified urban regions into five major developmental trajectories.

This classification framework moves beyond simplistic metrics of growth or decline. It functions as a diagnostic tool, allowing us to identify the symptoms of policy failure—such as investment in infrastructure that fails to attract residents ("Excess") or population growth that outpaces services ("Stagnation"). By mapping these patterns, we can pinpoint precisely where and why the compact city strategy is succeeding or failing.

The five classified patterns are as follows:

Growth: Regions characterized by synchronized increases in both PAI and accessibility, typically found in Induction Areas where the compact city policy is successfully promoting sustainable, transit-oriented development.

Excess: Defined by significant infrastructure improvements (increased accessibility) that fail to attract or retain residents (stagnant or decreased PAI). This exposes a critical misalignment where infrastructure investment fails to attract residents, leading to underutilized urban functions.

Stagnation: Regions where population increases (increased PAI) are concurrent with a decline in accessibility. This pattern highlights areas where demand outpaces facility provision, creating pressure and inefficiency, and requires urgent infrastructure upgrades.

Decline: Regions experiencing synchronized decreases in both PAI and accessibility, primarily located outside designated Urbanized Areas. This signals accelerating urban retreat and necessitates either managed shrinkage or targeted regeneration.

Stability: Regions showing consistent population density and stable accessibility levels, representing mature neighborhoods where planning measures are effectively maintaining a livable equilibrium.

Understanding these patterns is crucial, as they define where further, tailored interventions are necessary. The effectiveness of the compact city policy is highly dependent on addressing these diverse local contexts, moving beyond uniform goals to targeted, region-specific strategies.

2.4 Analyzing the Physical Dynamics: Measuring Urban Development and Retreat

To ground these macro-level patterns of accessibility and population change in the physical reality of the built environment, our analysis delves into the micro-dynamics of urban development and retreat (Ujihara et al. 2006). This is essential for understanding the tangible mechanisms driving the "Growth" or "Decline" observed in the previous section. Our focus narrows to the most critical zones targeted by the policy: the key transportation hubs.

The analysis quantifies these physical changes within an 800-m radius of each railway station—the designated focal points for consolidation under the compact city plan. We measure the rate of building development (P_d) and the rate of building retreat (P_r) using the following formulas:

$$P_d = \frac{N_d}{N_d + N_e} \times 100 \tag{12.6}$$

$$P_r = \frac{N_r}{N_r + N_e} \times 100 \tag{12.7}$$

In these formulas, P_d and P_r represent the respective rates of building development and retreat. These rates are calculated based on three key variables for each study area: N_d, which is the number of newly developed buildings; N_r, the number of retreated buildings that have been demolished or abandoned; and N_e, the existing stock of buildings at the start of the measurement period. These calculations provide a direct, quantitative measure of urban transformation at the nodal level. By tracking these rates, we can objectively assess whether designated centers are experiencing genuine consolidation through new development, or if they are instead facing fragmentation and hollowing-out through building retreat. This micro-level data offers critical insight into the on-the-ground success or failure of the city's urban consolidation efforts.

3 Anatomy of a Paradox: Evidence from Odawara City

To ground our analytical framework in empirical reality, this study focuses on Odawara City. Located approximately 100 km southwest of Tokyo, Odawara embodies the quintessential duality of a Japanese suburban city: it possesses a strong foundation for transit-oriented development with its extensive network of 18 stations, yet simultaneously faces the severe challenges of low urban density, high car dependency, and accelerating population decline and aging. This makes it an ideal laboratory for a critical, on-the-ground examination of the compact city policy in practice, a vision operationalized through its official Location Optimization Plan (Fig. 12.2).

3.1 Spatial Shifts in Facility Location and Accessibility

An examination of the trends in the number and accessibility of four key facility types reveals divergent spatial trajectories that challenge the core tenets of the compact city policy. The first step in dissecting Odawara's paradox is to map the physical changes in its urban services. Our analysis from 2018 to 2023 reveals a landscape

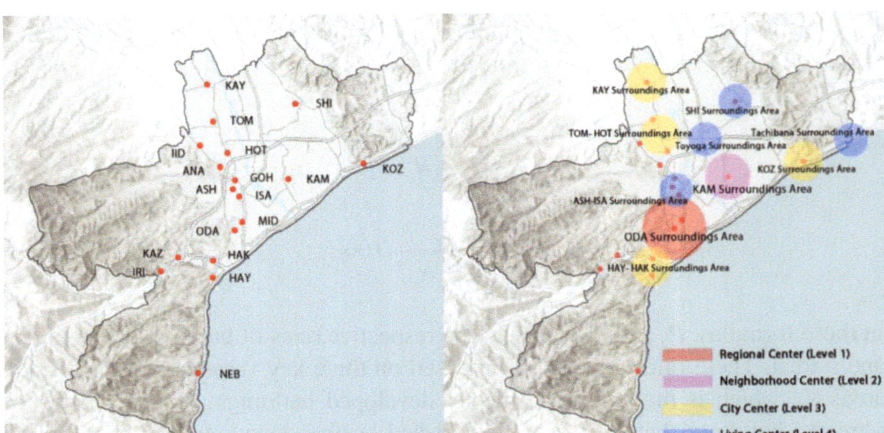

Fig. 12.2 Odawara official's original plans for 18 stations (Yang et al. 2023)

in flux, where facility changes follow a logic often at odds with the compact city's transit-oriented goals. As visualized in the maps presented in Fig. 12.3, this period saw not a simple consolidation but a nuanced and often contradictory restructuring of the city's service network.

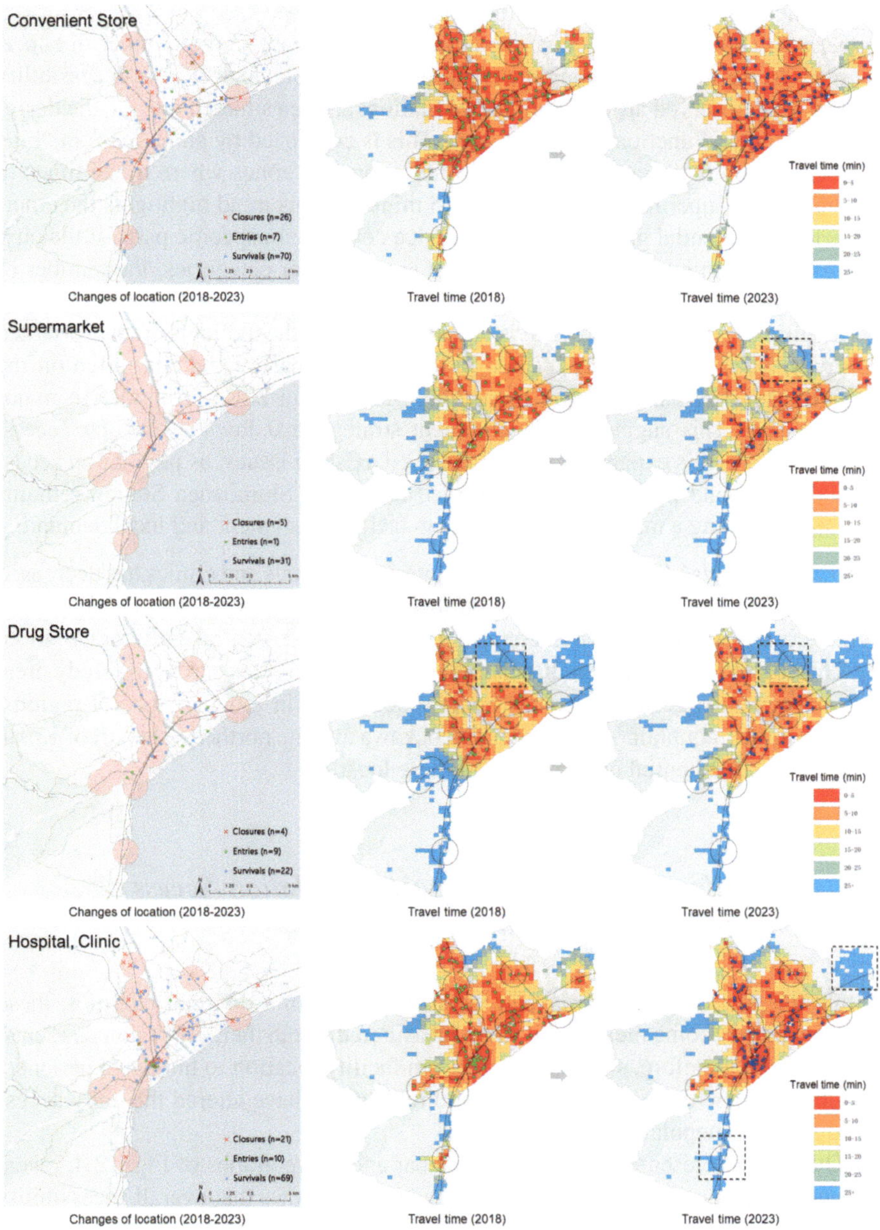

Fig. 12.3 Changes in facility locations and travel times

Convenience stores: The total number of convenience stores has decreased by approximately 20%, with this decline predominantly occurring along the Odakyu Line and in the city's eastern urban area. Despite this significant numerical reduction, the overall impact on accessibility has been minimal. The effective spatial distribution of the remaining stores ensures that most residents still have access to a convenience store within a 15-minute travel time, indicating a resilient network that has absorbed the consolidation.

Supermarkets: In contrast, the approximate 10% decrease in the number of supermarkets has led to more tangible consequences. A notable decline in accessibility has been observed around Shimosoga Station, an area situated outside the designated Urban Function Induction Area. This is evidenced by an increase in "blue meshes" on the accessibility map, which represent zones where travel times to the nearest supermarket now exceed 15 minutes. This trend highlights the emergence of potential gaps in essential service coverage in specific parts of the city.

Drugstores: Contrary to the consolidation seen in other categories, the number of drugstores has increased by about 20%. However, the placement of these new facilities reveals a logic that is misaligned with the policy's transit-oriented goals. They are primarily located in the southern part of the city, often on the outskirts of Urban Function Induction Areas and in close proximity to major roadways. This suggests a development strategy that favors vehicular access. Meanwhile, this expansion has not resolved existing issues, as population accessibility around persistently underserved areas like Shimosoga Station remains low, indicating a mismatch between new facility placement and local demand.

Internal hospitals/clinics: The number of internal hospitals and clinics has decreased by about 12%. A clear trend of centralization is visible, with new facilities concentrating within the commercial complex near the central Odawara Station. This has led to a less even distribution of medical services across the study area. Consequently, accessibility has notably decreased in more peripheral regions, such as the mountainous areas near Nebukawa and the northeastern part of Kozu, reflecting a potential underservice in these locations.

3.2 *Quantifying the Human Impact: Population Access and the Eroding 15-Minute Sphere*

While the previous section mapped the spatial shifts of individual facilities, these changes only become meaningful when measured through their impact on residents. This section, therefore, shifts the focus from facility location to human experience, quantifying how these consolidations and relocations have altered the daily accessibility for the population.

The results, presented in the combined bar and curve graphs of Fig. 12.4, reveal a deteriorating accessibility landscape. The analysis shows that overall accessibility to convenience stores, supermarkets, and medical facilities has declined between

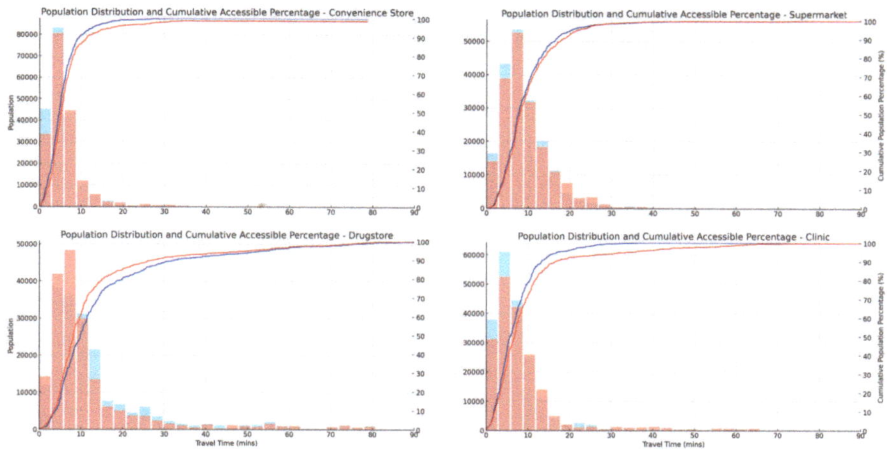

Fig. 12.4 Distribution of travel time and cumulative accessible population

2018 and 2023. This trend suggests that facility consolidation, likely aimed at cost reduction, is negatively impacting residents' convenience. The only exception is drugstores, which have seen improved accessibility due to an increase in their numbers, particularly in high-density areas near major roads.

A closer look at the cumulative accessibility curves in Fig. 12.4 and the combined overlay in Fig. 12.5 provides further nuance. While a high level of basic retail access is maintained—with approximately 90% of residents still able to reach a convenience store or supermarket within 15 minutes—the underlying trends are concerning. Accessibility to supermarkets is weakening, with a noticeable shift toward the 20-minute range for a larger portion of the population. Ultimately, although a significant majority (approximately 80%) currently live within the idealized 15-minute living sphere, this metric masks a gradual erosion of convenience for most essential services.

3.3 The Accessibility–Population Paradox and Divergent Development Patterns

The analyses of facility locations and population access reveal a troubling picture: while a nominal 15-minute city remains for the majority, the quality of that access is eroding. However, the most critical question remains: Are these changes in accessibility influencing where people choose to live? To answer this, we now integrate our accessibility analysis with population distribution trends, uncovering a fundamental paradox that challenges the core logic of the compact city policy.

This integrated analysis reveals a deeply counterintuitive trend. As visualized in the differential maps in Fig. 12.6 and quantified in the correlational analysis of Table 12.1, areas with previously high levels of accessibility are experiencing a

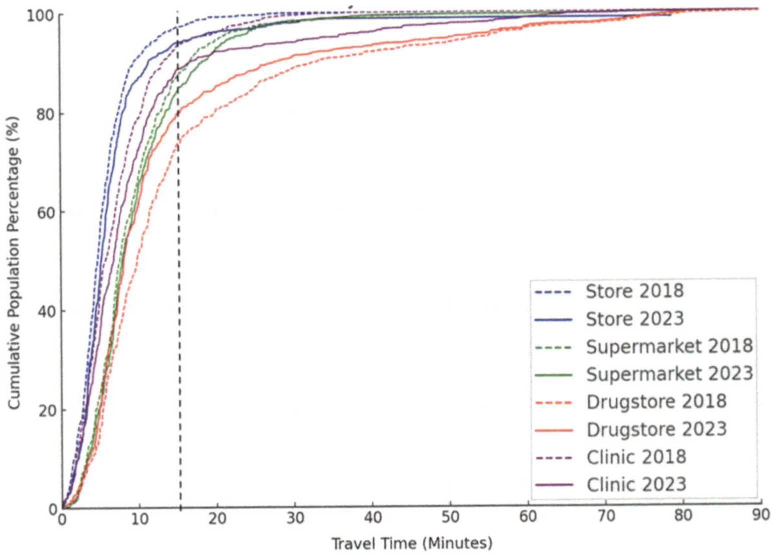

Fig. 12.5 Overlaid cumulative population accessibility curves

more pronounced decline in population than less accessible areas. This finding directly refutes the central premise of the Location Optimization Plan—that enhancing accessibility will naturally induce residential concentration. It suggests that more powerful forces, such as housing affordability and car-dependent lifestyles, continue to shape the urban landscape, trends which the current policy has been unable to reverse.

Based on this integrated analysis, we have classified the urban landscape into five distinct development patterns (Fig. 12.7). This typology serves as a diagnostic tool to understand how different parts of Odawara are responding to policy interventions, exposing critical misalignments between planning intent and actual outcomes.

Growth: Characterized by an increase in both PAI and accessibility (↑ PAI, ↑ accessibility). These areas, found predominantly within designated Induction Areas, represent zones where the compact city policy is achieving its intended effects.

Excess: Defined by a decrease in PAI despite an increase in accessibility (↓ PAI, ↑ accessibility). In these zones, urban functions are retained, but they are failing to attract or retain population.

Stagnation: Marked by an increase in PAI concurrent with a decrease in accessibility (↑ PAI, ↓ accessibility). This dangerous mismatch indicates that population is outpacing the provision of facilities, signaling an urgent need for infrastructure upgrades.

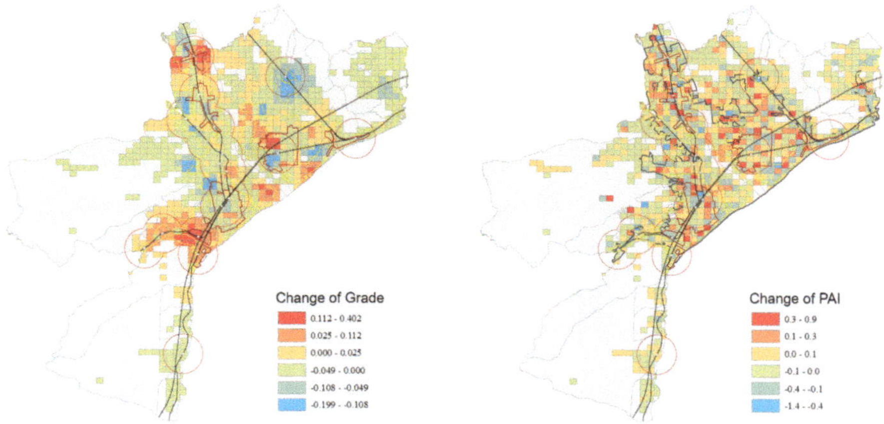

Fig. 12.6 Changes in accessibility (left) and PAI (right) over 5 years

Table 12.1 PAI in different accessibility grades

Accessibility grade	PAI						Ration of total area	
	2018 (%)	2023 (%)	2018–2023 (%)	2018 (65+) (%)	2023 (65+) (%)	2018–2023 (65+) (%)	2018 (%)	2023 (%)
0.8 ~ 1.0	216.7	225.5	8.8	234.0	220.8	−13.3	0.9	0.9
0.6–0.8	206.4	205.2	−1.1	195.2	205.9	10.7	3.1	2.9
0.4–0.6	220.0	211.6	−8.4	203.8	196.1	−7.7	12.3	12.3
0.2–0.4	132.3	136.9	4.5	129.0	131.6	2.6	35.2	32.4
0.0–0.2	37.4	43.4	6.0	46.2	52.6	6.5	39.8	42.8

Decline: Characterized by a decrease in both PAI and accessibility (↓ PAI, ↓ accessibility). This pattern is concentrated outside the main Urbanized Areas and points to regions in need of targeted regeneration or managed retreat strategies.

Stability: Areas with no significant changes in either PAI or accessibility (≈ PAI, ≈ accessibility), where conditions have reached an equilibrium.

These classifications, when mapped spatially across the city, provide a powerful visual framework for identifying where policy is succeeding and where additional, more targeted intervention is essential for achieving balanced urban growth.

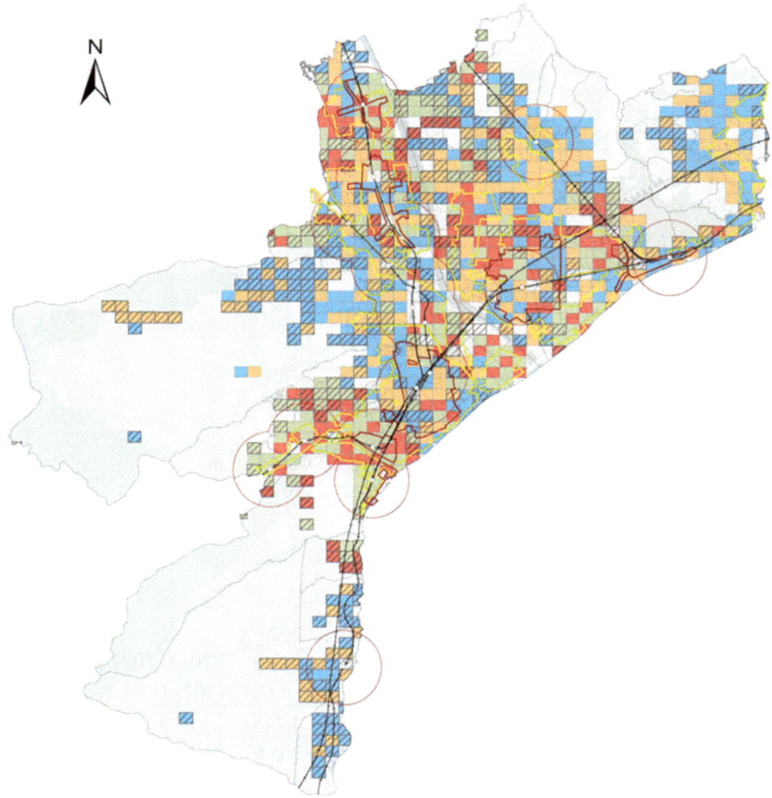

Fig. 12.7 Spatial distribution of meshes with different development pattern

4 Discussion

Our findings reveal a systemic failure: the Location Optimization Plan is not working as intended. This section moves from what is happening to why. We argue that this failure is rooted in two critical disconnects: first, a mismatch between the plan's hierarchical node strategy and on-the-ground development realities; and second, the persistent dominance of a car-oriented market logic that the voluntary policy framework is too weak to overcome.

4.1 Assessing Node-Level Performance

To directly evaluate the policy's impact, our analysis focuses on the 800-m radius surrounding each of Odawara's 18 railway stations. By calculating the area proportions of our five development patterns within these critical TOD zones (Fig. 12.8),

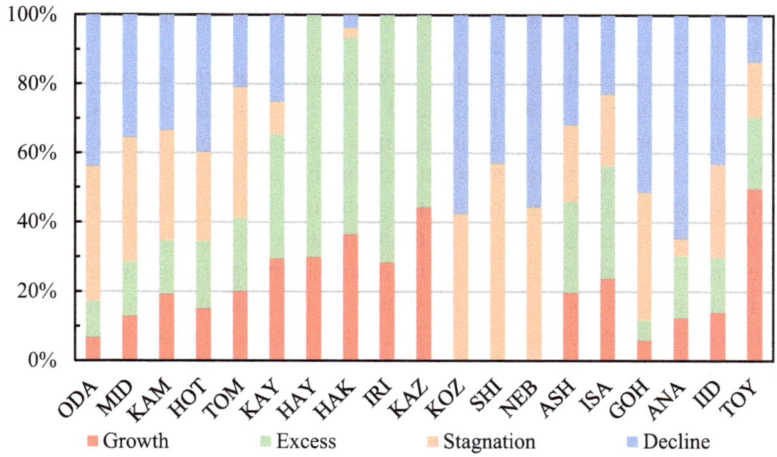

Fig. 12.8 Development pattern distribution by node

we can compare the on-the-ground reality with the strategic roles assigned to each node in the official plan. The findings reveal a significant and troubling disconnect between planning intent and actual development trajectories.

Stagnation at the core: The city's two most important nodes, Odawara (ODA) and Kamonomiya (KAM)—designated as the sole Regional and Neighborhood centers—are surprisingly dominated by patterns of Decline and Stagnation. Instead of acting as engines of consolidation, these high-level centers appear to have reached their developmental capacity or are failing to adapt to evolving urban needs.

Success of alternative anchors: In contrast, areas designated as City Centers, such as the node around Hayakawa (HAY), exhibit more balanced development. More strikingly, the Toyogawa (TOY) area, a Living Center anchored by a school rather than a station, has demonstrated significant growth, highlighting the potential of non-transit community anchors (Tasaki et al. 2018).

Decline on the periphery: Nodes located on the outskirts of the main urbanized area, such as Kozu (KOZ), Shimosoga (SHI), and Nebukawa (NEB), are overwhelmingly characterized by Stagnation and Decline, indicating a clear failure to meet development expectations in the city's most vulnerable zones.

4.2 A Granular Analysis of Development Realities

While the node-level statistics provide a broad overview of these failures, a deeper understanding requires a granular examination of specific areas. We selected three representative nodes for detailed analysis, each with a distinct planning goal:

Kamonomiya (KAM) (Neighborhood Center), Tomizu (TOM) (City Center), and Shimosoga (SHI) (Living Center).

A GIS-based analysis overlaying development patterns on the urban fabric reveals critical insights. Both KAM and TOM, despite differing planning goals, exhibit a development logic that undermines the transit-centric policy: patterns of Growth and Excess are clustered not at the core but at the edges of induction zones, adjacent to major roadways where new pharmacies have located. Meanwhile, Stagnation and Decline are dispersed throughout the residential zones. The case of SHI is even more stark, exhibiting the highest proportion of Stagnation, particularly within the residential zones on the west bank of the Morito River, while areas near the station front are predominantly in Decline. This spatial analysis also highlights how natural barriers, such as railways and rivers, contribute to the fragmentation of development.

To understand these patterns at the building level, we analyzed the rates of building development and retreat over the past two decades. The data reveals a critical disconnect between planning and reality. As illustrated in Fig. 12.9, the overall building retreat rate has been on a concerning upward trend, indicating a hollowing out of the urban fabric. This trend is particularly telling at the node level. For instance, despite Kamonomiya (KAM) being designated as a high-level Neighborhood Center, its retreat rate has remained persistently high, while its development rate has stagnated. This graphically demonstrates a lack of internal renewal momentum and challenges the effectiveness of its strategic designation. While Tomizu (TOM) experienced a temporary surge in development from 2014 to 2019, this momentum has since subsided, and its retreat rate is also rising. These divergent behaviors underscore the need for tailored policy approaches. For major nodes like KAM, intervention must focus on curbing building retreat and stimulating functional replacement. For city centers like TOM, collaboration with railway operators is crucial, and for peripheral living centers like Shimosoga (SHI), leveraging natural resources to counteract decline is paramount.

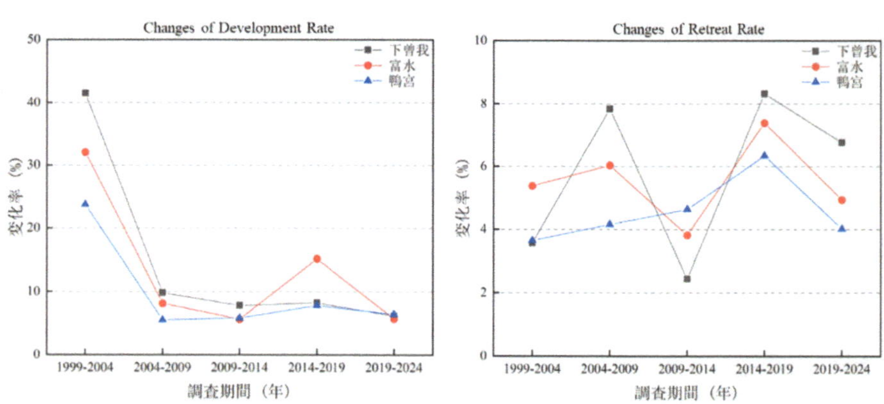

Fig. 12.9 Changes in development rate and retreat rate

The case of SHI is particularly illustrative of the challenges on the urban fringe. As a living center on the outskirts of the urbanized area, its development does not align with its intended function. A chronological analysis of building developments and retreats over three distinct periods reveals a disordered pattern of change, with widespread retreats coupled with unsustainable developments leading to a proliferation of fragmented open spaces. To counter this, future spatial planning around the station must be transformative. Leveraging SHI's untapped natural and tourism resources—by converting vacant plots into community assets such as supermarkets integrated with public spaces and implementing place-based strategies such as linear landscapes and community events—could revitalize the area, enhance its life-service capabilities, and foster a unique, sustainable identity.

Ultimately, the divergent realities of KAM, TOM, and SHI provide the ground-level explanation for the node-level failures identified previously. They demonstrate that the top-down designation of a nodal hierarchy is insufficient to overcome the bottom-up logic of car-dependent development, existing building stock, and fragmented land ownership. The policy's vision is being systematically undermined by a thousand small, uncoordinated decisions on the ground.

5 Conclusions

This study of Odawara City reveals a stark disconnect between the vision of Japan's compact city policy and its outcomes. Our analysis exposes a fundamental paradox: improved accessibility in urban cores has not induced population growth; instead, the most accessible areas are often the ones declining fastest. The "15-minute city" is becoming a statistical illusion, masking an erosion of convenience as development follows the logic of arterial roads, not railway stations. In short, the Location Optimization Plan is failing to steer the market, resulting in key urban nodes that stagnate or decline.

These findings point to a clear policy imperative. In areas with limited planning influence, the proliferation of unorganized building developments and demolitions is creating fragmented landscapes with considerable vacant land. To counter this, a paradigm shift is required. The future of the compact city cannot rely on voluntary measures alone; it necessitates timely policy revisions and stronger, more direct administrative interventions.

The challenge for Japan's compact city strategy, therefore, is not one of vision, but of execution. The case of Odawara serves as a critical lesson for shrinking cities globally: top-down spatial plans are insufficient when they ignore the powerful, bottom-up logic of the existing urban fabric and market behavior. True urban resilience in an era of contraction will not be achieved through zoning maps alone. It demands robust policy interventions that can actively shape markets, leverage unique local assets, and bridge the ever-widening gap between planning intent and human reality.

It should be noted that this study has certain limitations. For instance, the 5-year time span is relatively short for assessing the full impact of a long-term urban policy. Additionally, while the weighting of facilities was based on established norms, future research could achieve greater precision by incorporating surveys of Odawara residents to reflect their actual needs and preferences more accurately. Despite these limitations, the "paradox" phenomenon and its underlying mechanisms revealed by this research still hold significant policy reference value.

Acknowledgments The authors thank Odawara City for providing the essential GIS datasets. We acknowledge Professor Yan for his supervision and Professors Shaw and Miyamoto for their feedback. The collaborative environment and computational facilities of the Yan Laboratory at Keio University were instrumental to this research.

References

Akimoto N (2014) Implications of a cluster-type compact city design for accessibility in the suburbs of Toyama City: simulations in the Fuchu Area. Geogr Rev Jpn 87:314–327

Hoshi T, Umehara K, Hachiya Y, Maruoka T (2019) Study on relations with life convenience and the population distribution change in local cities with declining population: case study of Hakodate, Aomori, and Hachinohe. J Archit Plan 84:1393–1400

Kaido K (2007) Planning and design of compact cities. Gakugei Shuppansha Co, Ltd. https://book.gakugei-pub.co.jp/gakugei-book/9784761531607/

Ministry of Land, Infrastructure, Transport and Tourism (2015) About the revised urban regeneration special measures law. Retrieved from https://www.mlit.go.jp/common/001091253.pdf

Ministry of Land, Infrastructure, Transport and Tourism (2020) Guidelines for creating the location optimization plan. Retrieved from https://www.mlit.go.jp/toshi/city_plan/content/001379331.pdf

Odawara City Hall (2019) The Location Optimization Plan in Odawara City. Retrieved from https://www.city.odawara.kanagawa.jp/field/municipality/report/compactcity_network.html

Tasaki T, Manabe R, Murayama A, Okata J (2018) Study on the school-centered reorganization of public facilities in the Tokyo metropolitan suburbs: case study in Matsudo City, Kasukabe City, and Higashimurayama City. J City Plan Inst Jpn 53:1283–1288

Ujihara T, Taniguchi M, Matsunaka R (2006) The reality of urban retreat: reverse sprawl and its relation with residential zone types. J City Plan Inst Jpn 41:977–982

Wang J, Matsunaka R, Oba T, Tanaka K (2022) Analysis of the relationship between the location of large-scale retail stores and the distribution of urban population in local cities. J City Plan Inst Jpn 57:1086–1093

Yang W, Yan W, Chen L, Wei H (2023) Developing a TOD assessment model based on node–place–ecology for suburban areas of metropolitan cities: a case in Odawara. Environ Plan B: Urban Anal City Sci 51(4):839–853. http://dx.doi.org/10.1177/23998083231202880

Chapter 13
Nexus Model of Suburban Industrial Communities for Urban Sustainability

Ruiyi Zhang and Wanglin Yan

Abstract Suburban industrial parks (SIPs) drive urbanization in metropolitan peripheries. They aim to boost local economies while reducing environmental impacts. However, their contributions to sustainable urbanization remain unclear. This chapter introduces the suburban industrial community (SIC) nexus model. The model examines interdependencies between SIPs and surrounding urban systems across multiple scales. We apply this framework to Tokyo's metropolitan area to assess sustainability performance. Key findings include that neighborhood areas improve energy performance by 30% when integrated with SIPs. The model reveals three developmental phases over 50 years: growth (0–20 years), stabilization (20–35 years), and renewal (35+ years). Each phase requires specific policy interventions. Policy implications recommend transition from isolated SIP management to integrated SIC governance. This approach could benefit both emerging and developed economies pursuing sustainable industrial development.

Keywords Suburban Industrial Communities (SIC) · Industrial agglomeration · Energy efficiency · Metropolitan peripheries · Industrial policy

1 Introduction

Suburban industrial parks (SIPs) shape metropolitan peripheries worldwide, driving economic growth while facing sustainability challenges. We lack frameworks to assess their systemic impacts on urban sustainability.

R. Zhang (✉)
School of Media and Governance, Keio University, Fujisawa, Kanagawa, Japan
e-mail: zry1030@keio.jp

W. Yan
Faculty of Environment and Information Studies/Graduate School of Media and Governance, Keio University, Fujisawa, Kanagawa, Japan
e-mail: w_yan@keio.jp

W. Yan (ed.), *Urban Nexus Approach for Sustainable Cities*,
https://doi.org/10.1007/978-981-95-6156-8_13

1.1 *Suburban Industrial Parks (SIPs): Evolution and Global Context*

SIPs represent a distinctive urban development instrument that has shaped metropolitan peripheries worldwide since the mid-nineteenth century. These purposefully designed zones for industrial agglomeration have evolved from isolated manufacturing enclaves to integrated components of urban systems (Ian 2005), fundamentally altering the spatial organization and economic dynamics of metropolitan regions. The concept of planned industrial development originated during the Industrial Revolution, with early examples including Chicago's "company towns" (Minnery 2012) and the industrial zones within Howard's Garden City model (Clark 2003). However, the modern SIP emerged in the early twentieth century as governments recognized the need to manage industrial location and growth systematically. The prominent industrialized nations commenced the establishment of SIPs surrounding their large cities. Japan established its first SIP in 1934 at Shimomaruko, a Tokyo suburb, marking the beginning of Asia's engagement with planned industrial suburbanization (Kikuchi 2000).

The post-World War II period witnessed accelerated SIP development globally, driven by rapid industrialization and urban reconstruction in OECD nations (Kim and Gallent 1998; Scott 1986). In Japan, the "Capital Region Development Act" of 1956 and the "Industrial Park Development Program" explicitly promoted SIP establishment in the Suburban Development Zone (SDZ; 20–60 km to Tokyo station; Zhang et al. 2024), creating a controlled peripheral development model that differs markedly from the sprawl patterns observed in many Western cities. Throughout this period, SIPs evolved in response to global industrial development trends and emerged as a significant policy tool for governments to direct industry location and expansion (UNIDO 2019a). By 2010, the Tokyo Metropolitan Area contained 133 inland industrial parks, with development peaks during the 1960s–1980s rapid growth period. And three distinct developmental phases of SIPs were observed in past five decades, with Phase I (under 20 years) showing strongest agglomeration, Phase II (20–35 years) stabilization, and Phase III (over 35 years) diversification (Zhang and Yan 2024).

The implementation of SIPs has been facilitated by numerous governmental initiatives, including financial incentives, land-use rules, and infrastructural development. In Japan, the "Industrial Park Development Project" and related national policies implemented in the 1950s directly targeted the phenomenon of industrial suburbanization in metropolitan outskirts (Fig. 13.1). These regulations facilitated the conversion of industrial property to SIPs by providing tax incentives to developers. In recent decades, SIPs have assumed a diverse role in the urbanization process. They have facilitated the decentralization of industrial operations from overcrowded urban centers, generated employment prospects in suburban regions, and stimulated the development of housing and infrastructure nearby. SIPs have served as catalysts for the development of satellite cities and new urban centers, transforming the

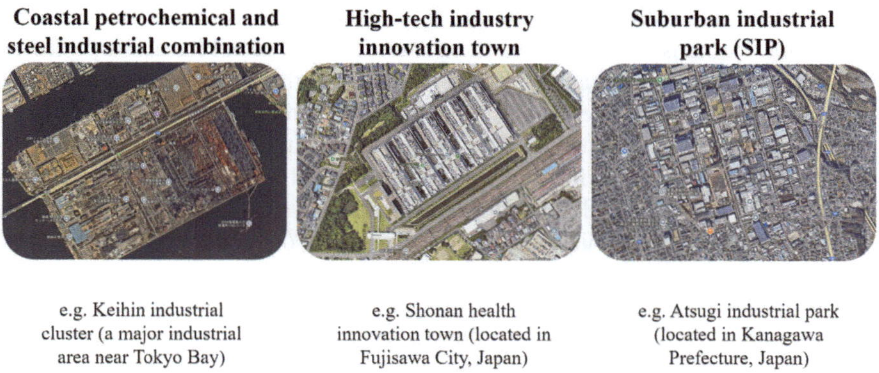

Fig. 13.1 Three types of industrial clusters in the Tokyo metropolitan area (Zhang 2025)

spatial configuration and socioeconomic dynamics of metropolitan areas (UNIDO 2019b).

Nevertheless, the swift advancement of SIPs has introduced obstacles, including urban sprawl, environmental damage, and issues of social equality. The discord between SIPs and their adjacent environments has resulted in significant industrial spatial transformations, with certain industrial areas being supplanted by alternative urban land uses and polluting industries migrating to new SIPs located further from urban centers or to developing nations (Hossain and Huggins 2021; Hurley et al. 2023). In recent decades, the emergence of green industrial parks and eco-industrial development projects has sought to tackle these difficulties by advocating for cleaner manufacturing, industrial symbiosis, and integrated urban–industrial planning (Lyu et al. 2022). Certain conventional SIPs have evolved into mixed-use developments, seamlessly merging with the urban fabric of suburban communities, while others have been enhanced to support new generations of industries and technologies (Evans 2009).

Despite of the positive role of SIPs in industrial suburbanization, job creation, and infrastructure development, challenges remain, such as urban sprawl, environmental degradation, and social inequity. The transition from heavy industries to mixed industrial–commercial–residential developments, the integration of green infrastructure, and the pursuit of circular economy principles represent fundamental shifts in SIP planning and management paradigms.

1.2 Rethinking SIP's Role in Sustainable Urbanization

Despite growing recognition of SIPs' importance in sustainable urbanization, significant knowledge gaps persist in understanding their systemic impacts and evolution. Existing research has examined multiple facets of SIP development, emphasizing their social, economic, and environmental advantages (Okata and

Murayama 2011; Yoshida et al. 2002). This encompasses the mitigation of urban congestion by industrial relocation, the advancement of social fairness by generating opportunities in urban and suburban regions, and the decrease of carbon emissions through reduced travel lengths for goods and commuters (Sakai et al. 2020).

Research on SIPs so far have predominantly focused on individual parks or single-dimension assessments, overlooking the critical spatial and functional relationships between SIPs and their surrounding urban systems. Limited research has incorporated the geographical and temporal aspects of industrial land transition within the larger urban systems that encompass SIPs. The long-term and multi-scale effects of SIPs on urban sustainability are yet ambiguous. Although certain studies have examined industrial land agglomeration at national and prefectural levels (Kuang et al. 2016; Li et al. 2022), there is a deficiency of research that connects regional and local dimensions. This constrains our comprehension of the spatial dynamics between SIPs and their surrounding areas, along with their aggregate impacts on urban development patterns. Therefore, three critical gaps motivate this research:

- *Spatial integration*: existing frameworks treat SIPs as isolated entities rather than components of broader urban systems, missing crucial park-neighborhood synergies.
- *Temporal transition*: limited understanding of how SIPs transform over multiple decades and their long-term contributions to sustainable urbanization.
- *Multidimensional assessment*: absence of integrated evaluation methods that simultaneously capture economic agglomeration, environmental performance, and social development.

To address these gaps, this chapter introduces a nexus model of suburban industrial community (SIC)—a conceptual and methodological framework to integrate SIPs with their surrounding neighborhoods to form coherent analytical units, enabling multi-scale temporal analysis across developmental phases, and providing quantitative tools for assessing sustainability performance across economic, environmental, and social dimensions.

2 The Nexus Model of SIC: Conceptual Framework

2.1 Theoretical Foundation

The SIC model builds upon balanced sustainable city concept, associated with multiple theoretical traditions in each subsector of economic, environmental, and social dimensions (Fig. 13.2). The sustainable development theory integrates the three pillars of sustainability through specific approaches:

- Economic sustainability through industrial agglomeration analysis
- Environmental sustainability through energy performance assessment

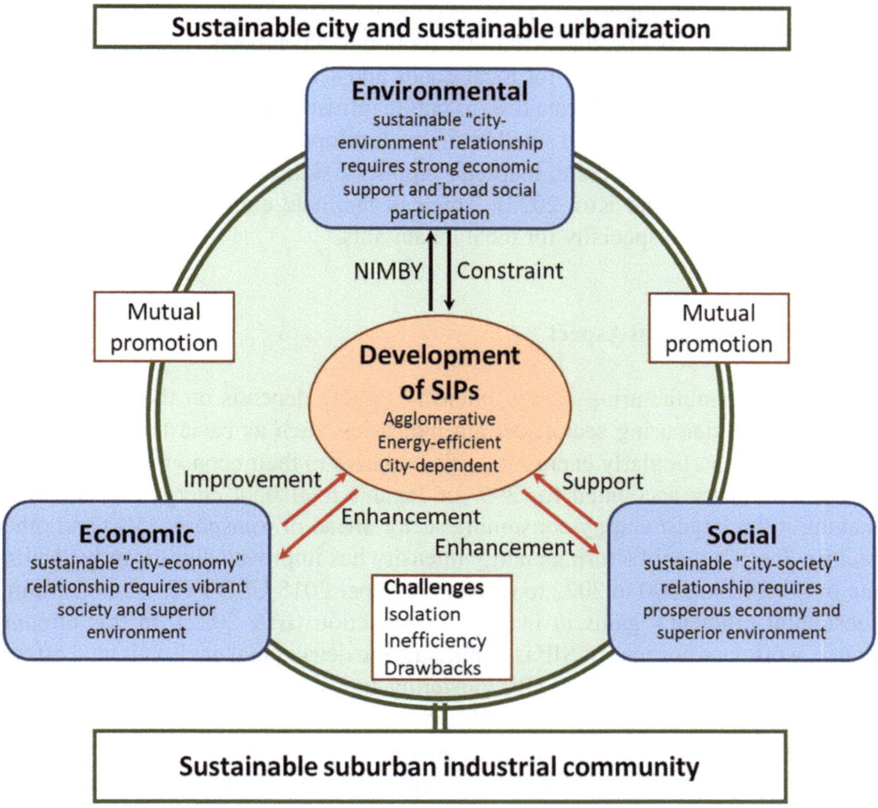

Fig. 13.2 Nexus framework of suburban industrial communities showing interdependencies, feedback loops, and synergies between economic, environmental, and social dimensions

- Social sustainability through urban function diversification metrics

These three dimensions are not independent but form a nexus of mutual dependencies: economic agglomeration influences energy consumption patterns, which in turn affects social development opportunities. A nexus model of SIC is expected to explicitly captures these cross-dimensional linkages and feedback mechanisms.

2.1.1 Economic Aspect

The implementation of SIPs bolsters the local economy by fostering a conducive climate for business growth. The concentration of industries within and around SIPs allows enterprises to use economies of scale, common resources, and knowledge spillovers. The regional transportation infrastructure could attract industrial agglomeration (Botelho 2019). The clustering effect can enhance productivity, innovation, and competitiveness among the local enterprises. Furthermore, SIPs frequently draw a variety of businesses, fostering economic diversification and resilience. The existence of SIPs also produces indirect economic advantages for adjacent towns, including heightened demand for local goods and services, along with tax revenue for local governments to allocate toward public infrastructure and services. SIPs can promote entrepreneurship and small business development by creating a supporting ecosystem that offers access to markets, suppliers, skilled labor, and business support services (Park and Kim 2022). This can facilitate employment creation and income generation, especially for local inhabitants.

2.1.2 Environmental Aspect

A country's manufacturing energy intensity largely depends on the makeup of a country's manufacturing sector: certain industries, such as basic metals and pulp and paper, are particularly energy intensive relative to their economic contribution. In 2022, industry accounted for 29% of Japan's total final energy consumption, making it the largest energy-consuming sector ahead of transport (25%) and other sectors. Japan's manufacturing energy intensity has improved significantly, declining by 43% from 2000 to 2022 to reach 3.5 MJ per 2015 USD PPP,[1] demonstrating substantial efficiency gains in industrial production (IEA 2022). In this circumstance, a primary purpose of SIP is to mitigate the detrimental environmental effects linked to industrial operations. By clustering industries in a certain region, SIPs enable the establishment of centralized environmental management systems, including waste treatment facilities, energy-efficient infrastructure, and pollution control initiatives. This facilitates enhanced oversight and control of industrial emissions, effluents, and waste management. Moreover, SIPs frequently encourage the implementation of cleaner production technologies and practices among local enterprises via incentives, regulations, and knowledge dissemination. The closeness of enterprises within SIPs fosters industrial symbiosis, allowing the waste or by-products of one company to be utilized as raw materials by another, thus diminishing overall resource consumption and environmental impact. SIPs may integrate green infrastructure, including parks, green roofs, and sustainable drainage systems, to alleviate the urban heat island effect, boost air quality, and promote biodiversity. By exemplifying optimal practices in industrial sustainability, SIPs can operate as paradigms for eco-industrial advancement and motivate the environmental enhancement of industries outside their confines.

[1] Purchasing Power Parity, an economic adjustment that accounts for differences in price levels between countries, allowing for more accurate international comparisons of economic data like energy intensity.

2.1.3 Social Aspect

Central place theory from Christaller's hierarchical concepts is used to understand SICs as specialized nodes within a polycentric metropolitan area (Martyniuk-pęczek et al. 2020). Despite the apparent advantages, evaluating the impact of SIPs on sustainable urbanization continues to pose challenges. A significant difficulty is the absence of comprehensive and consistent measures to evaluate the social success of SIPs. The intricate relationships between SIPs and adjacent urban systems, including transportation networks, housing supplies, and social dynamics, complicate the evaluation process. The diversity of SIPs regarding size, industrial makeup, and management systems complicates the generalization of their impacts on sustainable urbanization. To tackle these problems, it is essential to develop more comprehensive and cohesive assessment frameworks that account for the multifaceted dimensions of sustainability and the circumstances of each SIP. This may entail the creation of context-specific indicators, participatory evaluation methodologies, and multi-stakeholder engagement to encompass the many viewpoints and experiences of SIP stakeholders, including enterprises, employees, communities, and local authorities.

2.2 *From SIP to SIC: A Paradigm Shift*

The SIC (Fig. 13.3) represents a fundamental reconceptualization through a nexus lens—recognizing that SIPs function within a web of spatial, functional, and temporal interdependencies that extend well beyond their formal boundaries. The SIC nexus reveals how changes in one component cascade through the system, creating synergies or trade-offs across sustainability dimensions.

The SIC consists of three spatial components:

- *SIP:* the designated industrial park with defined boundaries (typically 26–67 ha in the TMA)
- *Fringe area:* a 160-m buffer zone immediately surrounding the SIP
- *Neighborhood area:* the broader influence zone extending 626–1831 m from the SIP boundary, determined by gravity modeling based on park size and daytime population density

2.3 *Progressive Quantification Approaches*

The nexus model of SIC is a multi-scale analytical framework that conceptualizes suburban industrial parks as integrated components of broader urban systems. It examines the synergies and trade-offs between:

- *Spatial nexus:* SIP, fringe area, and neighborhood area

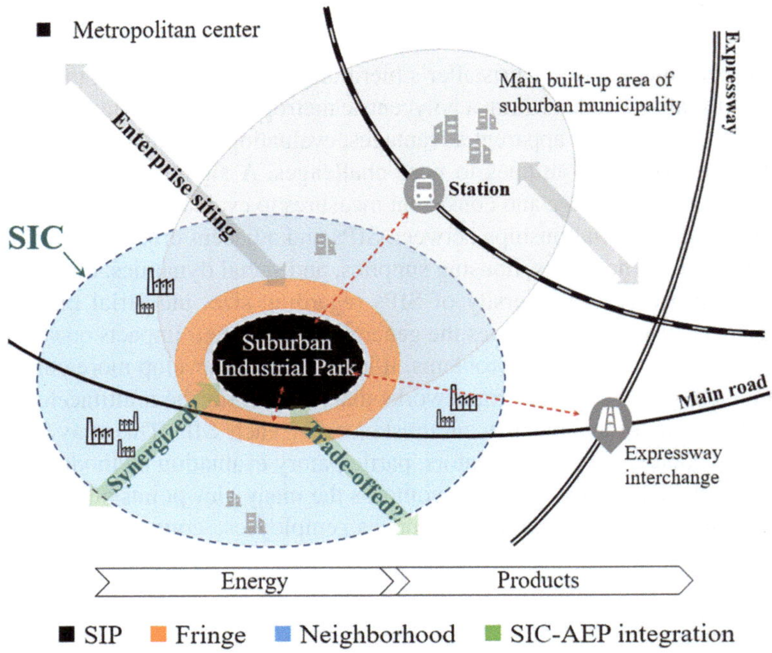

Fig. 13.3 Nexus model of SIC and SIC-based Area Energy Performance framework

- *Functional nexus:* industrial agglomeration, energy efficiency, and urban diversification
- *Temporal nexus:* developmental phases and their feedback loops

The SIC model employs three interconnected quantification methods that reveal the nexus dynamics of sustainability performance (Fig. 13.4), where each approach progressively informs and reinforces the others through feedback loops: (1) industrial agglomeration (Zhang and Yan 2024); (2) energy efficiency (Zhang and Yan 2025); and (3) urban function diversification.

Among them, industrial agglomeration approach employed Temporal Pattern of Spatial Features categorization through agglomeration indices at three scales—contrasting SIC to municipality; SIP to SIC; and fringe to neighborhood area—to reveal four agglomeration strength grades (A-strongest to D-weakest) in SIC samples. And energy efficiency approach developed Area Energy Performance (AEP) indicator, combining industrial energy productivity and local industrial structure, as well as realizing comparative assessment between SIP, SIC, and municipality levels. Urban function diversification represents the level of urbanization and capability of improvement in residents' quantity of life.

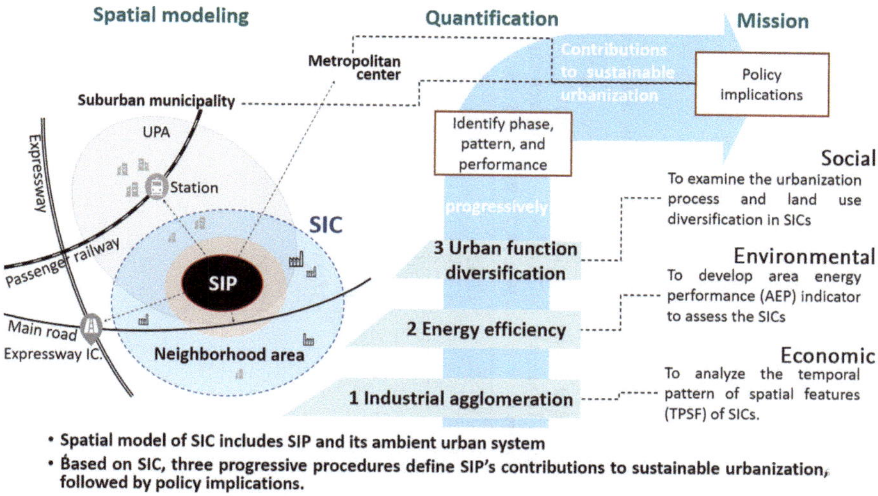

Fig. 13.4 Progressive SIC sustainable approaches (Zhang 2025)

3 Empirical Studies in the Tokyo Metropolitan Area

3.1 Study Area and Context

The Tokyo Metropolitan Area (TMA) provides an exemplary case for examining SIC evolution. It has systematic industrial suburbanization policy since the 1950s (Zhang et al. 2024), comprehensive data availability across multiple decades, and diverse industrial composition and developmental stages (Table 13.1 and Fig. 13.5). Peripheral prefectures show higher manufacturing concentration (31.23% of GDP) than Tokyo (12.45%). Machinery manufacturing dominates peripheries, especially in Ibaraki (14.86%) and Kanagawa (9.92%). Chemical industries concentrate in Chiba (6.70%), while transport/logistics distribute evenly across regions (4.5–7%). Employment concentration (Fig. 13.5a) shows distinct clustering along transportation corridors, with 1000–5000 employees per firm in industrial zones. Manufacturing productivity (Fig. 13.5b) reveals a core–periphery gradient: Tokyo's center achieves 6000+ thousand yen per employee through high-value services, while peripheral SIPs show 2000–4000 thousand yen per employee in traditional manufacturing. The TMA's energy profile mirrors its economic composition, characterized by a substantial presence of manufacturing, logistics, and utility sectors, in conjunction with a predominant service economy. The industrial sector of the region, encompassing SIPs, constitutes a significant portion of this electricity consumption (Figs. 13.6 and 13.7).

The TMA has experienced a steady separation of industrial energy use from economic growth, propelled by enhancements in efficiency, structural transformations, and policy interventions. The rate and degree of this decoupling fluctuate

Table 13.1 Industries and economic shares in peripheries of the TMA (based on real gross prefectural domestic product classified by economic activities as of 2019, unit: billions of YEN)

	Manufacture of Machinery (%)	Manufacture of Chemical and Allied Products, Oil and Coal Products (%)	Transport and Postal Activities (%)	Electricity, Gas, Heat Supply and Water (%)	Manufacture of Food, Beverages, Tobacco and Feed (%)	11 main industries (Fig. 13.7) (%)	Total prefectural real GDP (%)
Tokyo	3.62	0.95	3.77	1.52	0.82	12.45	112,711
Kanagawa	9.92	4.28	4.99	2.99	2.37	27.57	34,604
Saitama	7.79	3.57	5.31	3.44	3.51	30.61	23,158
Chiba	2.82	6.70	6.99	4.88	3.54	30.55	20,816
Ibaraki	14.86	5.48	4.53	3.39	5.07	42.40	13,865
4 periphery prefectures	8.53	4.83	5.45	3.59	3.33	31.23	92,443
TMA	5.83	2.70	4.53	2.45	1.95	20.91	205,153

(a) Number of employee per manufacturing firm

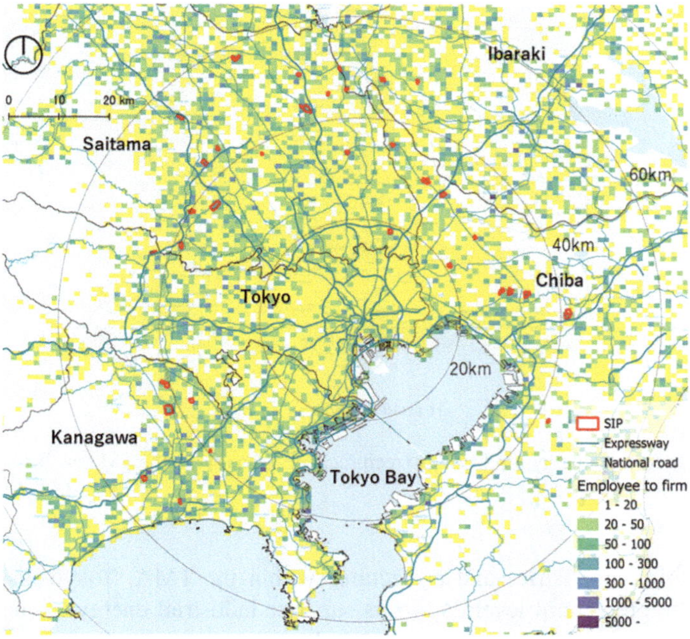

(b) Manufacturing shipping value per employ (unit: 10000 YEN)

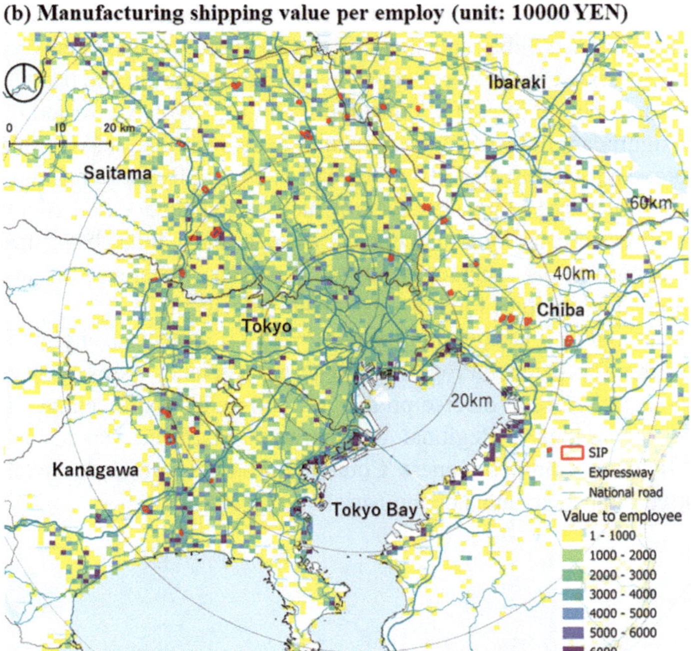

Fig. 13.5 Manufacturing and productivity in the TMA (based on National Statistics of Industries, mesh data, 2010)

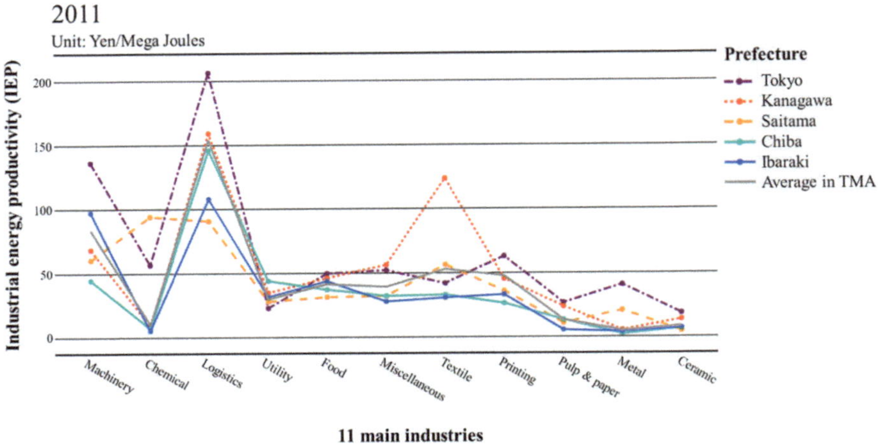

Fig. 13.6 Industrial energy productivity (reciprocal of intensity) in the TMA in 2011

among various industries and prefectures within the TMA. Tokyo distinguishes itself at the prefectural level due to its superior industrial energy intensity across several sectors, attributable to modern technologies, stringent regulations, and a predominantly service-oriented economic framework. Prefectures such as Kanagawa, Saitama, Chiba, and Ibaraki possess more energy-intensive industrial sectors, characterized by greater proportions of manufacturing and heavy industries. The spatial distribution of energy consumption in the TMA is affected by the positioning of significant industrial zones, such as SIPs, near the outskirts of the metropolitan core. The western regions of the TMA, especially in Kanagawa, exhibit a concentration of energy-intensive sectors such as machinery and chemicals. The northern and eastern regions, namely in Saitama and Chiba, have more diverse industrial compositions, characterized by a greater prevalence of less energy-intensive industries such as food processing and logistics.

Significant policy initiatives influencing TMA's industrial energy framework encompass the Top Runner Program for energy efficiency assessment, the Tokyo "HTT deco activity,"[2] and diverse prefectural and municipal incentives for the use of greener technologies. Regional collaboration platforms, such as the Kanto Regional Industry Competitiveness Council,[3] promote knowledge exchange and collaborative projects to enhance industrial energy efficiency.

[2] https://www.tokyo-co2down.jp/countermeasures/web-event/cool-choice

[3] https://www.meti.go.jp/policy/newmiti/mission/2016/pdf/2_3_3.pdf

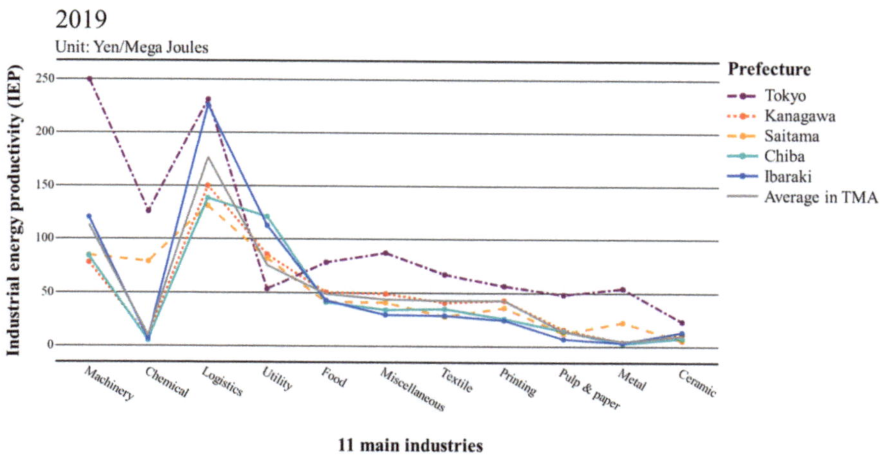

Fig. 13.7 Industrial energy productivity (reciprocal of intensity) in the TMA in 2019

3.1.1 Industrial Land Policy Framework

Administrative support for manufacturing industry land clusters primarily encompasses four main aspects: infrastructure development, soft administrative services, land provision, and supplementary assistance. Infrastructure development includes capital grants from higher administrative bodies to local municipalities, notably through programs such as the "Regional Industrial Structure Transformation Infrastructure Grant" for semiconductor-related industries and the "Comprehensive Social Capital Infrastructure Grant" for manufacturing.[4] Soft administrative services focus on developing industrial park information platforms, establishing investment promotion offices, and promoting industrial land advantages, supported by specialized fiscal budgets and digital implementation grants.

3.1.2 "Industrial Park Development Program"

The Industrial Park Development Program, established under the 1956 Capital Region Development Act, has been the primary mechanism for industrial land supply in Tokyo's metropolitan area.[5] By 2010, there were 133 inland industrial parks within the Suburban Development Zone of the Tokyo Metropolitan Area (Fig. 13.8). Development peaked during the rapid growth period of the 1960s–1980s, experienced some oversupply during the bubble economy, and saw renewed growth with highway network expansion in the new millennium. Figure 13.8 presents a regional configuration of the TMA, delineating various SIPs within a 20–60 km distance

[4] https://smartisland.mlit.go.jp/wp-data/wp-content/uploads/2024/03/001.pdf

[5] https://www.mlit.go.jp/hakusyo/syutoken_hakusyo/h18/images/12.pdf page 84–86

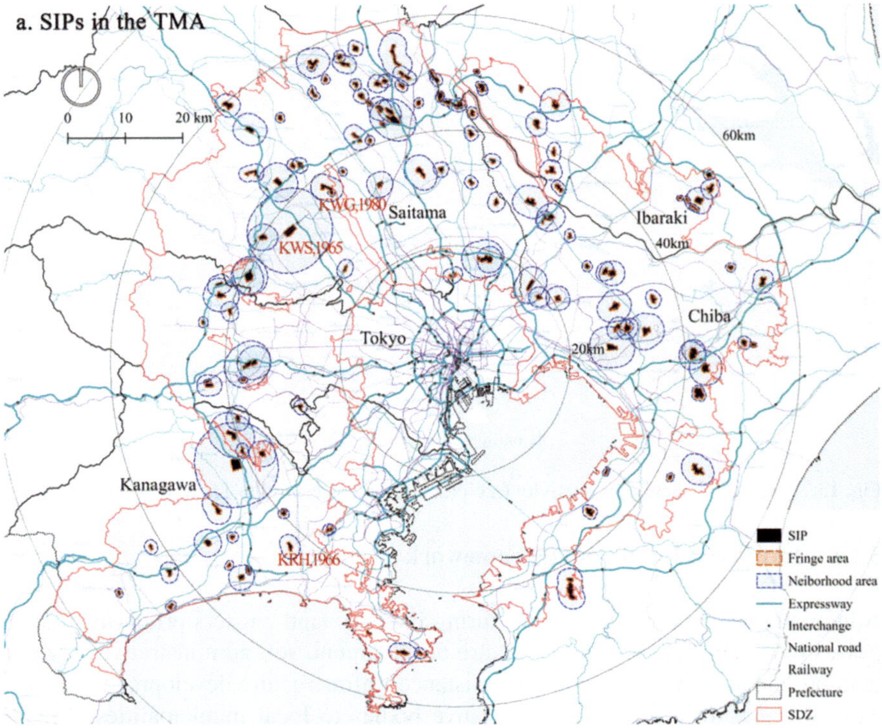

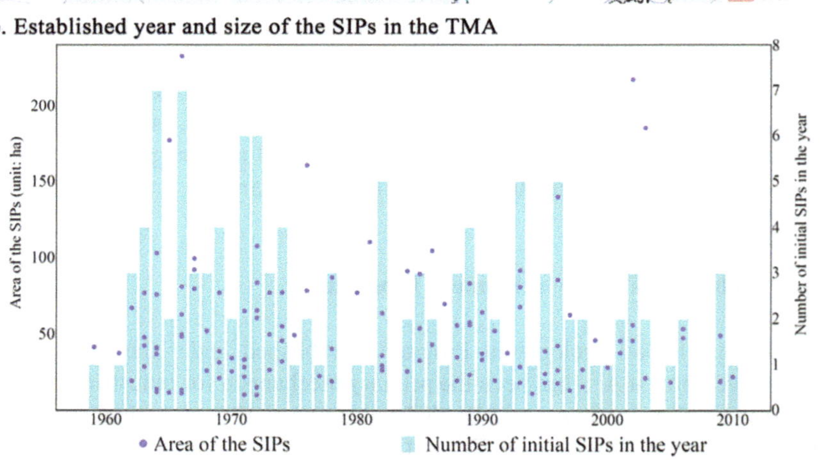

Fig. 13.8 (**a**) SIPs in the TMA, with their 160-m fringe areas and neighborhood areas; (**b**) year of establishment and size of the SIPs

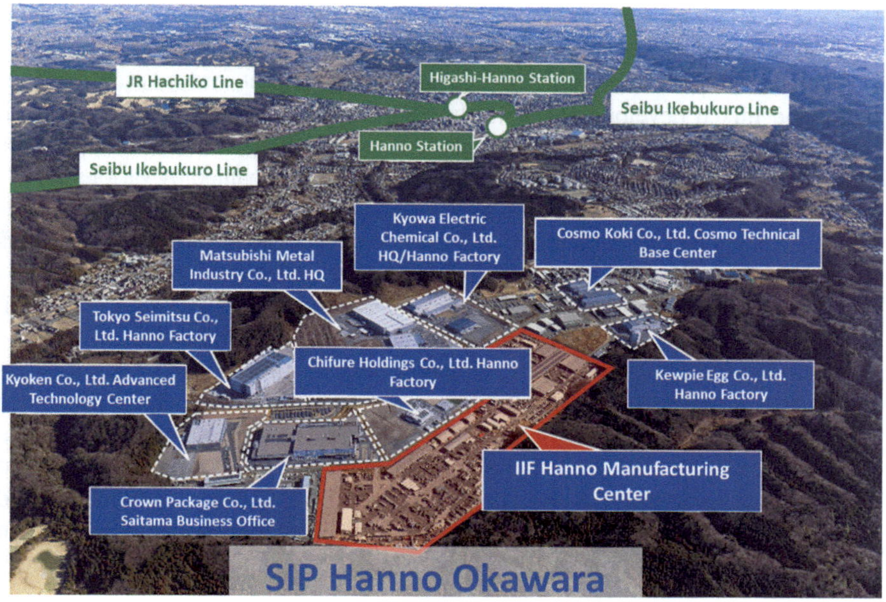

Fig. 13.9 SIP Hanno Okawara (areas in red indicate plots acquired through CRE/PRE financing methods, translated based on source from Industrial Investment Corporation, 2023, https://www.iif-reit.com/portfolio/guq72t0000010jn.html)

from the metropolitan center. The figure highlights the spatial distribution and developmental extent, with black marking inland SIPs. Notably, these SIPs tend to cluster around transportation corridors, and mainly in the SDZ[6] of the TMA. The program encompasses comprehensive infrastructure development, including buffer forests, retention ponds, park areas, roads, and utility systems, though building construction remains the responsibility of land-purchasing companies (Fig. 13.9).

3.1.3 Development Entities, Land Acquisition, and Models

Development entities primarily consist of various administrative levels and their managed development corporations, including the national-level Urban Renaissance Agency (UR), prefecture-level entities such as the Saitama Prefecture Enterprise Bureau, and municipal-level administrative organizations. Private development accounts for less than 10% of industrial parks in Tokyo's suburbs, primarily through local landowner committees or major real estate developers such as Mitsui Real Estate. The land acquisition process involves careful coordination with existing

[6] Established in 1956, the SDZ aims to direct the TMA's rapid suburban expansion in line with broader objectives such as decentralization and infrastructure enhancement. Well-connected to the metropolitan core via railways and expressways, the SDZ incorporates a mix of natural, residential, commercial, and industrial zones, promoting inclusive and sustainable growth.

land-use regulations, agricultural and forest protection policies, and various administrative procedures, which typically takes 1–2 years to complete.

Three development models exist: advance development, order-based development, and mixed development, with the mixed approach becoming predominant after the bubble economy. Land distribution is almost exclusively through sales rather than leasing in the Tokyo Metropolitan Area, accompanied by various preferential policies. Individual industrial park sizes range from 10 to 240 ha, with development closely tied to market demand to prevent land waste. Recent discussions have explored the possibility of lease-only industrial parks, though this approach is currently more common in nonmetropolitan areas such as Hokkaido, Aomori, Niigata, and Nagano prefectures, as well as parts of Ibaraki Prefecture.

Notwithstanding advancements, obstacles persist in reconciling TMA's industrial energy usage with enduring environmental objectives. This encompasses the necessity for enhanced decarbonization of energy supply, expedited dissemination of sophisticated efficiency technologies, and improved integration of energy and spatial planning at the SIP and regional levels. To effectively comprehend and address these spatial variances and issues in energy use, it is beneficial to envision SIPs as components of broader SIC.

3.1.4 Supply Chain Decarbonization Initiatives

The Japanese government has initiated a comprehensive program for promoting and advancing decarbonization management across entire supply chains, with a proposed budget of 1.5 billion yen for fiscal year 2023. The program aims to integrate climate change considerations into corporate management practices while strengthening businesses' competitiveness through decarbonization efforts. This initiative responds to the global expansion of ESG (Environment, Social, and Governance) financing, recognizing that supply chain emissions significantly impact corporate value. The program seeks to create a positive cycle between economic growth and environmental protection by attracting both domestic and international ESG investment through enhanced corporate decarbonization practices (Fig. 13.10).

The program's first major component focuses on supply chain decarbonization promotion, encompassing three key areas: organizational support for supply chain decarbonization, visualization and reduction of product/service emissions, and assistance with decarbonization strategy development and information disclosure. This component introduces the concept of Scope 3 emissions calculation, which allows companies to reflect the reduction efforts of their supply chain partners in their own emission calculations. Additionally, it supports the implementation of carbon footprint practices for products and services, while also incorporating TNFD[7]-aligned scenario analysis and the development of internal carbon pricing mechanisms to enhance corporate climate strategies.

[7] Taskforce on Nature-related Financial Disclosures

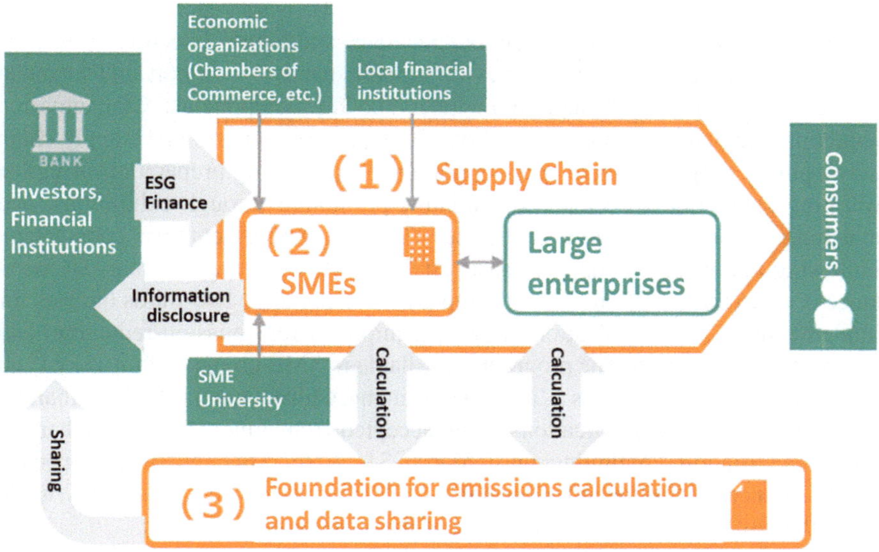

Fig. 13.10 Supply chain decarbonization management initiatives. (Translated based on source from Ministry of the Environment, Japan, https://www.env.go.jp/content/000156352.pdf)

The second significant element of the program specifically targets small and medium-sized enterprises (SMEs), addressing their unique challenges in implementing decarbonization management practices. This component includes the establishment of a comprehensive support system through three main initiatives: providing information and technical assistance for emissions calculations, developing regional support networks through collaboration with local financial institutions and chambers of commerce, and supporting the creation of emission reduction plans. The program recognizes that many SMEs lack the knowledge, resources, and technical expertise to initiate decarbonization efforts, making targeted support crucial for achieving supply chain-wide emissions reductions.

The final component of the program focuses on developing and maintaining the infrastructure necessary for emissions calculation and data sharing. This includes the maintenance and enhancement of the electronic reporting system for energy conservation, global warming countermeasures, and fluorocarbon regulations. The system will be upgraded to accommodate users beyond those legally required to report their emissions, enabling voluntary participation in emissions calculation and reporting. Furthermore, the program aims to establish a comprehensive platform for emissions data sharing, making it easier for both large corporations and SMEs to calculate, share, and report their emissions data. This infrastructure development is essential for creating a more efficient and accessible environment for corporate decarbonization efforts across the entire supply chain.

The Tokyo Metropolitan Government has launched an ambitious initiative for fiscal year 2024 called "HTT Deco Activities" to promote decarbonization across supply chains, particularly focusing on SMEs. The "H" is for "decrease" (electricity use) in Japanese and the two "T" are for "generate" and "store" (electricity), respectively. The program offers two types of support: hands-on support and financial subsidies.[8] The hands-on support includes six key components: CO_2 emission reduction planning, training of decarbonization promotion personnel, supply chain development support, effectiveness measurement (including preparation of global warming countermeasure reports), initiative promotion through exhibitions, and matching support with large enterprises (limited to representative companies). Following the hands-on support phase, eligible SMEs can receive financial assistance of up to three million yen per company, covering up to two-thirds of costs for implementing CO_2 visualization systems, equipment for emission reduction, and exhibition participation. The program is available to SME groups within Tokyo and eight surrounding prefectures (Kanagawa, Saitama, Chiba, Gunma, Tochigi, Ibaraki, and Yamanashi), with applications being accepted from June 21 to December 27, 2024, and support continuing until March 31, 2026. This initiative represents Tokyo's concrete step toward achieving its "Zero Emission Tokyo" strategy while strengthening the competitive advantage of local businesses.

3.2 *Application of SIC Model in the TMA*

Our analysis encompasses 133 SIP samples across 74 municipalities in peripheries of the TMA. The temporal coverage is 1960s–2010s establishment, analyzed at 1980s, 1990s, 2000s, and 2010s. Multiple datasets are used, including land use (10 m mesh), industrial statistics (1 km mesh), enterprise locations (point data), transportation infrastructure, and energy consumption. The utilization of the SIC model in the TMA facilitates the identification of geographical and functional links between SIPs and their adjacent regions, as well as the evaluation of their combined environmental and socioeconomic performance.

3.2.1 Industrial Agglomeration Patterns

Analysis of 268 observations from 110 valid SIP samples reveals three distinct developmental phases (Zhang and Yan 2024):

- *Phase I* (under 20 years): strongest agglomeration at all spatial scales, with 74% of observations showing positive agglomeration indices, dominated by machinery, chemical, and metal industries.
- *Phase II* (20–35 years): stabilization period, mixed agglomeration patterns with 55% positive indices, with industrial land transition beginning in fringe areas.

[8] https://www.my.metro.tokyo.lg.jp/w/000-20240621-41728558

- *Phase III* (over 35 years): diversification and renewal, with significant land-use transformation in neighborhood areas, and emergence of mixed industrial–commercial functions.

Transportation infrastructure analysis reveals that SICs are "pulled by main roads" (average 499 m proximity), "pushed by passenger stations" (average 2721 m distance), with Grade A (strongest agglomeration) SICs show 18% higher main road density.

3.2.2 Energy Performance Dynamics

Area Energy Performance analysis of 89 SIC samples demonstrate neighborhood effects on AEP (Zhang and Yan 2025):

- 67% of SICs show improved AEP when neighborhood areas included.
- Average improvement of 15–30%.
- Optimal cases: SIP Mitsuhara in Ome City, Tokyo Pre. (Machinery and logistics concentration in neighborhoods).

The spatial distribution of energy consumption and production within the SIC constitutes a significant facet of AEP. The closeness of energy-intensive sectors to renewable energy sources, such as solar panels or biomass facilities, can facilitate the equilibrium of supply and demand while minimizing transmission losses. The aggregation of companies with complementary energy profiles can facilitate more efficient energy cascading and load balancing. Spatial planning instruments, such energy mapping and zoning, can enhance the energy performance of the SIC. The SIC model serves to evaluate the AEP of several SICs within the TMA and to discern the principal drivers and obstacles to energy efficiency and decarbonization. This can guide targeted policies and expenditures to foster cleaner production methods, renewable energy uptake, and industrial symbiosis. The model can assist in optimizing the location and land-use patterns of SICs by considering issues such as accessibility, resource availability, and compatibility, with adjacent urban functions. The socioeconomic examination of SICs in the TMA can yield significant insights into the dynamics of industrial agglomeration, employment, and innovation. By analyzing the expansion and contraction of various industries and clusters within SICs, planners can discern emerging opportunities and difficulties for regional economic development. The SIC model can aid in evaluating the social and distributional effects of industrial policy, ensuring that the advantages of industrial expansion are more equally distributed throughout the region.

The energy–economy–society nexus within SICs demonstrates that neighborhood integration creates synergistic effects: industrial symbiosis reduces energy intensity, which enhances economic competitiveness, ultimately improving social outcomes through job creation and urban amenities.

4 Policy Implications and Implementation

4.1 From SIP Management to SIC Governance

The nexus approach reveals that isolated interventions fail to capture systemic benefits. Policy must address the interconnected nature of SIC sustainability, where economic incentives, environmental regulations, and social development initiatives create reinforcing feedback loops. The TMA exemplifies how the SIC model may offer valuable insights and direction for planners, legislators, and practitioners aiming to foster more sustainable and resilient industrial communities.

- The SIC model can inform industrial policy and urban planning decisions by providing a holistic understanding of the impacts and potential of SIPs. It can guide the location, design, and management of SIPs to optimize their sustainability performance.
- The model supports the development of eco-industrial parks and networks by identifying synergies and complementarities among industries. It can facilitate the implementation of industrial symbiosis projects and the creation of closed-loop resource systems.
- SICs serve as units of analysis for assessing the sustainability of metropolitan regions. By comparing the performance of different SICs, regions can identify best practices, benchmark their progress, and prioritize interventions.
- The SIC model provides a platform for engaging multiple stakeholders, including industrial park managers, firms, local governments, and communities. It can foster dialogue, collaboration, and collective action toward sustainable urban–industrial development.

The findings from the application of SIC concept in the TMA necessitate a fundamental shift in industrial park policy. Traditional SIP-focused approach has limitations, such as isolated planning missing neighborhood synergies, single-dimension performance metrics, and static management ignoring temporal evolution. While our SIC-based integrated governance advantages coordinated planning across park–neighborhood boundaries, multidimensional sustainability assessment, and adaptive management responding to developmental phases.

4.1.1 Location of the SIPs

The location of SIPs is a crucial determinant of their environmental and socioeconomic effects. SIPs are generally situated in the outskirts of urban centers, where land is comparatively plentiful and economical. The precise placement of a SIP inside this periphery zone might fluctuate significantly, influenced by considerations such as accessibility, infrastructural availability, and political boundaries.

The placement of SIPs has considerable consequences on their energy and resource efficiency. Strategically positioned SIPs near significant transportation routes, including roads and ports, can see diminished logistics expenses and emissions. SIPs located in regions with ample renewable energy supplies, such as solar or wind power, can exhibit a reduced carbon footprint. The closeness of SIPs to residential zones and other urban functions might impact their environmental performance by influencing waste heat recovery, industrial symbiosis, and eco-efficient land-use patterns. Decisions regarding the siting of SIPs are frequently affected by regional and municipal land-use rules, zoning regulations, and economic development initiatives. These regulations can either promote or inhibit the aggregation of industrial operations, the diversity of land uses, and the development of supporting infrastructure. For instance, certain places may possess distinct incentives or prerequisites for the creation of eco-industrial parks, whereas others may exhibit more permissive environmental rules or diminished enforcement. The alignment of siting policies with sustainability objectives is essential for the long-term viability and efficacy of the SICs.

4.1.2 Socioeconomic Structure

The socioeconomic structure of SICs encompasses the composition of industries, enterprises, and employment inside the community, along with their interactions with the broader regional economy. This structure is influenced by the agglomeration effects of the SIP, which draw associated sectors and support services to the adjacent regions. The diversity and specialization of the industrial base, the scale and ownership of enterprises, and the skill composition of the workforce are all critical components of the socioeconomic framework. An essential benefit of SICs is their capacity to promote industrial symbiosis and intersectoral collaboration. By clustering several industries within a confined geographic region, SICs can promote the interchange of materials, energy, and information among enterprises. For instance, the waste heat generated by one industry can be utilized to energize another, or the by-products of one process can function as inputs for another. These symbiotic interactions can enhance resource efficiency and competitiveness within the community, while simultaneously mitigating environmental concerns. The attainment of these advantages is contingent upon the composition and compatibility of industries within the SIC. Certain industries may exhibit complimentary resource flows and production processes, whereas others may present competing requirements or environmental consequences. The existence of anchor enterprises or industrial clusters might affect the dynamics of collaboration and rivalry within the SIC. The enterprise-oriented local industrial framework of the SIC is thus a crucial factor influencing its total energy performance and resilience.

4.2 Phase-Specific Policy Recommendations

Phase-specific policy recommendations could be acquired based on the industrial agglomeration assessment (Zhang and Yan 2024):

- Phase I SICs (Growth stage): accelerate industrial agglomeration through targeted incentives, strengthen infrastructure connections, particularly road networks, and reserve land for future mixed-use transition.
- Phase II SICs (Stabilization stage): promote industrial symbiosis and resource efficiency, begin selective mixed-use development in fringe areas, and enhance public transportation accessibility.
- Phase III SICs (Renewal stage): implement comprehensive redevelopment strategies, encourage vertical mixed-use and increased density, and transform portions for residential and commercial use.

4.3 Energy Performance Optimization

SIPs and their surrounding areas are substantial energy consumers in metropolitan peripheries. The aggregation of manufacturing, logistics, and utility firms within SIPs results in significant electricity and fuel consumption. Notwithstanding enhancements in efficiency within specific industries, the aggregate industrial energy consumption of the community remains elevated due to their magnitude and the energy-intensive characteristics of several industrial processes. The energy consumption profile of SIPs comprises a combination of electricity, fossil fuels (coal, oil, natural gas), and, in certain instances, renewable sources. Electricity energizes machines, illumination, HVAC[9] systems, and office functions. Fossil fuels are utilized for process heating, steam production, and on-site electricity generation. The proportion of renewable energy sources, including solar photovoltaic, wind, and biomass, fluctuates but is often minimal. Manufacturing sectors such as equipment, chemicals, metals, and ceramics are typically the most energy-intensive. Logistics and utilities substantially influence SIP energy consumption. The specific energy consumption patterns are contingent upon the industrial composition and technological advancement of each SIP. Principal obstacles in regulating SIP energy usage encompass outdated infrastructure, suboptimal equipment, absence of cohesive energy management systems, and insufficient implementation of cleaner technology. Addressing these obstacles necessitates focused policies, investments, and engagement among stakeholders.

Based on AEP analysis, policy priorities include (Zhang and Yan 2025):

[9] Heating, Ventilation, Air Conditioning

- *Industrial composition optimization*: prioritize high-productivity industries (machinery, logistics, chemicals); phase out energy-intensive low-productivity sectors; and promote industrial symbiosis for waste heat utilization.
- *Neighborhood integration strategies*: coordinate industrial development across SIC boundaries; create shared infrastructure and utilities; and implement district energy systems.
- *Digital and circular economy integration*: smart park management systems; real-time energy monitoring and optimization; and material flow tracking and exchange platforms.

Various factors influence the energy consumption patterns of SIPs, functioning at diverse scales. At the macro level, national and regional energy policy, fuel costs, and technical developments establish the overarching context. Regulations, incentives, and infrastructure provision at the prefectural and municipal levels also affect SIP energy use. The composition of sectors inside SIPs and their corresponding energy intensity are the principal factors influencing consumption trends. The existence of energy-intensive industries such as chemicals, metals, and machinery can substantially elevate a SIP's energy footprint. The technological advancement and age of industrial equipment also influence efficiency levels. Spatial elements, including the configuration and concentration of industrial operations within SIPs, can influence energy consumption via transportation demands, infrastructure necessities, and opportunities for waste heat recovery. Compact and integrated designs of SIPs may exhibit reduced energy intensity. Firm-specific factors, including operational processes, energy management competencies, and investments in efficiency initiatives, also influence consumption trends. Large corporations with specialized energy management teams and financial resources may be more adept at optimizing their energy consumption. Individual and organizational behavioral factors, including the awareness, abilities, and priorities of managers and employees, can impact energy usage via operational decisions and compliance with efficiency measures. Training and engagement initiatives can facilitate the alignment of habits with efficiency objectives. External factors such as market demand, supply chain constraints, and stakeholder expectations can influence SIP energy use by modifying production levels or necessitating efficiency enhancements for competitive and environmental performance.

4.4 Additional Recommendations

4.4.1 Industrial Policy Integration

Embedding SIC principles into national and regional industrial development plans is essential for advancing sustainable industrialization. These policies should explicitly recognize spatial integration and environmental sustainability as core components of industrial planning frameworks. By incorporating SIC concepts at the policy level, governments can create consistent regulatory environments that promote resource efficiency and industrial symbiosis while aligning with broader sustainable development goals.

4.4.2 Zoning Reform for Infrastructure Sharing

Reforming land-use and zoning regulations is critical to enable the clustering of compatible industries and development of shared infrastructure. Such revised regulations should incentivize the colocation of complementary businesses that can benefit from resource exchange, while removing barriers to establishing common utility systems, waste management facilities, and transportation networks. These zoning modifications would create the physical and regulatory conditions necessary for industrial symbiosis to flourish, reducing both infrastructure costs and environmental impacts.

4.4.3 Digital Optimization of Resource Flows

Leveraging digital technologies and data-driven approaches is key to maximizing resource efficiency within SIC. Implementation of eco-industrial park information systems can enable real-time monitoring and management of material flows, energy consumption, and waste generation. These digital tools support decision-making for optimizing industrial processes, identifying new symbiotic opportunities, and measuring environmental performance improvements. Advanced analytics can further enhance resource circulation by predicting potential exchanges and evaluating their economic and environmental benefits.

5 Conclusions

Traditional approaches treat industrial parks as isolated entities, missing critical interdependencies between parks and their surrounding urban systems. The proposed nexus model of SIC in this chapter addresses this gap by recognizing that sustainable industrial development requires understanding the interconnections between spatial scales (park–fringe–neighborhood), temporal phases (growth–stabilization–renewal), and sustainability dimensions (economic–environmental–social). It was applied in the TMA, offering a significant framework for comprehending and regulating the intricate relationships between industrial parks and their adjacent urban systems. The model provides insights into the principal drivers and difficulties of industrial suburbanization in metropolitan peripheries by incorporating spatial, economic, and environmental elements. The findings in the TMA illustrates its capacity to inform policy and planning decisions. The analysis indicates considerable regional disparities in the local industrial composition and energy performance of SICs. These variances are affected by factors including the establishment period of the SIPs, the location choices of firms, and the agglomeration impacts of associated industries. This chapter advances sustainable urbanization theory and practice through:

- *Conceptual innovation*: the nexus model provides a new lens for understanding industrial parks as integrated urban systems rather than isolated entities.
- *Methodological advancement*: the progressive quantification approach enables comprehensive sustainability assessment across multiple dimensions and scales.
- *Empirical insights*: the 50-year analysis of Tokyo's SIPs reveals systematic patterns of evolution and sustainability performance previously undocumented.
- *Policy framework*: phase-specific recommendations and SIC-based governance strategies offer actionable guidance for metropolitan planning.

This study acknowledges several limitations, including neighborhood boundary definitions based primarily on population and size factors and the requirement for social quality indicators beyond urban function and land use. Future research should address these gaps while exploring applications across global metropolitan contexts and investigating how Industry 5.0 will impact sustainable industrial clustering patterns.

The nexus model of SIC offers valuable insights for sustainable urbanization in the metropolitan peripheries worldwide. For emerging economies, the framework for balanced industrial–urban development helps avoiding sprawl patterns. For developed countries, the strategies of neighborhood improvement promote industrial park renewal and sustainable transformation. For global sustainability, concrete pathways for achieving SDG 9 (Industry, Innovation, Infrastructure) and SDG 11 (Sustainable Cities and Communities) can be learnt from the SIC approaches. The transition from SIP to SIC represents more than semantic evolution—it embodies a fundamental shift toward integrated, sustainable industrial–urban development essential for twenty-first-century cities and metropolitan areas.

Acknowledgments This chapter presents a key concept from research supported by JST SPRING, Japan (Grant Number JPMJSP2123), the Taikichiro Mori Memorial Research Grants, and Keio University's Doctoral Student Grant-in-Aid Program through the Ushioda Memorial Fund. We express our sincere gratitude for this support.

References

Botelho R (2019) Does Transit-Oriented Development TOD Attract Private Investment? Exploring the Investment Strategies of Horeca Businesses in Station Areas Radboud University. https://doi.org/10.13140/RG.2.2.14501.04326

Clark B (2003) Ebenezer howard and the marriage of town and country. Organ Environ 16(1):87–97. https://doi.org/10.1177/1086026602250258

Evans G (2009) Creative cities, creative spaces and urban policy. Urban Stud 46(5–6):1003–1040. https://doi.org/10.1177/0042098009103853

Hossain MA, Huggins R (2021) The Environmental and Social Impacts of Unplanned and Rapid Industrialization in Suburban Areas: The Case of the Greater Dhaka Region, Bangladesh. Environ Urbaniz Asia 12(1):73–89. https://doi.org/10.1177/0975425321990319

Hurley J, Grodach C, Martin D, Taylor E (2023) Do industrial firms follow zoning? changing firm location and the introduction of metropolitan zoning. Urban Policy Res:1–18. https://doi.org/10.1080/08111146.2023.2236133

Ian D (2005) Peri-urban ecosystems and societies: transitional zones and contrasting values. In: Duncan M, David S (eds) The peri-urban interface: approaches to sustainable natural and human resource use, 1st edn. Routledge, pp 17–28. https://doi.org/10.4324/9781849775878

IEA (2022) Japan—Countries & Region. Available Online at: https://www.iea.org/countries/japan/efficiency-demand, Accessed on 9 Dec 2024

Kikuchi I (2000) The change of industrial structure and the development of industrial parks in Kanagawa prefecture. Kyoikugakubukiyo 34. https://bunkyo.repo.nii.ac.jp/record/342/files/BKK0000214.pdf

Kim KS, Gallent N (1998) Regulating industrial growth in the South Korean Capital region. Cities 15(1). https://doi.org/10.1016/S0264-2751(97)10002-6

Kuang W, Liu J, Dong J, Chi W, Zhang C (2016) The rapid and massive urban and industrial land expansions in China between 1990 and 2010: A CLUD-based analysis of their trajectories, patterns, and drivers. Landsc Urban Plan 145:21–33. https://doi.org/10.1016/j.landurbplan.2015.10.001

Li M, Li L, Li X (2022) Restructuring urban outskirts industrial areas from the industrial clustering perspective: a case study in Shunde, China. Land 11(11). https://doi.org/10.3390/land11112004

Lyu Y, Liu Y, Guo Y, Sang J, Tian J, Chen L (2022) Review of green development of Chinese industrial parks. Energ Strat Rev 42. https://doi.org/10.1016/j.esr.2022.100867

Martyniuk-pęczek J, Martyniuk O, Parteka T (2020) Entrepreneurship nests in a polish edge city. GUT Publishing House. https://pbc.gda.pl/publication/92005

Minnery J (2012) Model industrial settlements and their continuing governance. Plan Perspect 27(2):309–321. https://doi.org/10.1080/02665433.2012.654980

Okata J, Murayama A (2011) Tokyo's urban growth, urban form and sustainability. In: Sorensen A, Okata J (eds) Megacities: urban form, governance, and sustainability. Springer Japan, pp 15–41. https://doi.org/10.1007/978-4-431-99267-7_2

Park J-I, Kim J-O (2022) Does industrial land sprawl matter in land productivity? A case study of industrial parks of South Korea. J Clean Prod 334. https://doi.org/10.1016/j.jclepro.2021.130209

Sakai T, Kawamura K, Hyodo T (2020) Logistics facilities for intra and inter-regional shipping: Spatial distributions, location choice factors, and externality. J Transp Geogr 86. https://doi.org/10.1016/j.jtrangeo.2020.102783

Scott AJ (1986) Industrialization and urbanization: a geographical agenda. Ann Assoc Am Geogr 76(1):25–37. https://doi.org/10.1111/j.1467-8306.1986.tb00101.x

UNIDO (2019a) International Guidelines for Industrial Parks. Available Online at: https://www.unido.org/sites/default/files/files/2020-05/International_Guidelines_for_Industrial_Parks_EN.pdf, Accessed on 9 Dec 2024

UNIDO, World Bank Group, Deutsche Gesellschaft für Internationale Zusammenarbeit, Ministry of Trade, Industry and Energy of the Republic of Korea (2019b). A Practitioner's Handbook for Eco-Industrial Parks: Toolbox. Available Online at: https://openknowledge.worldbank.org/entities/publication/d7c1b724-3c60-5dfa-91d8-fe0100c852be, Accessed on 9 Dec 2024

Yoshida N, Morioka T, Fujimura H, Takebayashi M (2002) Evaluation of eco-industrial park as societal demonstration site in Japan [Article]. J Glob Environ Eng 8:115–127. https://www.scopus.com/inward/record.uri?eid=2-s2.0-33745670430&partnerID=40&md5=2f94004e18d556910450fa477bc52b81

Zhang R (2025) Rethinking the role of industrial parks for sustainable urbanization in the peripheries of Tokyo metropolitan area [Research Thesis, Keio University]. Tokyo. https://koara.lib.keio.ac.jp/xoonips/modules/xoonips/detail.php?koara_id=KO90001001-20256471-0003

Zhang R, Yan W (2024) The 50-year spatial transition of suburban industrial parks and impacts on sustainable urbanization in the Tokyo metropolitan area. Sustain Cities Soc 113. https://doi.org/10.1016/j.scs.2024.105679

Zhang R, Yan W (2025) Evaluating the area energy performance of suburban industrial parks in the Tokyo metropolitan area. Clean Eng Technol 24. https://doi.org/10.1016/j.clet.2025.100888

Zhang R, Yan W, Shaw R (2024) Land use planning and conservation policy in the Tokyo metropolitan area. In: Yan W, Galloway W, Shaw R (eds) Resilient and adaptive Tokyo: towards sustainable urbanization in perspective of food-energy-water Nexus. Springer Nature Singapore, pp 79–103. https://doi.org/10.1007/978-981-99-3834-6_5

Chapter 14
Urban Nexus in Municipal Solid Waste Management

Qiannan Zhuo

Abstract As one of the world's largest and most densely populated cities, the effective municipal solid waste management is critical for Tokyo for sustainable urban living. This chapter introduces current situation of waste generation and treatment in Tokyo, reviews the existing problems of its waste management system and released policy plans, and discusses the expected effects to further reduce environmental impacts and promote circular economy practices from the Urban Nexus point of view. The findings offer a brief understanding of the current state and future policy direction of waste management in Tokyo and insights into how other megacities can optimize their waste management systems within the Urban Nexus context to improve the interconnectedness among various urban systems, including water, energy, food, and waste.

Keywords Municipal solid waste (MSW) · Waste management system · Tokyo · Urban Nexus

1 Introduction

1.1 History and Legal Background of Waste Management in Japan

Since waste is inherently a by-product of human production and consumption activities, the goals and challenges associated with its management have continuously evolved in response to shifts in societal needs, economic development, and environmental awareness. These changes in waste management priorities not only reflect the practical demands of each era but also serve as a mirror of broader societal

Q. Zhuo (✉)
College of Policy Studies, Tsuda University, Tokyo, Japan

W. Yan (ed.), *Urban Nexus Approach for Sustainable Cities*,
https://doi.org/10.1007/978-981-95-6156-8_14

transformations. Consequently, the study of waste management offers a valuable perspective on understanding the dynamics of society itself, as the way waste is handled, regulated, and perceived often indicates deeper social, economic, and environmental trends.

In the early twentieth century, waste management in Japan was primarily driven by the need to improve public health conditions and combat the spread of infectious diseases. At that time, the lack of adequate sanitation infrastructure posed significant risks to the well-being of urban populations. This was especially evident in the rapidly growing cities where improper waste disposal contributed to outbreaks of diseases such as cholera and dysentery. Recognizing the urgent need to address these public health threats, the Japanese government introduced the Cleanliness Law (Seisouhou) in 1954.

As Japan's economy grew rapidly in the postwar period, so too did the volume and complexity of waste. Industrialization and urbanization led to the emergence of new types of waste, including hazardous industrial by-products and chemical pollutants. By the mid-twentieth century, the nation faced an unprecedented pollution crisis, with air and water quality deteriorating due to unregulated industrial activity. In response, the objectives of waste management shifted from merely maintaining public health to addressing the more complex and pervasive issue of pollution control.

Moving into the latter half of the twentieth century, Japan's waste management policies underwent another significant transformation as environmental issues gained global prominence. By this time, the limitations of a purely pollution-control-based approach became apparent. Simply mitigating the harmful effects of waste was no longer sufficient; instead, there was a growing recognition of the need to prevent waste generation at its source and to promote the efficient use of resources. This shift in focus was closely tied to the broader global movement toward sustainable development, which emphasized the importance of balancing economic growth with environmental stewardship and resource conservation. In response to these changing priorities, Japan began to adopt more comprehensive policies that integrated waste management into the broader framework of environmental protection and resource efficiency.

One of the most significant milestones in this evolution was the enactment of the Basic Act for Establishing a Sound Material-Cycle Society in 2000. This law marked a fundamental shift in how Japan approached waste management by emphasizing the need to create a "sound material-cycle society," in which resources are used efficiently and sustainably throughout their life cycle. Specifically, the Basic Act establishes a dual approach: the proper treatment of waste through the Waste Management and Public Cleansing Law, and the promotion of the 3Rs—Reduce, Reuse, and Recycle—through the Act on the Promotion of Effective Utilization of Resources.

The Waste Management and Public Cleansing Law categorizes waste into two types: municipal waste and industrial waste. Municipal waste refers to all waste that is not classified as industrial waste, while industrial waste includes materials such as burnt ash, sludge, waste oil, and other waste which are generated as a result of business activities as specified by Cabinet Order. Industrial waste also includes specially controlled industrial waste, which is hazardous in nature, such as explosive, toxic, or infectious materials, and must be managed by the business generating it.

On the other hand, municipal waste is further subdivided into household waste and business-related waste. It also includes specially controlled municipal waste, which, like its industrial counterpart, consists of hazardous substances such as explosives, toxins, or infectious materials. Local municipalities are responsible for the collection, transportation, and disposal of municipal waste, ensuring that it is managed safely and sustainably.

In summary, the evolution of waste management in Japan reflects the country's changing social, economic, and environmental priorities over time. From its early focus on public health and sanitation, to pollution control during the height of industrialization, and finally to the promotion of a sustainable, resource-efficient society in the twenty-first century, Japan's waste management policies have continually adapted to meet the challenges of each era (Ministry of the Environment 2014).

1.2 FEW Nexus in Waste Management

Although food, energy, and water have been treated as separate policy domains, recent research and policy discourse increasingly recognize these sectors as part of a complex, interdependent system, where actions in one domain inevitably affect the others. However, there are still very limited attention paid to waste management.

Within the food–energy–water (FEW) nexus, waste management should not be viewed as a subordinate function but rather as a central hub that interconnects all three systems. For example, food waste not only reflects inefficiencies in the food system but also contributes to energy and water cycles when repurposed through composting or anaerobic digestion. From an energy perspective, waste incineration generates electricity and heat, feeding directly into urban energy infrastructure and aligning with broader renewable energy strategies. In parallel, the treatment of leachate and wastewater generated through waste processes plays a crucial role in urban water management and environmental protection.

In megacities like Tokyo—characterized by high population density and advanced infrastructure—waste management is not merely a downstream operation. Rather, it functions as a critical entry point for systemic connectivity. The generation and processing of waste are deeply intertwined with the production, consumption, and circulation of food, energy, and water. Managing waste appropriately and strategically offers a key leverage point for improving urban sustainability.

1.3 Objectives and Scope of This Chapter

This chapter aims to provide an integrated analysis of Tokyo's municipal waste management practices within the broader Urban Nexus framework, focusing on the interplay between waste management and other critical urban systems. The objectives of this chapter are:

1. *Analyze current trends and practices and highlight interconnections in the FEW nexus*: To examine the trends in waste generation and to explore the interdependencies of waste management with food, energy, and water systems.
2. *Address key challenges and policy directions*: To identify the challenges faced by Tokyo's waste management system and the future policy directions to further enhance within the FEW nexus context.
3. *Provide strategic recommendations*: To propose operational recommendations that align with Tokyo's long-term sustainability goals and lessons from Tokyo's practices to other global cities.

The scope of this chapter includes an in-depth evaluation of Tokyo's waste management system, its alignment with the FEW nexus principles, and its relevance for other global megacities. By addressing these areas, this chapter aims to provide actionable insights and serve as a reference for policymakers and urban planners striving for sustainable urban living.

2 Current State of Waste Management and Practices Within Urban Nexus Context

2.1 Trends in Waste Generation

Waste generation in Japan has undergone significant changes over the past several decades, reflecting shifts in the nation's economy, policies, and social behaviors. During the late twentieth century, particularly in the 1990s, Japan experienced a peak in waste generation, driven by rapid economic growth and heightened consumer activity during the bubble economy. Annual waste production reached over 50 million tons nationwide. This increase was accompanied by a rise in per capita waste generation, which exceeded 1 kg per day.

However, since the enactment of progressive waste management policies and public awareness campaigns, Japan has achieved remarkable reductions in waste generation. The introduction of the Containers and Packaging Recycling Law in 1995 and the Basic Environment Law in 1993 played pivotal roles in shifting societal attitudes and practices toward waste reduction and recycling. Currently, total waste generation has steadily declined, with the average daily waste output per person now hovering around 900 g. These reductions highlight the effectiveness of legislative frameworks and community-level participation in mitigating environmental impacts.

Tokyo's waste generation trends have mirrored broader national patterns, but with unique characteristics reflecting its role as Japan's largest metropolitan area (Fig. 14.1). In 1989, the waste volume for the 23 central wards of Tokyo peaked at 4.9 million tons annually, with landfill disposal reaching 2.4 million tons. This peak was driven by rapid urbanization, economic growth, and high levels of consumption during the bubble economy period. Following this, Tokyo implemented a series of aggressive waste reduction policies and resource recovery initiatives that have drastically reduced waste generation. By 2023, total waste generation in Tokyo had decreased by nearly 50%, with landfill disposal volumes reduced by 92% compared to 1989, now standing at just 0.18 million tons.

The significant reductions in Tokyo's waste generation can be attributed to a combination of innovative policies, technological advancements, and public engagement. In 1991, the city introduced a policy requiring payment for the collection of large-sized waste. This incentivized residents to reconsider their waste disposal habits and encouraged waste reduction at the source. Similarly, by 1996, all business-generated waste was made subject to fees, compelling businesses to adopt better waste separation practices and reduce overall volumes. The establishment of intermediate processing systems in 1996 marked another milestone, ensuring that all types of waste—combustible, noncombustible, and large-sized—underwent proper processing before final disposal. This significantly enhanced resource recovery rates and reduced environmental burdens. In 2008, Tokyo fully integrated thermal recycling of waste plastics, transforming what was previously a disposal challenge into

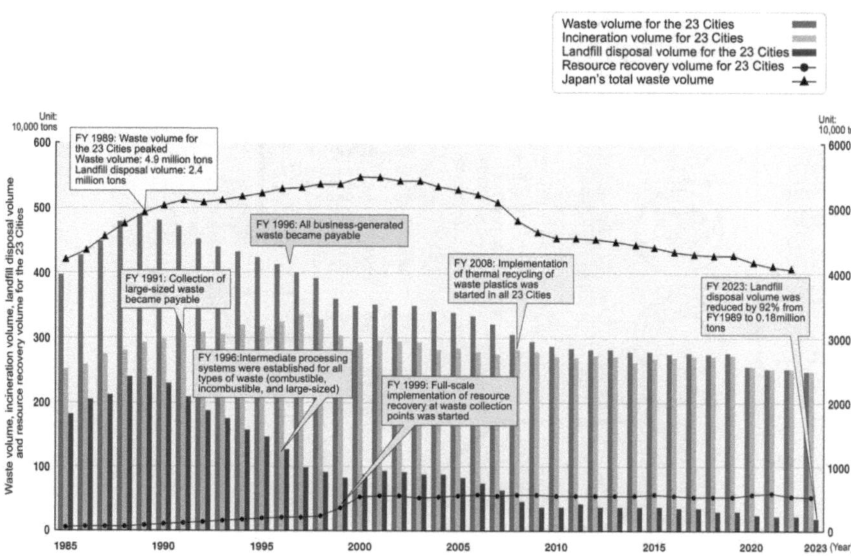

Fig. 14.1 Transition in waste volume, incineration volume, landfill disposal volume, and resource recovery volume. (Clean Authority of TOKYO 2024)

an energy resource. These measures, along with the introduction of resource recovery points at collection sites in 1999, allowed for greater separation of recyclables, thereby reducing the reliance on incineration and landfill facilities while simultaneously supporting renewable energy production through waste-to-energy systems.

Tokyo's waste processing flow is an intricate system designed to maximize resource recovery, minimize landfill dependency, and promote environmental sustainability (Fig. 14.2). The process begins at the household, commercial, and municipal levels, where residents and businesses are required to separate waste into combustible, noncombustible, large-sized items, and recyclable materials in adherence to strict guidelines.

Combustible waste is transported to one of the 22 incineration plants in the 23 central wards of Tokyo. These plants are equipped with state-of-the-art technology to ensure efficient incineration while minimizing environmental impacts. The heat generated during incineration is captured and converted into electricity or district heating. Incineration ash is treated and often repurposed for use in construction materials, further contributing to resource efficiency.

Noncombustible waste is processed in specialized facilities, including two primary noncombustible waste processing centers in the city. These centers employ advanced sorting and recovery technologies to extract valuable materials such as metals and glass. Recovered metals are sent to smelting facilities, while glass is processed into aggregates for construction or repurposed into new products. Ceramics and other inert materials are typically pulverized and used in infrastructure projects.

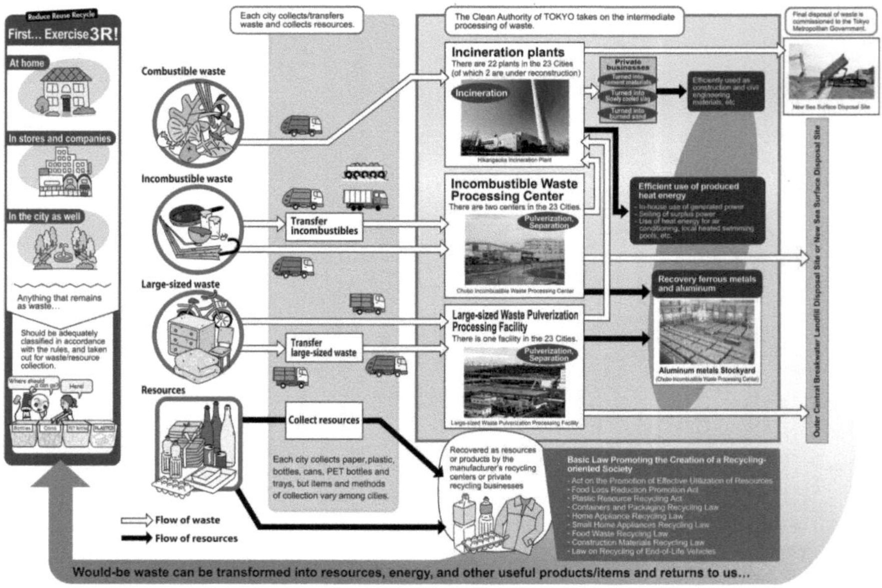

Fig. 14.2 Flow of waste and resources in the 23 cities. (Clean Authority of TOKYO 2024)

Large-sized waste, including furniture and appliances, undergoes a separate pulverization and sorting process at a dedicated facility. Metals and other recoverable components are extracted for recycling, while remaining materials are sent for incineration or repurposing.

Recyclable resources are collected through a network of resource recovery points and curbside programs. Items such as paper, plastics, and PET bottles are transported to recycling facilities, where they are processed into raw materials for new products. The city's robust recycling framework is supported by legislative measures, including the Basic Law Promoting the Creation of a Recycling-Oriented Society and specific recycling mandates for plastics, food waste, and home appliances.

Tokyo's waste composition reflects a sophisticated and well-managed system that addresses the city's diverse waste streams. Data from fiscal year 2024 provides detailed insights into the materials transported to incineration facilities and non-combustible waste processing centers, as shown below (Tokyo 23-Ward Cleaning Association 2025).

Waste processed at incineration facilities predominantly consists of combustible materials, accounting for a significant portion of the city's waste output. Paper waste is the largest contributor, comprising 41.64% of this category. This is followed by food and organic waste, which constitutes 22.15%. Efforts to minimize food waste through awareness campaigns and composting programs have aimed to reduce this percentage. Plastic waste represents 20.14% of incinerated materials, highlighting the need for enhanced recycling initiatives and the reduction of single-use plastics. Smaller fractions include wood (7.63%) and textiles (6.02%), which pose challenges for energy recovery during incineration but are still processed efficiently.

Noncombustible waste, handled at specialized processing centers, showcases a distinct material composition. Stone and ceramics represent the largest portion, accounting for 23.80%, largely stemming from construction and demolition activities. Glass (23.80%) and metals (15.83%) also make up significant shares of this category. Both materials are successfully recycled, with glass often reused in construction aggregates and metals reintroduced into industrial production chains. Additionally, household appliances and electronics (19.77%) highlight the growing need for robust e-waste recycling programs, ensuring the recovery of valuable components such as rare earth metals. Finally, plastics (11.49%) and miscellaneous materials reflect the ongoing challenges of sorting and processing waste streams with diverse characteristics.

2.2 Food Loss and Food Waste

Food loss and food waste represent critical challenges within Tokyo's broader waste management framework. In the Japanese context, these two terms are defined distinctly. Food loss refers to food that is still edible but is discarded for various reasons, such as unsold items from retailers or untouched leftovers from consumers. Food waste, on the other hand, encompasses all discarded food, including inedible parts such as vegetable stems, fish bones, and shells. In this framework, food loss is considered a subset of food waste, focusing specifically on avoidable waste of edible food.

In fiscal year 2021, Tokyo generated approximately 368,000 tons of food loss. Among this, households contributed 143,000 tons (38.8%) due to common issues such as over-purchasing, improper storage, and misinterpretation of expiration dates. This demonstrates a need for more awareness and practical measures to reduce waste at the consumer level. The business sector accounted for the remaining 225,000 tons (61.2%), with significant contributions from retail operations (74,000 tons), food service industries (130,000 tons), and food manufacturing (13,000 tons). In retail, unsold inventory and aesthetic standards for produce are significant drivers, while in the food service industry, excess portions and spoilage during storage are key contributors (Tokyo Metropolitan Government, Environmental Bureau 2024).

Tokyo's food waste management system faces distinct challenges due to the lack of widespread food waste separation. Unlike some other major cities, Tokyo has not yet implemented a city-wide program for separating food waste from general combustible waste. Although centralized recycling or composting of food waste is not common, some wards in Tokyo have introduced measures to encourage household-level management of organic waste. For example, certain districts offer subsidies for the purchase of food waste composting machines, promoting at-home composting as a practical solution. These localized initiatives, however, remain limited in scale and impact. As a result, food waste is primarily disposed of as combustible waste and transported to incineration facilities for thermal processing. For the majority of residents, food waste is treated as part of combustible waste, contributing to its high proportion of the total combustible waste stream. This reliance on incineration underscores the need for more comprehensive and scalable solutions to manage food waste effectively.

2.3 *Thermal Recycling*

Tokyo's management of combustible waste is centered around its advanced incineration facilities, which form the backbone of the city's waste-to-energy and resource recovery system. Combustible waste, which includes paper, plastics, and food waste, accounts for the majority of the waste stream and undergoes a meticulously designed processing flow.

Combustible waste is collected from households and businesses according to strict separation guidelines. Residents are required to separate combustible waste from other waste categories, ensuring that materials intended for incineration are not contaminated. The collected waste is then transported to one of Tokyo's 22 incineration plants, strategically located across the city to minimize transportation distances and logistical challenges. Upon arrival at an incineration plant, the combustible waste undergoes thermal treatment in highly controlled environments. Key steps in the process include:

1. *Pre-incineration sorting*: In some cases, bulky or inappropriate items are removed to optimize the efficiency of the incineration process.
2. *High-temperature combustion*: Waste is burned at temperatures exceeding 800 °C. This ensures the complete breakdown of organic materials, significantly reducing the volume of waste.
3. *Energy recovery*: The heat generated during combustion is captured and used to produce steam. This steam drives turbines to generate electricity, which is fed back into the local power grid.

The heat generated during incineration is utilized primarily in two ways. First, electricity is produced by powering turbines, contributing significantly to Tokyo's energy network. Second, surplus heat is distributed through district heating systems to nearby residential and commercial buildings, enhancing energy efficiency and reducing carbon emissions.

Thermal recycling provides Tokyo's waste management system with substantial benefits. Environmentally, it significantly reduces the volume of waste by up to 90%, conserving valuable landfill space. Additionally, the incineration process mitigates greenhouse gas emissions that would otherwise result from organic waste decomposition in landfills. Modern incineration plants equipped with advanced emissions control technology further minimize pollutants, ensuring compliance with stringent environmental standards.

Economically, thermal recycling contributes to Tokyo's energy independence by converting waste into a renewable energy source. In FY2023, Tokyo's incineration plants generated over 1226.00 MWh of electricity, with 718.37 MWh sold to power companies. The heat supplied from these facilities, amounting to 466.00 GJ, also generated significant revenue while offsetting energy demands from conventional fossil fuels. These achievements highlight the dual environmental and economic benefits of Tokyo's thermal recycling framework (Clean Authority of TOKYO 2024).

Beyond energy generation, these facilities incorporate technologies to capture and utilize residual heat effectively. For example, the surplus heat not used directly for power generation is often employed to heat swimming pools, botanical gardens, and other public amenities. This multifaceted utilization underscores Tokyo's commitment to sustainable energy practices.

2.4 Leachate and Wastewater

Leachate and wastewater management are critical components of Tokyo's comprehensive waste management system, particularly in waste processing facilities and landfill. These processes involve strict adherence to environmental standards and the implementation of advanced treatment technologies to mitigate potential ecological risks.

In recycling facilities, wastewater is generated from processes such as cleaning, cooling, and material separation. This wastewater undergoes sedimentation, chemical coagulation, and filtration to remove suspended solids and recover valuable materials such as metals. These efforts contribute to resource efficiency and reduce environmental impacts.

Incineration plants, on the other hand, produce wastewater through wet scrubbers used to clean flue gases. This wastewater is treated in multiple stages, including neutralization and biological processing, to remove contaminants. Treated water is often reused within the plants for cooling and cleaning operations, reducing reliance on freshwater. Advanced reverse osmosis systems are also employed to ensure compliance with strict discharge standards.

At landfill sites, rainwater infiltrates through layers of waste, forming leachate—a liquid containing pollutants such as organic compounds, heavy metals, and other hazardous substances. To manage this, Tokyo employs a multilayered approach, starting with impermeable liners and leachate collection systems to prevent groundwater contamination. Collected leachate is stored in reservoirs with capacities of up to 150,000 cubic meters before undergoing advanced treatment.

The outcomes of Tokyo's advanced leachate treatment processes demonstrate significant achievements in meeting environmental and regulatory standards:

1. *pH stabilization*: The treated leachate achieves a pH of 7.4, which falls well within the sewage discharge standards of 5–9.
2. *Nitrogen reduction*: Nitrogen levels are reduced to 17 mg/L, significantly below the discharge standard of 120 mg/L.
3. *COD reduction*: Chemical oxygen demand (COD) is lowered to 64 mg/L, effectively meeting the agreed threshold of 150 mg/L with the Bureau of Sewerage.

These results underscore Tokyo's commitment to reducing the environmental impact of landfill leachate and maintaining water quality in compliance with stringent regulatory requirements (Tokyo Metropolitan Government, Environmental Bureau 2022).

3 Challenges and Policy Direction

3.1 Policy Frameworks and Regulations

The Waste Management Plan demonstrates Tokyo's waste policy direction as a specific area plan under the Tokyo Environmental Basic Plan (Tokyo Metropolitan Government, Environmental Bureau 2021). It is formulated under Article 5–5 of the Waste Management and Public Cleansing Law (Act No. 137 of 1970), serving as a detailed plan to address Tokyo's waste management challenges. It was initially established in FY2001 as a unified plan for both municipal and industrial waste. Since then, it has evolved to address emerging challenges and societal demands for every 5 years.

The 2021–2025 Waste Management Plan is a sub-plan within the framework of broader initiatives, including the "Future Tokyo" Strategy (March 2021) and the Tokyo Environmental Basic Plan (March 2016). This plan outlines key policies and strategies aimed at achieving long-term sustainability. The plan's timeframe spans 5 years, from FY2021 to FY2025, with an eye toward the vision for 2030 (FY2030) and a long-term goal of carbon neutrality by 2050 (FY2050). It embodies a clear and structured approach to addressing the city's waste challenges. It is formulated through careful consideration of current trends, societal needs, and future projections, ensuring alignment with national and global environmental goals.

The 2021–2025 Waste Management Plan is built upon three strategic pillars that address Tokyo's unique challenges while fostering sustainable development:

1. *Sustainable resource use*: Tokyo aims to reduce natural resource consumption and environmental impacts by promoting efficient material use and the 3Rs (Reduce, Reuse, Recycle) in alignment with the Basic Act on Establishing a Sound Material-Cycle Society. This addresses both local and global sustainability concerns.
2. *Advanced waste management systems*: Recognizing the pressures of a super-aging society and reduced workforce, the plan emphasizes digitization and process improvements across waste disposal and management systems to enhance efficiency and infrastructure resilience.
3. *Tackling social challenges*: The plan ensures the continuity of waste management during crises, such as pandemics and disasters, while contributing to net-zero CO_2 emissions by 2050 through targeted waste sector interventions.

To establish its goals along with the three pillars, the 2021–2025 Waste Management Plan integrates data from existing "Plastic Reduction Program" and "Food Loss Reduction Promotion Plan" and considers the advances in digital transformation (DX) and developments in waste processing technologies. The resulting projections provide a foundation for setting specific targets. A series of ambitious goals have been set. For instance, the plan aims for a 40% reduction of plastic waste incineration and a reduction to 380,000 tons of food loss by FY2030. Besides, the recycling rate is targeted to 31% by 2025 and 37% by 2030 from 23% in 2018.

The following five key measures have been identified by 2021–2025 Waste Management Plan to achieve these goals:

1. *Reduction of resource loss*: Promote programs such as the "Plastic Reduction Program" and the "Food Loss Reduction Promotion Plan"; encourage collaboration among consumers, manufacturers, and businesses to reduce waste generation.
2. *Enhancement of recycling systems*: Expand household recycling programs for plastics and other materials; facilitate 3R (Reduce, Reuse, Recycle) initiatives for business waste in collaboration with municipalities.

3. *Strengthening waste processing systems*: Optimize waste collection and transportation systems in response to demographic and social changes; enhance treatment methods for difficult-to-process waste.
4. *Development of a reliable circular economy business*: Foster a business environment that supports innovative solutions aligned with the circular economy.
5. *Addressing social challenges*: Build a resilient waste management system capable of maintaining operations during disasters and crises; advance policies for achieving net-zero emissions through targeted initiatives and enhanced corporate participation.

3.2 Reduce Food Loss

To address the critical issue of food loss, the Tokyo Metropolitan Government (TMG) has undertaken a series of comprehensive surveys aimed at quantifying and deeply understanding the sources, causes, and magnitude of food loss across various sectors (Tokyo Metropolitan Government, Environmental Bureau 2022). According to a 2017 analysis, Tokyo's annual food loss amounted to 510,000 tons, with approximately 125,000 tons originating from households and 385,000 tons from businesses. In households, the primary causes of food loss include overpurchasing, cooking excessive amounts, and food leftovers. In the commercial sector, food loss stems from excessive inventory in the supply chain, the "1/3 rule" in retail requiring products to be delivered within the first third of their shelf life, and food waste during meal preparation or customer leftovers in the dining industry.

Tokyo's approach to food loss reduction follows a hierarchical framework that prioritizes prevention, effective utilization, and recycling. The primary focus is on minimizing food waste generation at the source by sharing effective practices among households and businesses. For effective utilization of food resources, Tokyo's approach is to strengthen partnerships with food banks and other organizations to redirect surplus but edible food to those in need, and utilize disaster stockpiles efficiently before expiration to reduce waste while benefiting communities. For recycling, Tokyo is going to provide subsidies and incentives for businesses to adopt in-house recycling systems and composting technologies.

In details, the Tokyo Metropolitan Government (TMG) has outlined comprehensive initiatives to address food loss as part of its "Zero Emission Tokyo Strategy." This strategy envisions achieving net-zero food loss by 2050 through a multipronged approach that includes prevention, effective utilization, and recycling of food waste. Recognizing the critical importance of the next decade, TMG has set an ambitious interim goal to halve food loss by 2030, compared to 2000 levels. Achieving these goals requires not only enhanced measures but also the cultivation of public awareness and collaboration with all stakeholders, including businesses and consumers.

The most effective strategy for reducing food loss is to prevent it from occurring in the first place. In households, this involves raising awareness about the importance of proper food management, such as mindful purchasing, regular stock checks in refrigerators, and cooking only the necessary portions. Efforts are also being made to promote smart consumption habits aligned with modern lifestyles. Public education campaigns are key to encouraging these behaviors and making them routine.

In the business sector, food loss prevention focuses on sharing successful case studies of waste reduction initiatives. For example, companies are encouraged to revise traditional supply chain practices, such as the "1/3 rule," which mandates that products be delivered within the first third of their shelf life. By relaxing such restrictions and fostering collaboration across the supply chain, businesses can significantly reduce unnecessary waste.

TMG is actively promoting the use of ICT and AI to reduce food loss. These technologies enable businesses to optimize inventory management, predict demand more accurately, and minimize overproduction. For instance, AI-driven demand forecasting systems can help retailers and manufacturers better align supply with consumer needs. Additionally, innovations in packaging and freezing technologies are being supported to extend the shelf life of food products, reducing spoilage and waste. New business models that integrate these advanced technologies are also being developed to drive systemic change.

Improving practices within the food supply chain is another critical focus area. TMG is advocating for the reevaluation of commercial customs, such as adjusting delivery deadlines and adopting expiration date labeling that includes year and month rather than exact dates. By publicizing and promoting business efforts to reduce food loss, TMG aims to create a ripple effect that encourages widespread adoption of these practices across industries.

Another key initiative involves the effective use of surplus food. Disaster stockpile foods are being proactively utilized before their expiration dates, ensuring that these resources do not go to waste. TMG is also working closely with food banks to redistribute edible surplus food to those in need. Sharing best practices and successful models in this area helps to expand and solidify these efforts.

For food waste that cannot be prevented, TMG is advancing recycling initiatives to ensure proper resource utilization. This includes converting inedible food waste into animal feed, fertilizers, or energy. By supporting businesses in adopting these practices, TMG aims to establish a more circular economy for food waste.

Looking toward 2050, TMG has identified several innovative approaches to further reduce food loss and contribute to a sustainable future.

1. *Optimizing food supply and demand*: High-precision demand forecasting systems using AI are being developed to better match food production with consumption needs, thereby minimizing overproduction.
2. *Innovative preservation technologies*: Research and development are focused on creating new methods for long-term food preservation that maintain quality while reducing CO_2 emissions. These technologies include advanced storage and processing methods.

3. *Expanding food sharing platforms*: Food-sharing services are being promoted to turn surplus food into economic opportunities. These platforms aim to normalize the concept of sharing food resources among households and businesses.
4. *Encouraging environmentally friendly eating habits*: Efforts are underway to introduce carbon footprint labeling for food products, enabling consumers to make informed choices that minimize environmental impacts. Technologies that help households reduce food loss are also being developed and promoted.

Through these comprehensive measures, TMG is committed to transforming Tokyo into a leader in sustainable food systems, contributing significantly to global efforts to combat food loss and achieve a circular economy.

3.3 Waste-to-Energy and Wastewater Management

The Tokyo Metropolitan Government (TMG) recognizes that waste-to-energy (WtE) systems, while historically significant, are becoming less central to its waste management strategy due to evolving priorities and emerging challenges. As Tokyo's population is projected to peak by 2030, large-scale increases in waste generation are unlikely. Besides, TMG is shifting its focus toward waste reduction and recycling, aligning with its broader objectives of promoting a circular economy and achieving carbon neutrality by 2050. The decline in waste available for incineration, coupled with the increasing emphasis on reducing greenhouse gas emissions, has led to a strategic deprioritization of WtE facilities.

Another critical challenge for Tokyo's WtE strategy is the reduction of plastics, which contribute significantly to the calorific value of waste and is a key objective under Tokyo's waste management plan. This naturally leads to a decrease in the energy potential of combustible waste, further diminishing the reliance on WtE facilities.

Moreover, incineration facilities have a finite lifespan and require long-term planning for refurbishment or replacement. The aging infrastructure of some of Tokyo's WtE plants highlights the need for significant investments to maintain operational efficiency and environmental compliance. However, the declining volume and calorific value of waste raise questions about the cost-effectiveness of continuing to rely on WtE systems as a cornerstone of waste management.

While Tokyo is deprioritizing WtE facilities, the city remains committed to optimizing their operation during the transition period. This includes modernizing existing plants to ensure they meet the highest environmental standards and exploring opportunities to integrate renewable energy sources into the waste management system (Tokyo Metropolitan Government, Environmental Bureau 2021).

Tokyo's wastewater management in the context of waste processes is highly efficient, demonstrating advanced planning and environmental protection measures. Currently, there are two active landfill sites in operation. As part of its future waste management strategy, Tokyo aims to further reduce landfill volumes and extend the lifespan of these final disposal sites. This proactive approach minimizes the risk of water pollution associated with large-scale landfilling.

The city's commitment to environmental preservation is evident in the long-term nature of its measures. These protective actions continue even after landfilling operations have ceased, ensuring that surrounding environments remain unaffected. For example, Tokyo's 15th landfill site in Koto Ward, Wakasu, ceased operations in 1974, yet its leachate continues to be treated at the dedicated wastewater treatment facility located within the Central Breakwater Inner Landfill area. This ongoing management underscores Tokyo's dedication to mitigating environmental impacts and maintaining water quality.

By maintaining robust wastewater management practices, Tokyo not only safeguards its urban environment but also sets a precedent for sustainable waste management. These efforts reflect the city's long-term vision for creating a resilient and eco-friendly urban system that balances waste processing with environmental preservation (Tokyo 23-Ward Cleaning Association 2023).

4 Conclusion

4.1 Summary of Key Findings

Tokyo's waste management system is, overall, highly effective and well-structured, reflecting its role as a leading global megacity. As one of the most populous urban areas in the world, Tokyo has long relied on incineration as its primary waste treatment method. This approach has been instrumental in managing waste efficiently while mitigating environmental impacts. Key findings from this analysis highlight Tokyo's strengths and areas for improvement:

First, Tokyo's alignment with the Zero Emission Tokyo strategy underscores its commitment to sustainability. The city has set ambitious goals to reduce plastic waste by 40%, halve food loss by 2030, and extend the lifespan of its landfill sites. It is actively working to reduce waste generation and promote recycling as part of its ongoing efforts to achieve a circular economy. Current initiatives focus on the reduction of single-use plastics, food loss prevention, and robust recycling systems.

Second, thermal recycling is being utilized to convert waste into energy, and the resulting heat and electricity are integrated into urban infrastructure, powering homes and facilities. However, the amount of electricity generated from waste-to-energy (WtE) facilities is expected to decline in the future due to reductions in waste generation and plastic usage. While this trend may decrease energy outputs, it aligns positively with environmental impact assessments, reflecting progress in reducing overall waste and carbon emissions. These measures represent Tokyo's forward-looking approach to sustainable urban management, with long-term success still contingent on continuous adaptation and innovation.

Third, advanced wastewater treatment processes have also been successfully implemented, ensuring that environmental risks from leachate are minimized even at older landfill sites, such as the 15th landfill in Wakasu.

Although challenges persist, such as shortages in collection personnel and limitations in the recycling industry, the aging infrastructure of incineration plants necessitates strategic investment and planning to ensure continued efficiency and environmental compliance, despite these challenges, Tokyo remains focused on achieving its long-term goals, including reducing greenhouse gas emissions and transitioning to a circular economy.

Finally, Tokyo's governance and collaborative approach have been pivotal to its success. By engaging stakeholders across government, businesses, and communities, Tokyo has fostered a culture of shared responsibility and innovation.

4.2 Policy Recommendations for Sustainable Waste Management

To build on Tokyo's successes and address its challenges, the following policy recommendations aim to provide actionable strategies that align with Tokyo's long-term sustainability goals and offer insights for other global cities:

1. *Expand integration with the FEW nexus*: While Tokyo has made significant efforts in reducing food loss, it should also focus on promoting the segregation of food waste and its appropriate treatment. This would enhance the efficiency of incineration facilities, improve combustion rates and contribute further to WtE electricity generation.
2. *Modernize infrastructure for long-term resilience*: Prioritize the refurbishment and modernization of waste-to-energy facilities to develop flexible infrastructure capable of adapting to changes in waste composition and volumes, particularly as plastic usage declines.
3. *Promote technological innovation and adoption*: Invest in research and development of emerging technologies. These innovations should, for instance, address the pressing issue of workforce shortages in waste collection, streamlining operations and ensuring continuity in waste management services, or decentralize composting, or enhance food-sharing platforms.
4. *Strengthen community engagement and education*: Expand public awareness campaigns to normalize sustainable practices, such as proper waste segregation and mindful consumption. Engage with schools, businesses, and local communities to foster a sense of shared responsibility for waste reduction.
5. *Foster knowledge exchange with global cities*: Share Tokyo's best practices and lessons learned through international collaborations and forums. Establish partnerships to develop standardized metrics for measuring waste reduction and resource efficiency, enabling benchmarking and shared progress tracking.

Tokyo's approach offers a comprehensive blueprint for global cities aiming to improve waste management systems and enhance sustainability. Key lessons from Tokyo's experience include:

1. *Establish comprehensive and data-driven plans*: Effective waste management begins with meticulous surveys and assessments that incorporate both quantitative and qualitative data. Cities should craft ambitious yet feasible short-term, mid-term, and long-term goals that align local needs with global sustainability trends. For example, Tokyo's detailed analysis of waste composition and its integration into broader strategies demonstrates the importance of data-driven decision-making. These plans must remain flexible and adaptable to evolving societal and environmental conditions, ensuring their continued relevance and effectiveness.
2. *Ensure long-term commitment to pollution control*: Pollution control, particularly in wastewater management, requires a steadfast and long-term approach. Tokyo exemplifies this commitment by maintaining leachate treatment systems at decommissioned landfill sites, such as the 15th landfill in Wakasu, where water quality monitoring and treatment continue decades after its closure. This practice underscores the critical need for sustained environmental stewardship to protect ecosystems and public health.
3. *Target specific waste streams through collaborative efforts*: Identifying and prioritizing specific waste streams, such as plastics and food waste, is essential for impactful waste reduction. Tokyo's targeted initiatives, informed by comprehensive waste composition analyses, highlight the effectiveness of focusing on high-impact materials. Collaborative efforts between government, industry, and the public ensure the successful implementation of these initiatives. For instance, the city's push to reduce food loss and single-use plastics demonstrates how a unified approach can drive meaningful change.
4. *Adapt to future demographic and societal changes*: Waste management strategies must anticipate and adapt to changes in population dynamics, economic trends, and societal behaviors. Tokyo's proactive planning to address an aging population, workforce shortages, and declining waste calorific values sets a strong precedent for other cities. Developing resilient systems that can accommodate these shifts ensures that waste management practices remain effective and sustainable over time.

By embracing these strategies and tailoring them to their specific contexts, global cities can address the multifaceted challenges of waste management while contributing to global sustainability goals. Tokyo's experience underscores the importance of aligning local actions with global trends, fostering collaboration among stakeholders, and maintaining a long-term vision for sustainable urban living.

Acknowledgments The author acknowledges that the research related to this chapter is supported by the Taikichiro Mori Memorial Research Grants.

References

Clean Authority of TOKYO (2024) "Waste Report 23 2025". https://www.union.tokyo23-seisou.lg.jp/seiso/seiso/pamphlet/report/index.html

Ministry of the Environment (2014) "History and Current State of Waste Management in Japan." Policy Planning Division, Office of Sound Material-Cycle Society, Waste Management and Recycling Department, Minister's Secretariat, Ministry of the Environment. http://www.env.go.jp/content/900453392.pdf

Tokyo 23-Ward Cleaning Association (2023) "Final Disposal of Waste." Tokyo 23-Ward Cleaning Association. https://www.union.tokyo23-seisou.lg.jp/shiro/saishu.html

Tokyo 23-Ward Cleaning Association (2025) "Waste Composition Survey Results, Fiscal Year 2024." Waste Composition Survey Results. Tokyo 23-Ward Cleaning Association. https://www.union.tokyo23-seisou.lg.jp/gijutsu/kankyo/toke/seijyou.html

Tokyo Metropolitan Government, Environmental Bureau (2021) "Tokyo Resource Circulation and Waste Treatment Plan 2021–2025." Resource Circulation and Waste Treatment Plan. Tokyo Metropolitan Government, Environmental Bureau. https://www.kankyo.metro.tokyo.lg.jp/documents/d/kankyo/resource-waste_treatment-files-r3-cwplan

Tokyo Metropolitan Government, Environmental Bureau (2022) "Tokyo Waste Landfill Sites." Tokyo Metropolitan Government, Environmental Bureau. https://www.jwrf.or.jp/news/upload/3a16262e380da006ca25018c8e09b3cbfd3a99ba.pdf

Tokyo Metropolitan Government, Environmental Bureau (2022) "Formulation of Tokyo Food Loss Reduction Promotion Plan." Tokyo Food Loss Reduction Promotion Plan. Tokyo Metropolitan Government, Environmental Bureau. https://www.kankyo.metro.tokyo.lg.jp/resource/recycle/tokyo_torikumi/keikaku

Tokyo Metropolitan Government, Environmental Bureau (2024) "Food Waste Survey Report, Fiscal Year 2023: Status and Trends in Tokyo." Food Waste Survey Report, Fiscal Year 2023. Tokyo Metropolitan Government, Environmental Bureau. https://www.kankyo.metro.tokyo.lg.jp/documents/d/kankyo/2024-07-11-091228-432

Chapter 15
Understanding Food-Related CO_2 Emissions by Food Design Tools

Amane Fujita, Shun Nakayama, and Wanglin Yan

Abstract Food-related activities account for 12% of Japan's greenhouse gas emissions, making consumer-level reduction essential for achieving carbon neutrality by 2050. While various initiatives have attempted to raise awareness through databases, educational materials, and outreach events, these efforts often lack systematic integration and fail to address consumers' primary concerns of taste and health alongside environmental impact. This study developed an integrated approach combining a comprehensive CO_2 emission database with two educational tools: "Food Design Cards" that enable calculation of both emissions and nutritional balance and a "Tangible Food Design Game" that translates invisible emissions into physical weight. Through five workshops engaging 176 participants from preschool children to adults, we demonstrated that over 80% of participants improved their understanding of food-related CO_2 emissions, with 90% finding the tools engaging. However, only 33% expressed immediate intention to change purchasing behaviors, revealing a persistent awareness–action gap. Our findings validate the effectiveness of integrating environmental data with health and taste considerations while highlighting the ongoing challenge of translating awareness into behavioral change. This research provides a framework for developing engaging environmental education tools that can contribute to consumer-driven emission reductions in the food sector.

Keywords Food design · Database · Development of tool · CO_2 emissions · Cognition

---

A. Fujita
SFC Research Institute, Keio University, Fujisawa, Kanagawa, Japan
e-mail: pontatta9012@keio.jp

S. Nakayama (✉)
School of Network and Information, Senshu University, Kanagawa, Japan

Graduate School of Media and Governance, Keio University, Fujisawa, Kanagawa, Japan
e-mail: shunnkym@keio.jp

W. Yan
Faculty of Environment and Information Studies/Graduate School of Media and Governance, Keio University, Fujisawa, Kanagawa, Japan
e-mail: w_yan@keio.jp

W. Yan (ed.), *Urban Nexus Approach for Sustainable Cities*,
https://doi.org/10.1007/978-981-95-6156-8_15

1 Introduction

To achieve carbon neutrality by 2050, reducing CO_2 emissions from food consumption has become an urgent priority at the consumer level, as the food sector accounts for 12% of Japan's total greenhouse gas emissions. Enabling consumers to reduce the environmental impact of their dietary choices requires raising awareness about the carbon footprint associated with their food consumption.

Various approaches have been developed to enhance this awareness through creating CO_2 emission datasets for food products, developing educational materials and tools, and implementing outreach activities such as events and classroom instruction (Ueno et al. 2015). The National Institute for Environmental Studies, for instance, has compiled the Environmental Load Basic Unit Data Book, which documents energy consumption across different life cycle stages and CO_2 emissions per purchase price for 50 food items (Nansai et al. 2012). Other initiatives include the development of an application that teaches users about food miles and CO_2 emissions per meal through the process of making lunch boxes, as well as daily bus tours focusing on food-related decarbonization organized by TabeKifu Corporation.

Some of these initiatives adopt a comprehensive approach that connects the entire food chain from ingredient selection through cooking to disposal. The integration of ingredient selection and cooking processes, in particular, enables the development of coherent narratives that can effectively enhance consumer awareness of food-related CO_2 emissions. However, several challenges currently limit the full integration of these elements. These include insufficient grounding in scientific evidence, inadequate consideration of effective communication strategies, and the absence of appropriate social implementation frameworks.

This study addresses these challenges by constructing a comprehensive database and developing "Food Design Cards" and a "Tangible Food Design Game" that incorporate considerations of taste and health, which are primary concerns for consumers. The research aims to enhance awareness of food-related CO_2 emissions through this integrated approach. We conducted workshops targeting diverse consumer groups, including preschool children and adults, to evaluate the effectiveness of these tools in raising environmental awareness. This research reveals how the systematic integration of database construction, educational tool development, and event implementation influences awareness outcomes, thereby contributing to the development of innovative tools for educating diverse consumer populations about food-related CO_2 emissions.

2 Technical Elements of Food Design

2.1 Conceptual Framework for Food-Related CO_2 Emissions

The calculation of food-related CO_2 emissions has traditionally involved constructing databases that trace the entire food chain from ingredient selection through cooking to disposal (Ueno, Ueda, and Seno 2015). This approach draws on established methodologies such as life cycle assessment and material flow analysis. In recent years, these conceptual frameworks have gained increasing application in environmental education, energy education, and consumer education.

Current trends in database construction have moved beyond calculating environmental impacts for broad food categories to estimating CO_2 emission factors for individual food items across their entire life cycle. These CO_2 emission factors can be classified into two primary categories based on their calculation method. Monetary-based factors determine emissions using production data per unit of currency, while weight-based factors calculate emissions using physical measurements such as transaction weight units or production volume per unit area.

2.2 Development of the Food Environmental Impact Database

This study organized CO_2 emissions data based on four key stages of the food life cycle: production, distribution, access, and cooking. The disposal stage was excluded from consideration as it does not directly influence consumer decision-making at the point of food selection.

We compiled weight-based CO_2 emission data and emission factors for different food items across life cycle stages. This approach offers several advantages for general and educational applications. It provides intuitive understanding and enables calculations across various scales, from macro to micro level analyses. When data gaps existed, we converted monetary-based CO_2 emission factors to weight-based values using statistics from the Statistics Bureau of the Ministry of Internal Affairs and Communications. Recognizing that production-stage CO_2 emissions vary significantly based on location and production methods even for identical food items, we organized the data by regional blocks. Our primary data sources included the Ministry of Agriculture, Forestry and Fisheries, research by Oda et al., and the National Institute for Environmental Studies (Nansai et al. 2012; Oda and Uchiyama 2008). For items heavily dependent on imports, we referenced Poore et al. (2018). To calculate CO_2 emissions for the distribution and access stages, which vary according to distance, transportation mode, fuel efficiency, and average speed, we utilized transportation statistics from the Ministry of Land, Infrastructure, Transport and Tourism, including automobile, domestic shipping, and railway transport data. For cooking-stage environmental impacts, we adopted Tsuda's framework, focusing on five cooking methods: grilling, boiling, deep-frying, rice cooking, and blanching (Tsuda et al. 2006).

2.3 Development of Food Design Tools

Based on our comprehensive food environmental impact database, we developed support tools for CO_2-conscious food design.

Existing food environmental impact tools include game-based educational materials incorporating food miles concepts and model menu CO_2 emission tools that calculate emissions by measuring actual energy consumption during cooking and applying emission factors. Research has demonstrated that these awareness-building materials effectively enhance knowledge specificity, stimulate interest and motivation, and promote problem recognition. However, consumer food consciousness primarily focuses on taste, health, and price, with environmental considerations remaining secondary. This hierarchy of priorities presents a significant challenge in encouraging tool adoption.

Our research addresses this challenge by recognizing health as a universal concern essential to human well-being. We developed food design tools that integrate environmental considerations with health aspects. Building on this foundation, we created two complementary tools. The Food Design Cards enable users to calculate CO_2 emissions and understand emission processes. The Tangible Food Design Game incorporates tangible interaction principles, giving physical form to information through direct sensory engagement. The game translates invisible CO_2 emissions into physical weight, making the concept accessible to diverse audiences including preschool children, elementary students, and adults who may be unfamiliar with CO_2 emission terminology.

3 Tool Interfaces

3.1 Food Design Cards

The cards were developed to combine CO_2 emissions data with health considerations that consumers prioritize, aiming to help users understand where and how much CO_2 is generated throughout the journey from production to consumption. To achieve this objective, we adopted an approach that calculates both CO_2 emissions and PFC balance.

The card system comprises 34 food ingredient cards, 24 recipe cards, CO_2 emission calculation sheets, and PFC balance sheets, as shown in Fig. 15.1. PFC balance refers to the ratio of three macronutrients that serve as primary energy sources for maintaining health: protein, fat, and carbohydrates. Optimal PFC balance varies among individuals and depends significantly on age, body weight, and desired activity level.

The food ingredient cards were created based on our food-specific CO_2 emission database and include 12 vegetables, 3 grains and legumes, 5 fruits, 7 processed foods, 3 seafood items, and 4 livestock products. Each card displays an illustration of the ingredient and its production method on the front, with CO_2 emissions from production listed on the reverse side.

Recipe cards were developed with reference to menus served in public elementary school lunches. The front of each card shows an illustration of the dish along

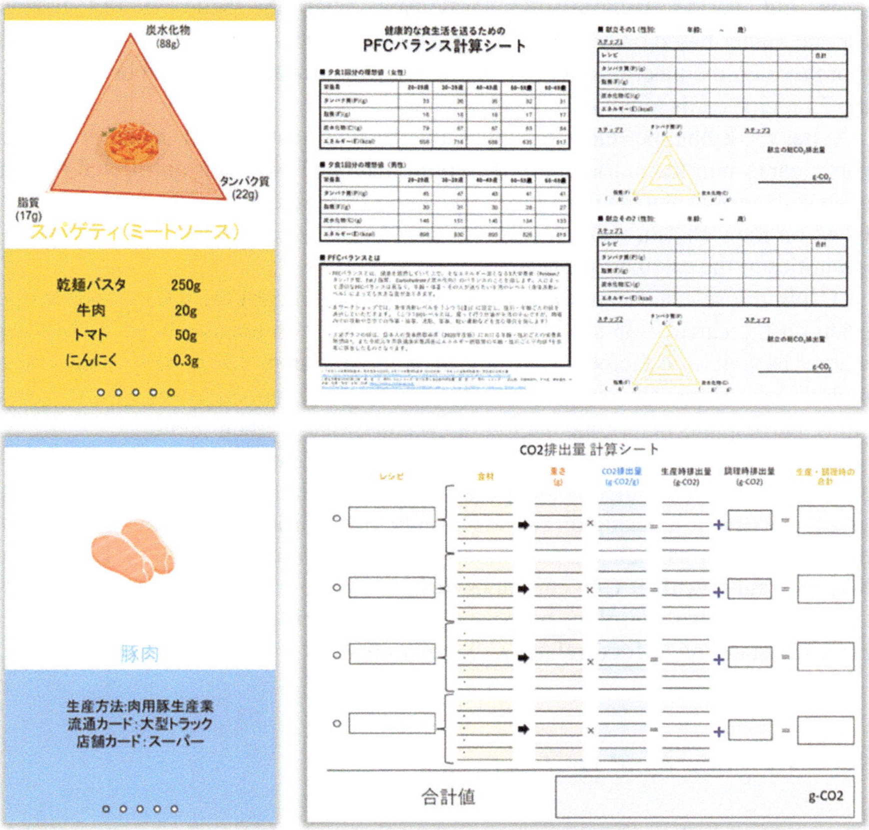

Fig. 15.1 Components of the card-based CO_2 emission and nutrition education system. The system consists of food ingredient cards (bottom left), recipe cards (top left), PFC balance sheets (top right), and CO_2 emission calculation sheets (bottom right)

with ingredient quantities and macronutrient content. The reverse side displays CO_2 emissions generated during cooking.

The CO_2 emission calculation sheet enables participants to calculate total emissions from production and cooking stages for their chosen menu. It includes sections for calculating production emissions by multiplying ingredient quantities by their emission factors, as well as spaces for recording cooking emissions. The single-sheet format allows users to track emissions from each ingredient and recipe throughout the process.

The PFC balance sheet displays ideal macronutrient values for one dinner serving based on gender and age for individuals with “normal” activity levels on the left side. The right side features a plotting diagram for comparing ideal values with the actual macronutrient content of selected recipes.

The card game proceeds through the following steps. First, each participant receives 34 food ingredient cards, 24 recipe cards, one CO_2 emission calculation sheet, and one PFC balance sheet. Participants determine their ideal PFC balance for one dinner based on their age and gender using the values on the PFC sheet and plot these on the macronutrient diagram. They then select three to four items from the 24 recipe cards to create a dinner menu.

After recording the selected recipes and their nutrient content on the PFC sheet, participants sum the nutrients for the entire meal and plot their nutritional balance against their personal ideal values. They then select ingredient cards matching those listed on their chosen recipe cards and transfer all recipe names, ingredients, and quantities to the calculation sheet.

Using the selected ingredient cards, participants record the CO_2 emissions per gram on the calculation sheet and calculate production emissions by multiplying weight by emission factors. They add cooking emissions from the reverse side of recipe cards to the calculation sheet. By summing all production and cooking emissions, participants determine the total CO_2 emissions for their menu.

This process is repeated for all selected recipes to calculate the complete meal's emissions, which are then recorded on the PFC sheet. The activity concludes with group discussion focusing on two key points: how participants' menus compare to the ideal nutritional balance recommended by the Ministry of Health, Labour and Welfare, and how their CO_2 emissions differ from those of other participants, identifying factors that contribute to these differences. This comparative analysis deepens understanding of food-related CO_2 emissions.

3.2 Tangible Food Design Game

The game challenges players to minimize CO_2 emissions for a curry meal while maintaining nutritional balance. It employs 39 food models, 4 shipping vessel models, curry recipe and CO_2 emission sheets, and a ranking chart as shown in Fig. 15.2.

Food models are shaped like actual ingredients and display illustrations, quantities, and countries of origin on the front. The reverse side indicates the primary nutrient category among the five major nutrient groups through color coding: green for vitamins, red for protein, blue for minerals, yellow for carbohydrates, and pink for fats. The collection includes 17, 11, 5, 5, and 1 variety of each nutrient type, respectively. Each model's weight represents the ingredient's production-related CO_2 emissions, eliminating the need for calculations as participants can simply place selected models on a scale to obtain total emissions.

Three sets of the 39 food models are available, with one set designated as imported products. Four additional shipping vessel models represent transportation emissions for imported items. For simplicity, all imports are assumed to originate from the United States.

The curry recipe and CO_2 emission sheet allows participants to record selected ingredients, quantities, total CO_2 emissions, and their curry's name. The front page illustrates the roles of the five major nutrients and indicates which models belong to each category, with spaces for recording ingredient details and scale-measured

Fig. 15.2 Components of the curry-making CO_2 emission game. The game consists of food models and shipping vessel models (top), CO_2 emission sheet (bottom left), and ranking chart (bottom right)

emissions. The reverse side provides information about factors contributing to food-related CO_2 emissions and potential reduction strategies.

The ranking chart enables comparison among participants by evaluating results on a five-tier scale. Rankings are assigned per 100 g of CO_2 emissions, with the best tier at 0–100 g and the lowest at over 400 g.

The game follows these procedures. Players select five food models based on three conditions: choosing two vitamin sources and one each from protein, minerals, and carbohydrates according to desired curry ingredients; selecting tasty ingredients with lighter model weights while limited to touching models a maximum of ten times; and completing selection within time limits of 90 s for middle school students and above, with no limit for elementary and preschool children.

Selected ingredient names and quantities are recorded on the curry sheet. Using a scale, players measure each selected model's weight individually and record values on the curry sheet as 1 g = 1 g-CO_2, subtracting 70 g for the model's base weight. All selected ingredients are then weighed together for total CO_2 emissions. For imported items, corresponding shipping vessel models are added based on nutrient categories.

The scale reading is recorded in the total CO_2 weight section of the curry sheet, again adjusting for base model weights. Players mark their emission level on the ranking chart with stickers, with lower emissions achieving higher ranks and determining winners.

3.3 *Implementation Methods*

The card system requires participants to calculate carbon emissions for creating a meal using ingredient and recipe cards, necessitating basic arithmetic skills. The venue should provide tables and chairs for comfortable card selection, arrangement, and calculation. One person can manage the activity while serving as both facilitator and support staff.

The game assigns CO_2 emissions as physical weight to models, which players select and weigh to determine total emissions, eliminating calculation requirements. Picture card presentations about global warming and CO_2 emissions are available, making the game accessible to preschool and elementary students encountering environmental issues for the first time. The venue must accommodate one to three sets of 43 models with adequate circulation space, requiring a certain minimum size. Each model set requires at least two staff members: one explaining rules and assisting with ingredient selection, while another records selected model information on curry sheets, ensuring smooth progression.

4 Case Studies and Results

To evaluate the cognitive effectiveness of our tools, we conducted five workshops in total: two using the Food Design Cards and three using the Tangible Food Design Game. Table 15.1 summarizes these workshops. We intentionally varied the participants, implementation methods, and baseline conditions across workshops to comprehensively assess cognitive impacts.

Table 15.1 Workshop implementation details

No.	Organizer	Date	Venue	Main participants	Tool used
1	Author's lab	December 10, 2021 10:00–12:30	PEOPLEWISE CAFÉ	8 local residents	Cards
2	Author's lab	August 4, 2022 14:00–16:00	WISE living lab	6 office workers	Cards
3	Author's lab	November 21, 2022 9:00–17:00	Keio University SFC	High school and university students	Game (Japanese version) × 1
4	Aoba Ward Office	December 3, 2022 9:30–15:00	Kodomo no Kuni	Preschoolers and elementary school students	Game (Japanese version) × 2
5	Next-generation suburban development	February 11, 2023 10:00–16:00	SPRAS AOBADAI	Elementary and high school students, local residents	Game (Japanese version) × 1, international version ×1

4.1 *Food Design Cards: Implementation and Cognitive Effects*

Fourteen participants experienced the card-based workshops, including local housewives and collaborative project partners. Participants expressed newfound realizations about food-related CO_2 emissions through the calculation process. One housewife commented that she could tangibly feel the magnitude of emissions by performing actual calculations, while another discovered that vegetarian options have relatively low environmental impact. The workshops also sparked meaningful discussions about emission reduction strategies. Participants grappled with dilemmas such as balancing protein intake from high-emission animal products against environmental concerns and questioned how to reduce emissions without drastically altering their familiar dietary patterns.

Post-workshop interviews with Next-Generation Suburban Community Development participants revealed several key insights about cognitive effects. The calculation-based approach proved particularly effective in helping participants understand emission sources in depth. Creating personalized menus facilitated meaningful comparisons with others, leading to significant personal insights. However, the complexity of simultaneous calculation and nutritional consideration made the activity challenging for younger students.

Most notably, the cards facilitated a shift from simple "recognition" of facts to deeper "awareness" and interest in understanding food's environmental impact. Participants reported gaining practical knowledge about emission reduction methods through discussions about high-impact ingredients and potential areas for improvement. Given that our target audience of middle school students through adults already possessed basic environmental knowledge, the tool succeeded in deepening rather than introducing understanding.

4.2 Tangible Food Design Game: Implementation and Cognitive Effects

The game workshops engaged 176 participants total, yielding 136 complete recipe datasets. Table 15.2 presents the workshop results, showing an average curry recipe emission of 238 g-CO_2 per meal across all sessions. Workshop No. 4 recorded the lowest average emissions while No. 5 showed the highest, with considerable variation in emission distributions across workshops.

Analysis of ingredient selections revealed interesting patterns. While vitamin and mineral choices remained consistent across workshops, protein selections varied significantly among eggs, chicken, shrimp, and other options. Workshop No. 3 showed overwhelming preference for eggs as the protein source, whereas later workshops demonstrated more balanced selections. Milk dominated mineral choices across all sessions, while onions and carrots were the preferred vitamin sources.

The game's tangible nature proved highly effective for engagement and learning. Participants appreciated the colorful, touchable models that attracted attention in open spaces, and the familiar curry theme provided an accessible entry point. Interview feedback highlighted specific learning outcomes, such as understanding the emission differences between beef and pork, and recognizing variations within food categories. One participant noted that while they had been environmentally conscious, the game introduced them to nutritional categorization for the first time.

Postgame questionnaires from Workshop No. 5 revealed compelling statistics about the tool's effectiveness. While 66% of participants admitted to either ignoring environmental concerns when food shopping or being concerned but inactive, over 90% found the game interesting and enjoyable enough to play again. More than 80% reported improved understanding of CO_2 emission sources, validating the game's effectiveness as an awareness tool.

However, the translation from awareness to behavioral intention showed more modest results. Only 33% expressed immediate intention to adopt low-carbon food purchasing practices, while 67% indicated gradual implementation plans. This gap between understanding and action intention suggests that while the game excels at

Table 15.2 CO_2 emissions by workshop

	No. 3	No. 4	No. 5
Number of samples	69	53	54
Data collection rate (%)	53.6	96.2	88.9
Average CO_2 emissions (g-CO_2)	229.8	192.9	293.1
0–100 g-CO_2 (participants)	14	16	8
100–200 g-CO_2 (participants)	20	16	13
200–300 g-CO_2 (participants)	17	10	13
300–400 g-CO_2 (participants)	14	8	11
400 g-CO_2 (participants)	4	3	9

building awareness, additional interventions may be necessary to drive behavioral change.

Both tools demonstrated success in achieving their primary objective of enhancing awareness about food-related CO_2 emissions, though through different mechanisms. The cards engaged participants through analytical thinking and calculation, while the game leveraged physical interaction and immediate feedback. These complementary approaches offer flexibility in addressing diverse learning styles and contexts, contributing to a comprehensive toolkit for environmental education about food systems.

5 Discussion

5.1 Target Users and Implementation Contexts

Building on the systematic approach outlined in the "Introduction"—integrating database construction, tool development, and event implementation—this discussion examines how our integrated framework enhanced cognitive effectiveness for food-related CO_2 emissions awareness. Through comparative analysis of the cards versus game and across different workshops, we interpret the variations in implementation outcomes and cognitive impacts.

The contrasting results between the Food Design Cards and Tangible Food Design Game reflect their fundamentally different approaches to embedding our CO_2 emission database into educational experiences. The game's colorful, visually appealing models proved ideal for public events and exhibitions, attracting spontaneous participation from diverse audiences. In contrast, the cards' requirement for seated, focused calculation made them better suited for structured educational settings such as classrooms. This distinction in implementation contexts directly influenced participant demographics and engagement patterns. While the cards' computational complexity limited accessibility to middle school students and above, the game's intuitive weight-based approach enabled participation from preschool children, thereby also engaging their accompanying parents—an unexpected benefit that broadened our reach beyond initial target audiences.

These implementation differences translated into distinct cognitive outcomes that validate our hypothesis about the importance of considering both taste and health alongside environmental impact. The cards facilitated deep "awareness" by requiring participants to methodically plot protein, fat, and carbohydrate values on nutritional balance triangular chart while calculating emissions. This dual focus on nutrition and environment, as evidenced by participant feedback about discovering protein sources and their environmental trade-offs, transformed vague understanding into concrete knowledge that sparked genuine interest in exploring reduction strategies. Conversely, the game achieved immediate "recognition" through multi-sensory engagement—the physical act of holding models of different weights made abstract CO_2 emissions tangible and memorable. Workshop observations revealed families and friends eagerly competing to minimize their curry's weight, demonstrating how the game successfully engaged even environmentally indifferent participants in understanding emission sources while having fun.

5.2 Design Specificity and Educational Outcomes

The significant variations in average CO_2 emissions across game workshops provide insights into how participant demographics and rule adaptations influence outcomes. Workshop No. 4's lowest average emissions resulted from strategic rule modifications for its predominantly young audience of preschoolers and elementary students who had limited exposure to environmental concepts. By removing time limits and restrictions on handling models, we allowed these young participants to focus purely on weight minimization, effectively prioritizing environmental learning over the nutritional balance constraints imposed in other workshops. This adaptation demonstrates the importance of calibrating educational tools to audience capabilities while maintaining core learning objectives.

Workshop No. 5's highest average emissions, despite having younger participants than No. 3, directly resulted from introducing imported food options with additional shipping vessel models. This finding validates our database's comprehensive approach to including transportation emissions and reveals how expanding choice complexity can lead to more realistic but higher-emission meal compositions. The 50% increase in emissions when imports were included mirrors real-world shopping scenarios where convenience often trumps environmental considerations.

The ingredient selection patterns across workshops revealed fascinating insights about how social dynamics influence environmental choices. Workshops No. 4 and 5, dominated by parent–child pairs, showed more diverse and realistic ingredient combinations compared to Workshop No. 3's student participants. Direct observations captured mothers guiding their children with comments such as "We usually put potatoes in our curry at home," resulting in familiar, practical recipes. However, when parents adopted a hands-off approach, children consistently prioritized weight minimization over culinary coherence—selecting eggs over chicken despite the unusual combination. This behavioral divergence highlights how environmental education must account for the tension between optimization goals and real-world constraints.

These findings illuminate both successes and limitations in our integrated approach. Our tools effectively achieved the primary objective of enhancing CO_2 emission awareness through scientifically grounded, engaging experiences. The 80% comprehension rate for emission sources and 90% enjoyment rate validate our strategy of combining environmental data with health and taste considerations. However, the modest 33% immediate behavioral change intention reveals the persistent challenge in translating awareness to action—a gap that even our comprehensive approach could not fully bridge.

The complementary nature of our tools suggests that effective environmental education requires multiple entry points. The cards' analytical approach suits formal educational settings and motivated learners seeking deep understanding, while the game's experiential design captures attention and builds foundational awareness across age groups and interest levels. This dual-tool strategy, grounded in our robust database and responsive to user preferences for taste and health, represents a significant advancement in food system environmental education. Future iterations might explore hybrid approaches that combine the game's accessibility with the cards' analytical depth, potentially through digital platforms that can adapt complexity to user profiles while maintaining scientific rigor.

6 Conclusion

This study addressed the urgent need to reduce food-related CO_2 emissions at the consumer level, recognizing that the food sector contributes 12% of Japan's greenhouse gas emissions. We developed an integrated system combining scientific database construction, user-centered tool design, and systematic implementation to enhance consumer awareness of food's environmental impact.

Our approach produced two complementary tools grounded in a comprehensive CO_2 emission database covering production, distribution, access, and cooking stages for 39 food items. The Food Design Cards enabled analytical exploration of emissions alongside nutritional balance, while the Tangible Food Design Game translated abstract emissions into physical weight. Both tools uniquely integrated environmental data with consumers' primary concerns of health and taste.

Testing with 176 participants across five workshops demonstrated strong effectiveness: over 80% improved their understanding of emission sources, and 90% found the experience engaging. The cards excelled at deepening knowledge among motivated learners, while the game successfully engaged all age groups through multisensory interaction. Average meal emissions varied significantly based on rule adaptations (186–279 g-CO_2), revealing how implementation choices influence outcomes.

However, the study revealed a critical gap between awareness and action. Despite high comprehension rates, only 33% of participants expressed immediate intention to change purchasing behaviors. This finding highlights the persistent challenge in environmental education: creating awareness does not automatically generate behavioral change.

Future work should focus on bridging this awareness–action gap through tools that incorporate real-world constraints like pricing and availability. The validated framework presented here—systematically integrating scientific data, engaging design, and adaptive implementation—provides a foundation for developing more effective interventions that can transform consumer awareness into meaningful contribution toward carbon neutrality goals.

Acknowledgments This study is done as part of The Moveable NEXUS: Design-Led Urban Food, Water and Energy Management Innovation in New Boundary Conditions of Change (No. 11314551), a Belmont Forum grant project under the Collaborative Research Area on Sustainable Urbanization Global Initiative (SUGI)/Food–Water–Energy Nexus.

References

Nansai K, Kondo Y, Kagawa S, Suh S, Nakajima K, Inaba R, Tohno S (2012) Estimates of embodied global energy and air-emission intensities of Japanese products for building a Japanese input-output life cycle assessment database with a global system boundary. Environ Sci Technol 46(16):9146–9154

Oda H, Yohji UA (2008) Analysis of energy and environmental burdens on food sector with input-output tables. Input-Output Analysis 16(2):27–41

Poore J, Nemecek T (2018) Reducing food's environmental impacts through producers and consumers. Science 360(6392):987–992

Tsuda T, Ohya C, Seto Y, Kubokura H, Inaba A (2006) Practical determination of life cycle CO_2 emission associated with domestic cooking in Japanese households. J Life Cycle Assess Jpn 2(3):288–297

Ueno M, Ueda S, Seno M (2015) An analytical investigation of descriptions of the environment in home economics textbooks. J Home Econ Jpn 66(2):39–53

Part V
Conclusion

Chapter 16
Policy Design-led Nexus Approach and Future Research Directions

Wanglin Yan

Abstract This volume advances the nexus approach as a practical framework for sustainable urbanization. Building on the SUGI M-NEX project in Tokyo, it integrates multimodal data, cross-sectoral analysis, and community-based practice to address interdependencies across food, energy, and water systems. The chapters demonstrate how nexus thinking informs urban governance, resilience to shocks, farmland protection, compact-city policy, and decarbonization strategies. By reframing the nexus through boundary objects, entry points, and multi-scalar perspectives, the book offers methodological innovations and policy insights. It positions nexus thinking—particularly the *policy design-led nexus approach*—as both an analytical tool and a governance strategy for resilient, equitable, and sustainable futures.

Keywords Spatial data integration · Nexus model of urbanization · Community-wide practice · Policy goal

1 The Structure of Nexus Approach for Sustainable Urbanization

The nexus approach to sustainable urbanization has often been criticized as a buzzword lacking methodological rigor. Much existing research relies on conventional statistical models to analyze interlinkages, reinforcing this critique. Such concerns are partly valid, given the conceptual ambiguity and complexity inherent in nexus thinking. In response, this volume seeks to clarify and operationalize nexus methods through *policy design-led* practical applications. Building on the SUGI M-NEX

W. Yan (✉)
Faculty of Environment and Information Studies/Graduate School of Media and Governance, Keio University, Fujisawa, Kanagawa, Japan
e-mail: w_yan@keio.jp

W. Yan (ed.), *Urban Nexus Approach for Sustainable Cities*,
https://doi.org/10.1007/978-981-95-6156-8_16

project, the editorial team focused on the Tokyo Metropolitan Area, translating conceptual debates into concrete policies for sustainable urbanization.

The chapters are organized to illustrate both theoretical advances and practical applications of nexus thinking—from nexus modeling and farmland conservation to design innovation and circular-economy practice—demonstrating how the *policy design-led nexus approach* bridges data analysis, governance, and community engagement.

Chapter 1 synthesizes recent advances in nexus assessment, modeling, governance, design, and policy integration, providing a roadmap for moving beyond sectoral silos toward a systemic approach to sustainability in the Anthropocene.

Chapter 2 underscores the role of farmland at the metropolitan periphery, examining how demographic change affects urban expansion, farmland use, and ecosystem services.

Chapter 3 reviews Urban Nexus methodologies and introduces the concept of the "Moveable Nexus" in Tokyo, demonstrating practical implications of the Food–Energy–Water (FEW) nexus for sustainable urban development.

Chapter 4 focuses on system dynamics modeling for nexus research, offering a framework for understanding cross-sectoral feedback.

Chapter 5 investigates the impacts of the Great East Japan Earthquake on energy structures and carbon emissions through detailed input–output analysis of direct and indirect consumption.

Chapter 6 evaluates the ecosystem services provided by urban agriculture in residential neighborhoods, where farmland coexists with urban land uses.

Chapter 7 quantifies CO_2 emissions from the transport and delivery stages of urban food retail operations, highlighting the hidden environmental costs of urban consumption.

Chapter 8 analyzes water utilities in Tokyo and Kanagawa, calculating CO_2 emissions from both utility-level processes and building-level water supply systems. It clarifies the interaction between water distribution infrastructure and urban form in shaping carbon emissions.

Chapter 9 employs GIS to quantify building-level CO_2 emissions across food, energy, and water sectors using a novel Service Point-Based Carbon Accounting (SPBCA) method, capturing nearly 70% of global emissions (compared to 50% with conventional approaches).

Chapter 10 addresses food-access challenges in suburban Tokyo, particularly in aging communities where demographic shifts exacerbate accessibility issues.

Chapter 11 assesses the compact city policy in the Tokyo Metropolitan Area, finding that transit-oriented development (TOD) can enhance quality of life near railway stations, though its benefits are less applicable to extra-urban regions such as Odawara region, where service distribution remains highly uneven.

Chapter 12 examines the conventional assumption that compact city may improve accessibility against demographic decline through the case in an extra-urban of Odawara City. Our analysis reveals a critical paradox: areas with the highest accessibility are, counter-intuitively, experiencing the sharpest population decline.

Chapter 13 explores interdependencies between suburban industrial parks (SIPs) and surrounding urban systems, analyzing sustainability performance across multiple scales in the Tokyo metropolitan context.

Chapter 14 discusses the potential of nexus thinking to further reduce environmental impacts and promote circular economy practices.

Chapter 15 introduces the "food design card," a practical tool for transforming environmental awareness into sustainable food consumption behaviors, while also considering the promise of compact city policies for improving accessibility in urban cores.

Together, these chapters demonstrate how the nexus framework—when rigorously applied—can provide both analytical depth and practical pathways for fostering sustainable urbanization.

2 Reframing the Nexus Approach for Urban Sustainability

Figure 16.1 presents a comprehensive yet distinctive policy design-led nexus framework for sustainable urbanization studies. The horizontal axis represents spatial evolution and multi-scale modeling for policy design; the left vertical axis shows multimodal spatiotemporal data; the right vertical axis depicts the transformation of values and policy goals associated with sustainable development; and the upper section reflects external environmental and social changes, including disaster risks. This framework illustrates the evolution of our understanding of the nexus—from development-first urbanization to higher dimensions of sustainability such as resilience, decarbonization, and energy efficiency. Nexus thinking thus provides powerful instruments for integrating diverse methods across disciplines and sectors. Our achievements are characterized by the following four main features.

2.1 Integration of Multimodal Geographic Information

We have developed methods that integrate spatiotemporal data such as satellite remote sensing from a wide area to the neighborhood level, and comprehensively analyzes the progress of urbanization, the use of land resources, the urban environment, and the deterioration of natural ecological systems using landscape ecology methods. The research on the urbanization of Tokyo Metropolitan Area has produced practical results that has been used in resilience and carbon neutral policies. It also contributes to policy recommendations on TOD and SIC (suburban industrial communities) for research and planning of sustainable urbanization.

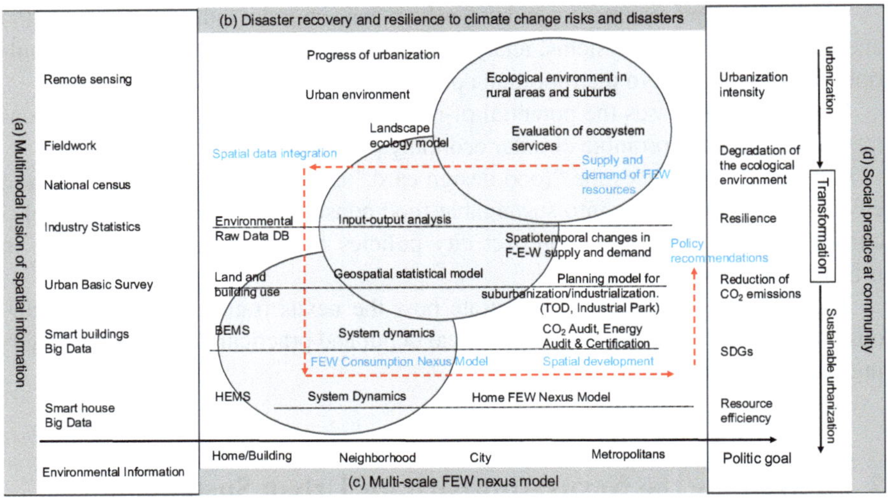

Fig. 16.1 Policy design-led nexus approaches for sustainable urbanization

2.2 *Disaster Recovery and Climate Change Resilience*

Regarding the reconstruction from large-scale disasters and shocks such as the Great East Japan Earthquake, we combined spatial analysis and environmental I–O analysis by GIS and contributed to the concept of resilience. Nexus approach can be also applied to the short caused by pandemic, the Great Pause caused during COVID-19 (Liu and Yan 2025). Their study successfully revealed a sharp decline in commercial energy use during the Great Pause, with reductions of up to 18% per month compared to the baseline. In contrast, household energy use increased drastically, surging by as much as 15%, primarily due to increased heating demands and high in-home food expenditures.

2.3 *The Multi-scale FEW Nexus Model*

M-NEX project is a public transportation-oriented development (TOD) in the Tokyo area and built a nexus model for food, energy, and water supply and demand on the dual axes of "urban center–suburbs" and "station front–station back living area." Through this project, we have integrated geographic base data and multi-scale spatiotemporal data (e.g., IO tables, big data from smart houses and buildings, etc.) to quantitatively elucidate the effects of nexus effects, resource-use efficiency, carbon emission effects, and unexpected events such as earthquakes and pandemics at multiple scales in households, buildings, blocks, and cities. Nexus principles and methods are introduced in the analyses of nexus effects and examination of policy goals, reflecting the origin of M-NEX-design-led urban food–energy–water management innovations.

2.4 Social Practice in Communities

Our research on sustainable urbanization in Tokyo is all rooted in the community and have been realized through co-creation with various actors and can be summarized as social practices at community level. Tools and concepts such as TOD and SIC, as well as the food design cards, are developed to deliver knowledge to the bottom of the society. These practices present the applicability of design-led nexus approach in urban planning and management. We believe this is not only beneficial to research but also to various actors.

3 Future Research Directions

To advance the practical utility of the nexus approach, several factors influence its success or failure in application—especially in public service delivery for essential FEW infrastructure, where limited budgets and fragmented governance persist. Future research should therefore refine scope, sectoral integration, and scale alignment while strengthening boundary objects, entry points, and evaluation of nexus effects.

3.1 Scope

Traditional nexus literature tends to focus on resource security—primarily from a supply-side perspective. Many attentions have been paid on the water–energy–food (WEF) nexus as a practical issue for assessing trade-offs, identifying synergies, and optimizing resource use and governance.

Kurian (2017) expanded the scope of nexus thinking by introducing three interrelated dimensions: resources, services, and risks. Kurian emphasized the importance of understanding how resources are transformed into services, and how these services interact with both institutional and environmental risks. This chapter proposed three critical considerations that define the operational scope of nexus thinking. The first is Nexus Nodes, or key variables that influence the structure and behavior of nexus systems—such as population growth, precipitation patterns, or income levels. These variables act as entry points for assessing system pressures and opportunities. The second is Nexus Critical Mass, which refers to the intensity and frequency of cross-sector interactions, indicating whether nexus approaches are likely to yield coordination benefits. Identifying clusters of related activity can help focus integration efforts. The third is Nexus Thresholds, where the interconnectedness of resources, services, and risks reaches a point that necessitates coordinated responses—for example, when energy used for water pumping affects access to equitable water services. This expanded scope enables a more grounded understanding of how resource systems function in real-world contexts and how governance strategies can be shaped to enhance resilience, equity, and long-term sustainability.

Therefore, scoping the nexus approach requires deep observations of a nexus system more than simply acknowledging the interlinkages among water, energy, and food (WEF) systems. It demands a systems-thinking mindset that embraces multi-scalar problem-solving, engages with complexity, and promotes participatory governance (Hogeboom et al. 2021). These attributes enable the nexus framework to serve not just as an analytical tool but as a strategic approach for integrated resource management and policy development.

3.2 Sector

A nexus-informed approach requires strengthened cross-sectoral dialogue, coordinated planning, and the inclusion of both trade-offs and co-benefits in policy formulation (Machell et al. 2015). Such integration encourages a move away from fragmented sectoral thinking and toward a more collaborative, multidimensional form of governance in urban systems.

Infrastructure systems are foundational in shaping urban food–energy–water (FEW) security, as they determine both access to and the availability of essential resources. Romero-Lankao et al. (2017) emphasize that infrastructure plays a critical role in the sustainability of urban resource systems. However, they also point out that urban infrastructure tends to be highly path-dependent, making transformational change difficult to achieve in the short term. This rigidity reduces the capacity of cities to respond effectively to challenges such as climate change, urban resilience, and integrated resource management. As a result, urban planning and infrastructure design must increasingly account for the interlinked nature of urban metabolism and FEW interactions in order to support sustainable and resilient development.

Recent research has begun to move beyond conventional siloed perspectives by exploring the nexus between green–blue infrastructure (GBI) and the FEW nexus. Cristiano et al. (2023) offer a qualitative assessment of green roofs through a food–water–energy–ecosystem nexus lens, showing that traditional silo-based evaluations often overlook the broader sustainability benefits of such infrastructures. Meng et al. (2023) note the growing academic interest in integrating GBI with nexus thinking.

Despite these integrative demands, resource management in practice remains largely sector-based and is often responsible for continuing resource insecurities (Taguta et al. 2022). These approaches frequently neglect the interdependencies that exist among water, energy, and food systems—interdependencies that are central to the strength of the nexus framework. Although some modeling efforts attempt to incorporate these connections, many still fall short of capturing the complexity and multidimensional nature of nexus interactions. More research is needed to evaluate how existing nexus tools represent these interlinkages across sectors.

3.3 Scale

One of the main obstacles to applying a common, reproducible nexus framework across different scales, domains, and regions lies in the sheer diversity of issues, challenges, and timelines that the concept seeks to encompass. Among them, the scale at which the food–energy–water (FEW) nexus is examined plays a critical role in shaping both understanding and implementation. This variability complicates efforts to standardize methodologies or create generalizable models. Shannak et al. (2018) reviewed FEW nexus literature and identified key strengths and limitations at various scales, drawing attention to persistent data and methodological gaps that hinder effective implementation.

Given the inherent complexity and multidimensional nature of the FEW nexus, a variety of frameworks have been developed across spatial scales—from regional and urban contexts to the scale of individual households. At the household or individual level, nexus studies often focus on footprint accounting and sustainable consumption of FEW resources (Liu et al. 2024). In contrast, urban-scale frameworks—such as the Urban Nexus approach—aim to enhance resource productivity and governance by encouraging collaboration among local and regional actors.

Although global challenges such as climate change, population growth, and resource depletion often drive the need for nexus thinking, practical interventions typically occur at local and regional levels. Ramaswami et al. (2017) proposed a method for calculating community-wide urban CO_2 emissions, showing how nexus modeling can incorporate both local and transboundary environmental impacts. McGrane et al. (2018) advocate for hierarchical frameworks that integrate across these levels, thereby enabling alignment between national policy objectives and local actions. Their work highlights how the nexus manifests differently depending on context—for example, national policy dynamics intersecting with local watershed conditions—emphasizing the necessity of contextual sensitivity in nexus planning.

Misalignments frequently arise across spatial and temporal scales, especially when human decision-making processes and biophysical systems operate on different timelines or geographic scopes. Temporal dimension is also equally important.

Scenario planning is a widely adopted tool for exploring long-term futures under conditions of climatic and demographic uncertainty (McGrane et al. 2018).

Therefore, successful implementation of the FEW nexus relies not only on interdisciplinary approaches but also on governance and design that are sensitive to scale. Bridging the gap between macro-level policy and micro-level realities—across both space and time—is essential for transforming nexus thinking into actionable, resilient, and equitable outcomes.

3.4 Boundary Objects

The food–energy–water (FEW) nexus provides a vital framework for addressing the interconnected challenges of urban sustainability. A central concept in understanding this interdependence is that of the nexus trigger or catalyst—an initial perturbation in one domain (such as demographic change, technological advancement, or an exogenous shock) that produces cascading impacts across the broader FEW system (Newell et al. 2019).

To address these challenges, the use of boundary objects—flexible, shared concepts that facilitate communication across disciplines—has gained prominence (Newell and Cousins 2015). Though interpretations may differ between fields, shared terminology enables collaboration and mutual understanding.

A prominent example is urban metabolism (UM), which functions as a boundary object with wide applicability across engineering, environmental sciences, social sciences, and urban planning. The concept of boundary objects has gained attention (Newell et al. 2019). Boundary objects are shared tools or representations that hold different meanings for different stakeholders but are flexible enough to allow mutual interpretation and collaboration across sectors and disciplines.

As another example, Andersson et al. (2021) suggest that traits—attributes related to ecological function, species dynamics, and biophysical constraints—can serve as boundary objects. While different communities may interpret these traits in diverse ways, they enable a common understanding of system behaviors, interactions, and thresholds within complex social–ecological systems. System dynamics modeling and group model-building approaches provide structured means for fostering shared comprehension of nexus challenges. These processes generate system maps that act as boundary objects, helping to reveal interdependencies and feedback between system components and stakeholder priorities (Getahun and Selassie 2017). By exposing the structure of complex problems across disciplines and sectors, these maps facilitate collaborative learning and strengthen cross-scale governance.

Therefore, boundary objects are crucial for advancing nexus governance. They help address mismatches in spatial and sectoral scales, support inclusive stakeholder engagement, and foster the development of adaptive and integrated strategies for sustainable resource management.

3.5 Entry Points

Entry points refer to the strategic starting positions—such as a particular resource, sector, or societal priority—through which nexus thinking can be embedded into management, planning, and governance frameworks. Identifying entry points is essential for operationalizing the water–energy–food (WEF) nexus. Al-Saidi and Elagib (2017) highlight water as a key entry point for integration within the nexus. They argue that water frequently acts as the "gateway resource" due to its pivotal role across sectors and its vulnerability to scarcity and crisis. Their analysis points to three main drivers for prioritizing water: growing interlinkages among resources driven by scarcity, recent crises in resource supply (such as droughts, water shortages, and energy–water conflicts), and the failures of traditional sector-based management strategies that ignore cross-sector dependencies. Nevertheless, the authors caution that there is no universal definition of "integration" in the nexus context, as entry points vary by circumstance.

Case studies further show how the choice of entry point can shape management and governance outcomes. Mcgrane et al. (2018) describe how a combination of participatory stakeholder workshops and data-driven regional analysis helped uncover context-specific nexus interactions and potential entry points. Likewise, Andersson et al. (2021) suggest that "traits"—easily understood characteristics tied to ecosystem services, such as urban cooling, clean air, or biodiversity—can serve as entry points that enhance ecological awareness and embed nature-based understanding into decision-making at both citizen and policy levels.

In some instances, entry points are defined by broader sociopolitical concerns. Globally, sustainable urbanization is increasingly seen as a powerful cross-cutting entry point for achieving multiple Sustainable Development Goals (SDGs). For instance, UN-Habitat (2020) emphasizes that initiatives focused on urban sustainability can drive integration across sectors, combining FEW systems with infrastructure and social services to deliver synergistic benefits. Mayer (2015) argues that once a problem is clearly articulated—such as climate-related migration—the selection of suitable entry points must consider both political feasibility and efficiency. According to his observation, while climate change is a major issue, it may not always be the most effective entry point, especially when other immediate priorities, such as economic growth or demographic pressures, take precedence.

The M-NEX project in Tokyo intentionally selected food as the entry point for introducing nexus thinking to communities (Yan and Nakayama 2021). By connecting themes such as urban agriculture, food waste reduction, and localized FEW cycles, the project engaged stakeholders using a familiar and relatable resource, thereby promoting awareness of system interlinkages in a meaningful and participatory manner.

3.6 Nexus Effects

While interconnections and interdependencies are abundant in nexus systems, the potential synergies and trade-offs are often unclear. This makes it difficult to evaluate consequences or craft effective policies. Most existing studies remain narrowly sectoral or isolated, which limits the development of comprehensive decision-support tools.

Understanding how resources transition into services is essential for evaluating individual behavior and institutional performance—two key factors influencing the sustainability of the nexus effects. Systems-based nexus assessments can reveal important synergies and feedback across environmental resources, both spatially and temporally, thereby enabling more efficient, resilient, and integrated urban development strategies.

Chang et al. (2020) provided a synthesis of FEW nexus theory with urban growth models (UGMs), highlighting how urban metabolism and ecology related to FEW systems can be embedded in UGMs. Their findings suggest that multi-criteria decision analysis, when paired with cellular automata (CA) modeling architectures, can effectively simulate the integration of FEW technologies and their trade-offs. These synergies and cascade effects underscore the potential of FEW-integrated UGMs to shape more adaptive and sustainable urban futures.

Successful nexus governance requires an iterative process grounded in stakeholder participation, reflection, and continual adjustment (Tye et al. 2022). The Moveable Nexus is designed to be flexible, participatory, and context sensitive. Rather than offering fixed solutions, it embraces the evolving and localized nature of FEW interactions and promotes codesigned responses that reflect the unique characteristics of specific social–ecological systems.

Therefore, new methodologies and decision-support tools are needed to better map, visualize, and analyze interactions across sectors—especially when these interactions cut across unclear or overlapping administrative boundaries.

4 Conclusions

This volume demonstrates that, when rigorously applied, the nexus approach can move beyond conceptual ambiguity to deliver practical, policy-relevant insights for sustainable urbanization. Drawing on experiences from the SUGI M-NEX project in the Tokyo Metropolitan Area, the chapters collectively advance methodological innovation and applied practice. They illustrate how nexus thinking integrates multimodal data, assesses cross-sectoral dynamics, and addresses pressing urban challenges such as farmland loss, urban metabolism, disaster resilience, and infrastructure decarbonization.

The reframed policy design-led nexus framework emphasizes multi-scalar analysis—from households to metropolitan regions—while accounting for external shocks such as earthquakes and pandemics. It also underscores community-level practices, co-creation, and the translation of scientific knowledge into actionable policy and design tools, including food-design cards and transit-oriented development models.

Looking ahead, the success of nexus applications depends on clarifying scope, strengthening cross-sectoral integration, and resolving scale mismatches that often hinder implementation. Boundary objects and strategic entry points can enable collaboration, while system-based methods will better capture nexus effects. Ultimately, the policy design-led nexus approach functions as both an analytical and governance framework—bridging disciplines, scales, and stakeholders to foster resilient, equitable, and sustainable urban futures.

References

Al-Saidi M, Elagib NA (2017) Towards understanding the integrative approach of the water, energy and food nexus. Sci Tot Environ 574:1131–1139. https://doi.org/10.1016/j.scitotenv.2016.09.046

Andersson E, Haase D, Anderson P, Cortinovis C, Goodness J, Kendal D, Lausch A, McPhearson T, Sikorska D, Wellmann T (2021) What are the traits of a social-ecological system: towards a framework in support of urban sustainability. Npj Urban Sustainability 1(1):1–8. https://doi.org/10.1038/s42949-020-00008-4

Bazilian M, Rogner H, Howells M, Hermann S, Arent D, Gielen D, Steduto P et al (2011) Considering the energy, water and food nexus: towards an integrated modelling approach. Energy Policy 39(12):7896–7906

Chang N-B, Hossain U, Valencia A, Qiu J, Kapucu N (2020) The role of food-energy-water nexus analyses in urban growth models for urban sustainability. Sustain Cities Soc 63:102486

Cristiano E, Farris S, Deidda R, Viola F (2023) How much green roofs and rainwater harvesting systems can contribute to urban flood mitigation? Urban Water J 20(2):140–157

Getahun M, Selassie YG (2017) Characterization, classification and mapping of soils of agricultural landscape in Tana Basin, Ethiopia. In: AESS interdisciplinary environmental studies and sciences series. Springer, pp 93–115

Hogeboom RJ, Borsje BW, Deribe MM, Hoekstra AY, Nelson AD (2021) Resilience meets the water–energy–food nexus: mapping the research landscape. Front Environ Sci 9:630395

Kurian M (2017) The water-energy-food nexus: trade-offs, thresholds and transdisciplinary approaches to sustainable development. Environ Sci Pol 68:97–106

Liu H, Yan W (2025) Exploring urban resilience through the food–water–energy nexus during the COVID-19 pandemic: insights from the great pause in Tokyo. J Clean Prod 523:146410

Liu H, Yan W, Kobayashi H (2024) Modelling the system dynamics of household food, water, and energy nexus effects. Heliyon 10:e25886

Machell J, Prior K, Allan R, Andresen JM (2015) The water energy food nexus-challenges and emerging solutions. Environ Sci: Water Res Technol 1(1):15–16. https://doi.org/10.1039/c4ew90001d

Mayer B (2015) Climate migration and the politics of causal attribution: a case study in Mongolia. Migration and Development 1–20. https://doi.org/10.1080/21632324.2015.1022971

McGrane SJ, Acuto M, Artioli F et al (2018) Scaling the nexus: towards integrated frameworks for analysing water, energy and food. Geogr J 185(4):419–431

Meng F, Yuan Q, Yuan Q, Bellezoni RA, Bellezoni RA, Yang Z, Seto KC, Seto KC (2023) Quantification of the food-water-energy nexus in urban green and blue infrastructure: a synthesis of the literature. Resour Conserv Recycl Adv 188(106658)

Newell JP, Cousins JJ (2015) The boundaries of urban metabolism: towards a political-industrial ecology. Prog Hum Geogr 39(6):702–728
Newell JP, Goldstein B, Foster A (2019) A 40-year review of food–energy–water nexus literature and its application to the urban scale. Environ Res Lett 14(7):073003
Ramaswami A, Boyer D, Nagpure AS, Fang A, Bogra S, Rao-Ghorpade B (2017) An urban systems framework to assess the trans-boundary food-energy-water nexus. Environ Res Lett 12:025008
Romero-lankao P, Mcphearson T, Davidson DJ (2017) The food-energy-water nexus and urban complexity. Nat Clim Change 7(4):233–235. https://doi.org/10.1038/nclimate3260
Shannak S'd, Mabrey D, Vittorio M (2018) Moving from theory to practice in the water–energy–food nexus: an evaluation of existing models and frameworks. Water-Energy Nexus 1(1):17–25
Taguta C, Senzanje A, Kiala Z, Malota M, Mabhaudhi T (2022) Water-energy-food Nexus tools in theory and practice: a systematic review. Front water 4. https://doi.org/10.3389/frwa.2022.837316
Tye MR, Wilhelmi OV, Pierce AL, Sharma S, Nichersu I, Wróblewski M, Goszczyński W et al (2022) The food water energy nexus in an urban context: connecting theory and practice for nexus governance. Earth Syst Gov 12:100143
UN-Habitat_Core_Team (2020) World cities report 2020: the value of sustainable urbanization (S/045/20E). UNHabitat. http://resp.llas.ac.cn/C666/handle/2XK7JSWQ/301181
Yan W, Nakayama S (2021) Redesigning the urban food life through the participatory living lab platform. In: TransFEWmation: towards design-led food-energy-water systems for future urbanization. Springer, pp 209–234

MIX
Papier aus verantwortungsvollen Quellen
Paper from responsible sources
FSC® C105338

If you have any concerns about our products,
you can contact us on
ProductSafety@springernature.com

In case Publisher is established outside the EU,
the EU authorized representative is:
Springer Nature Customer Service Center GmbH
Europaplatz 3, 69115 Heidelberg, Germany

Printed by Libri Plureos GmbH
in Hamburg, Germany